Michael Hoppert

Microscopic Techniques in Biotechnology

WILEY-VCH

WILEY-VCH GmbH & Co. KGaA

Author

Dr. Michael Hoppert
Georg-August-Universität-Göttingen
Abteilung Strukturelle Mikrobiologie
Grisebachstrasse 8
37077 Göttingen
Germany

Library of Congress Card No.:
Applied for.

British Library Cataloguing-in-Publication Data:
A catalogue record for this book is available from the British Library.

Bibliographic information published by Die Deutsche Bibliothek
Die Deutsche Bibliothek lists this publication in the Deutsche Nationalbibliografie; detailed bibliographic data is available in the Internet at <http://dnb.ddb.de>

Printed in the Federal Republic of Germany

Printed on acid-free paper

Composition Mitterweger & Partner, Kommunikationsgesellschaft mbH, Plankstadt
Printing betz-druck GmbH, Darmstadt
Bookbinding Litges & Dopf Buchbinderei GmbH, Heppenheim
ISBN 3-527-30198-4

Preface

Among the instruments used for the investigation of biological objects, only microscopes provide a real image and not just an abstract data set. Interpretation and analysis of the image allows in one determination the exploration of a variety of important features, such as the total quantity of the objects, cell volume and volume of compartments, the presence of contaminants or infective agents, the status of the cell cycle, and topography. Microscopic techniques are prerequisites for localization of subcellular fractions and proteins, differentiation of organisms, microspectrometry, and all micromanipulation techniques.

Over the past 50 years, electron microscopy has opened new pathways for biologists to understand structure-function interrelations at the subcellular level. New light-based optical techniques involving the detection of fluorescent molecules have introduced a new dimension of imaging into light microscopy, enabling cellular events to be studied in vivo down to molecular scales.

In fact, microscopes of any kind are now user-friendly, easy-to-operate instruments, and most of the preparation techniques are standardized and easily applicable. Though the book provides just a short glimpse into all these techniques, it is written for the reader seeking a quick start in key microscopic techniques.

Aim and Scope of this Book

This book is devoted to microscopic applications in biotechnology. Of course, most of the tools described do not differ from the instrumentation used in other fields of life sciences. The book seeks to convey the potential of microscopy for biotechnological applications.

The introductory sections give general overviews of the biological objects, which should also allow non-biologists (chemists, technical engineers, or computer software specialists) working in the field of biotechnology to understand the potential of microscopic imaging. The sections describing useful applications contain short and straightforward introductions into the structures of the objects described in terms of the view in the instrument used. The non-biologist will find a description of biological objects that is far from complete, but comprehensive enough to give an idea of the objects structures. In the "useful applications" sections, selected problems are described in more detail. These problems are illustrative for a broad range of biotechnological applications and demonstrate the possibilities of the techniques. These sections contain only incidental information concerning the technical procedures, which are described in greater detail in the "Preparation Techniques" sections. These contain short introductions to the apparatus used and describe the technical procedures in detail. Representative procedures are included as step-by-step protocols. Of course, a selection of protocols may be critical and the reader may miss a protocol for her or his special needs. Here, the extended reference list is helpful. In addition, some manufacturers provide excellent and detailed protocols especially adapted to their products. Those protocols are, of course, not included here. One has to consider that the number of protocols naturally widely exceeds the scope of a single book covering such a wide field of microscopic techniques.

Recently developed or developing techniques are summarized in a separate chapter. These techniques may come to be (or are already) of utmost importance for biotechnology, but are not (yet) available for a large range of users.

Dr. Michael Hoppert
Göttingen, February 2002

Acknowledgements

The protocols and figures for a book such as this are necessary to draw from a wider field than is encountered in our single laboratory. Thus, I would like to thank numerous colleagues for allowing me to take benefit from their helpful advice as well as from their published micrographs and illustrations. I would also like to thank all former and present members of the Structural Biology department for their diverse contributions, and Jan Hegermann (University of Göttingen) for his help during manuscript preparation. I am especially grateful to Prof. Dr Frank. Mayer for his essential foundations, without this book could not have been written.

Michael Hoppert

Concerning the contributions to Chapter 2.4.8.9, I would like to thank Drs. O. Bertrand (INSERM U76, Paris) and J.-L. Rigaud (Institut Curie, Paris) for helpful discussions and suggestions. Part of the ideas for this chapter were put on paper during my Visiting Professorship at the University of Marburg in the Institute for Virology, and I am grateful to Prof H.-D. Klenk and Dr. E. Bogner for inviting me. The collaboration with Dr. Bertrand was sponsored by The British Council (PN 99.121).

Andreas Holzenburg

Foreword

This book was initiated by Wiley-VCH after the remarkable success of the series BIOTECHNOLOGY. In volume 1 of this series, cell structure was described. The publishers felt that a more detailed treatment of this subject may be of interest for scientists engaged in research and development in biotechnology. They managed to convince one of the most competent scientists in the field of imaging of biological samples, Michael Hoppert, to do the writing. The outcome is presented here.

A first glimpse at the table of content gives an idea on the breadth of the work. Further inspection reveals that the book is indeed an in-depth compendium. The first part concentrates on representative examples of biological material to be investigated in the context of biotechnology, and on instructions for their preparation. This part is also combined with descriptions of the theoretical basis, applications, and limitations of multiple kinds of methodologies of light microscopical imaging, starting from the most basic light microscopy, and encompassing sophisticated light microscopical techniques, such as state-of-the-art fluorescence laser scanning microscopy in various modifications for qualitative and quantitative analysis of samples.

In the second part, imaging by application of electron microscopy is treated with the same accuracy. Descriptions of representative samples, ranging from isolated macromolecules and unicellular organisms, many kinds of differentiated cells of eukaryotic organisms, to complex tissue, are presented, together with careful but nevertheless straight-forward instructions on how to prepare these samples.

In the final section, Michael Hoppert describes specialized approaches of imaging not usually dealt with in conventional books on imaging techniques for biological samples, for example, total internal reflection microscopy, multiphoton excitation microscopy, fluorescence resonance energy transfer, fluorescence correlation spectroscopy, soft X-ray microscopy, microscopic imaging with photons and heavy ions, just to name a few of them.

Carefully selected illustrations, together with the references given at the end of each of the sections of the book, simplify the search for the technique of choice needed for a specific problem.

Michael Hoppert's book carries features that may make it the reference book *par excellence* in the field of imaging of biological samples, with a special emphasis on biotechnology.

I trust that this book will have the success it deserves.

Prof. Dr. Frank Mayer
Göttingen, Januar, 2003

Contents

1
Light Microscopy

1.1
Applications and Limitations in Light Microscopy

Light microscopes are the most widely used instruments for visualization of structures down to the scale of a few micrometers. The instruments are available in numerous variants, suitable for detection of phase contrast, fluorescence, changing polarity of light, etc. The big advantage of light microscopes is their mechanical stability, allowing operation under varying environmental conditions, and their ease of operation. The fact that fully hydrated specimens may be visualized under natural environmental conditions (room temperature, atmospheric pressure etc.) remains the most important feature of light microscopic applications in biomedical research. Visible light as an imaging medium does not usually induce artificial changes in the specimen (a big disadvantage of the shorter-wavelength radiation). With no other microscopic technique is such a broad range of well established staining procedures, additional observation facilities, manipulating tools, etc. available for visualization of fully hydrated specimens in a natural surrounding. Although alternative techniques may become applicable for routine research in the near future, only light microscopy currently allows the elucidation of dynamic processes such as growth, cell movement, vesicle transport, etc.. Biologists, though, have to take account of the fact that the resolution limit in light microscopes, essentially represented by the wavelength of the light, allows depiction of whole organisms and large subcellular compartments, but none of even the largest macromolecules. Although direct imaging of viruses, protein molecules, etc. is impossible, they may be *localized* by use of probes with specific binding affinities (antibodies, enzyme substrates, oligonucleotides, etc.). These techniques, especially in conjunction with fluorescent markers, allow the detection of small quantities of the specifically marked targets. Depending on the individual binding affinities, a minimum number of targets have to be present for a detectable signal to be emitted. Nevertheless, specialized fluorescent techniques allow detection down to one molecule. At the cellular level, it is – in principle – possible to detect one single cell in a specimen (e.g., a single pathogen in a wood sample), but this of course requires automated systems. The lower detection limit in a non-automated system is around 50 (bacterial) cells in a volume of 100 μl. The sample must be thin enough to allow transmission by the light beam or detection of the fluorescence signal (this is also true for confocal microscopes). Nevertheless, when the structures are large enough, and the sections are optically transparent (i.e., the multiple layers do not blur the final image), up to 0.5 mm are acceptably thin for an overview. This allows the cutting of, for example, wood samples with a simple razor blade. The lower limit of specimen thickness is around 200 nm.

The ease of specimen handling in the operation of the light microscope has facilitated the development of numerous microtechniques, from closed flow cells for circulation of liquid samples (for the observation of biofilm development, for example) to micromanipulators, optical tweezers, etc. that allow handling of single cells. The image information itself may be quantified by image processing, spectroscopic analysis of the transmitted light, volumetric measurements, and so on.

Essentially, all microscopes only reveal the properties of the specimen that produce alterations in the imaging medium (light, X-ray, electron beams, etc.). This is also true for light microscopy, and allows numerous intrinsic properties of the specimens to be detected. Upon interaction with the specimen, light may change its intensity – regularly at a specific wavelength – leading to contrast and color. Variations in the optical density of the specimen change the phase properties, leading to phase contrast. The arrangement of molecules in the specimen may turn the polarization plane. Visible light induces excitation of certain molecules to produce fluorescence, regularly applied in connection with fluorescent probes. Light absorption, induction of fluorescence, changes in the phase properties, and polarization are all used for construction of light microscopes. Raman imaging may be viewed as a microscopic technique that helps to elucidate the chemical composition of a specimen with the aid of visible light. However, the interaction of light with the specimen, unlike that of X-rays or electron beams, does not convey information on the elemental composition.

The essential drawback of optical light microscopy is the resolution limit, with 0.16 μm as a theoretical value, but no better than 0.3 μm under optimum conditions under the microscope. However, optical near-field microscopy is a microscopic technique also based on visible light as imaging medium that breaks the resolution limit. Instrumentation and applications are currently under development. This technique is presented in detail below (see Section 1.4.11).

1.2 Useful Applications

1.2.1 Cellular and Tissue Structure

1.2.1.1 Prokaryotic cell structures at the light microscopic level

All bacteria, irrespective of their degree of cellular complexity in various physiological or taxonomic groups, have an envelope consisting of a cytoplasmic membrane and, in most cases, a rigid wall. The envelope encloses the cytoplasm and interacts, at its exposed layer, with the environment. The cytoplasm contains, in most cases without obvious compartmentation, nucleic acids, ribosomes, storage materials, enzymes, and low-molecular weight components within the cytosol. Their size and structural simplicity restricts cytological investigation of bacterial structures by light microscopy to the detection of large cellular inclusions. Specialized staining methods allow localization of small intracellular targets, but do not help to elucidate their fine structures. Nevertheless, light microscopy for bacteria is indispensable when living cells or processes inside living cells have to be monitored.

The cell envelope

Simple negative staining allows detection and determination of the dimensions of spacious bacterial capsules, sheaths, or slimes (often termed "extracellular polymeric substances" or EPSs) as depicted in Fig. 1.1 a. When organized into a regular morphologically structured component, this is usually referred to as a capsule, glycocalyx, or sheath. When loosely organized, or of variable consistency, it is designated as a slime. Capsules, sheaths, or slimes represent the outermost layer of most free living Gram-positive and Gram-negative bacteria. Common sugars making up the polysaccharides are rhamnose, galactose, glucosamine, galacturonic acid, or hyaluronic acid; their relative amounts vary. Though usually composed of polysaccharides, EPSs composed of polypeptides or cellulose have also been described. EPSs are important virulence factors of pathogenic bacteria, because they counteract host defense mechanisms. Some of the more loosely attached EPSs (slimes) are of industrial interest; *Leuconostoc* converts sucrose into dextran gel. EPSs connect bacteria to each other and their substratum, thus allowing them to build up consortia, such as flakes or films. EPSs protect cells against sudden changes in environmental parameters (pH, water content, antibiotic agents). Under optimized conditions, in laboratory cultures, bacterial strains often lose the ability to synthesize EPSs. The EPSs appears to be bound to the cell wall not by covalent bonds, but by electrostatic interaction with the outer wall polymers.

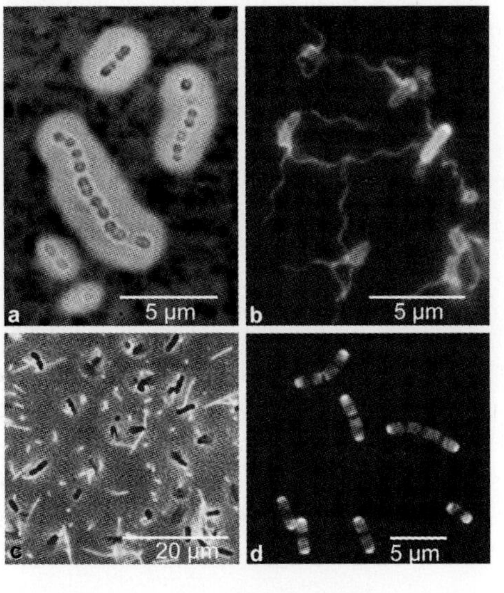

Fig. 1.1 Structures outside the bacterial cell wall: (a) India Ink preparation of encapsulated pneumococci. The cells appear slightly gray within the bright capsules. The dark ink particles do not penetrate the capsule material (Gillies and Dodds, 1977).
(b) *Escherichia coli* cells labeled with an amino-specific Alexa Fluor dye, examined in a fluorescence microscope. The wavy flagella are attached to the fluorescent cell surfaces (Linda Turner and Mary A. Nilsson, Rowland Institute for Science, Cambridge, Mass.).
(c) *Escherichia coli* K-12 Hfr H; visualization of the F-pili. A fluorescence marker is coupled to the pilus-specific bacteriophage MS 2 (Jarchau, 1985; Mayer, 1986).
(d) Immunofluorescence of D-cysteine residues in sacculi from *E. coli* cells labeled with D-cysteine and chased in the presence of cell division inhibitors (de Pedro et al., 1997)

Though the swimming of bacteria in liquid media is easily to observe under any standard light microscope, specific stains must be applied for visualization of bacterial flagella by bright-field microscopy. Recording of fluorescently labeled flagella also reveals details of their motion (Fig. 1.1 b). In addition, the localization of pili and fimbria by application of fluorescence techniques is possible (Fig. 1.1 c). These structures, together with the EPSs, are mainly involved in interactions between cells and their surroundings. They take part in the initial stages of biofilm formation, the primary attachment of a cell on a surface. During the attachment process, the cells often have to penetrate a layer of repulsive surface charges. Pili penetrate this layer, thus allowing contact between cell and surface. The contact between pili and other cell surfaces may be highly specific, during conjugation or in symbiotic and pathogenic cell-cell interaction, for example.

Since the thickness of the next, innermost layers of the cell envelope are below the detection limit of light microscopes, no direct imaging of these layers is possible. Differentiation between cell wall types by the use of specific stains (most commonly the Gram stain) is possible, though no direct information on the wall structure can be deduced from these methods. However, the light microscope also permits differentiation of larger labeled areas on the cell surface. Figure 1.1 d depicts the differentiation between old and newly synthesized parts of the murein sacculus in *E. coli*. Although, from the image alone, the cell wall itself is invisible, the intensities of the fluorescence signals represent the quantities of a cell wall component (here introduced artificially) and allow conclusions to be drawn on the distribution of old (brighter) and newly synthesized (darker) portions of the cell wall.

Structures inside the prokaryotic cell

Direct visualization of large intracellular inclusions (storage granules and endospores, Fig. 1.2 a, e, and f) is possible without any pretreatment. Endospores form typical, highly refractive inclusions (Fig. 1.2 a) and are important taxonomic markers for the genera *Bacillus* and *Clostridium*. A variety of other inclusions in bacteria (gas vacuoles, magnetosomes, chlorosomes, phycobilisomes, glycogen and polyphosphate granules, protein crystals) are at best very difficult to discriminate. Reliable structural analysis is only possible with electron microscopic techniques. Nevertheless, detection and quantification of the large inclusions may be possible with differential stains. Some bacterial polymeric inclusions are raw materials of biotechnological relevance. Polyhydroxyalkanoates (PHAs) form inclusion bodies, functioning as carbon and energy storage materials, accumulated in a variety of prokaryotes. These inclusions usually appear to be more or less spherical. The hydrophobic nature of these inclusions allows hydrophobic agents to be used as stains. Figure 1.2 b shows cells of *Pseudomonas* sp. stained with the fluorescent Nile Blue. The image shows a composite photograph of a phase contrast (i.e., bright-field) and fluorescence image. Though application of fluorescent stains is not necessary to visualize PHA inclusions, the two separate images (fluorescent and bright-field) allow easy quantification of the relative amount of PHA when the surface area of the whole cells (bright-field) is compared with the area of the fluorescence signal.

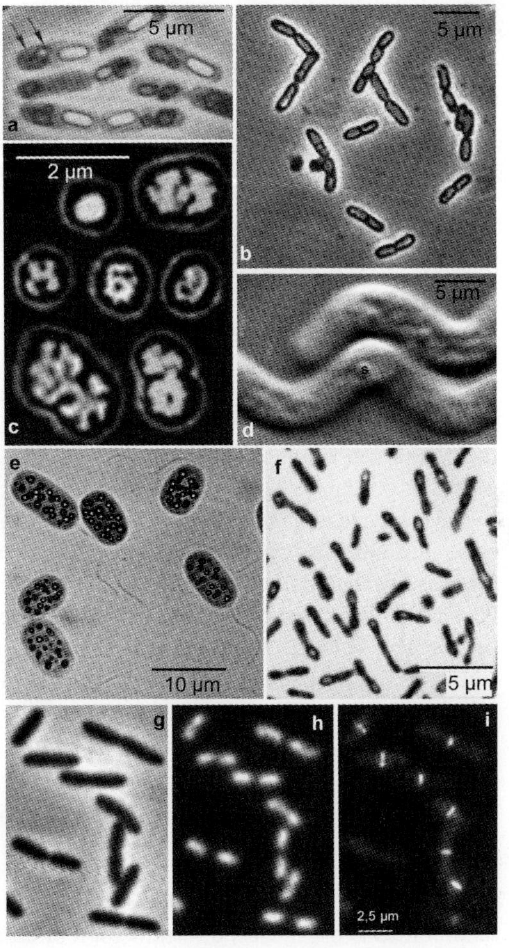

Fig. 1.2 Intracellular bacterial structures:
(a) *Bacillus cereus* var. schwetzova forms, besides a spore (bright, oval cellular inclusion), a diamond-shaped and a spherical parasporal body (arrows) (Gould and Hurst, 1969).
(b) Polyhydroxybutyrate (PHB) granules in *Pseudomonas*. Overlay of a bright-field and a fluorescence image of *Pseudomonas* cells. The phase contrast provides natural contrast between the cell and the background, but small PHB granules are not readily visible. The cells were therefore stained with Nile Blue A, a fluorescent stain that binds to PHB. The PHB can easily be seen as bright yellow granules (micrographs taken by William Ghiorse, Section of Microbiology, Cornell University).
(c) Nucleoid structure as revealed by DAPI staining of *Methanococcus jannaschii* cells (Malandrin et al., 1999).
(d) Two spirillum-shaped cells shown by differential interference contrast. Small sulfur globules (S) are visible (Guerrero et al., 1999).
(e) Large sulfur globules (here in *Chromatium okenii*) are highly refractive and have a characteristic appearance in the phase contrast image (Pfennig and Trüper, 1992).
(f) *Pseudomonas* cells containing aggregates of a recombinant protein. The aggregates deform the normal cellular shape (Carri and Villaverde, 2001).
(g, h, i) Localization of the FtsZ protein in *Escherichia coli*: (a) bright-field image (phase contrast), (b) DAPI-stained fluorescence image (staining of DNA), (c) immunofluorescence of FtsZ. The ring-like structure is depicted in side-view, as a short line at the site of cell division (Den Blaauwen et al., 1999)

Localization and the overall structure of the nucleoid may be visualized by fluorescence microscopy of DAPI-stained cells. The presented figure (1.2 c) implies a coralline shaped structure of the nucleoid. Since the preparation procedure requires a chemical fixation, artificial alterations of the nucleoid may not be excluded, but it becomes clear that the nucleoid shape varies during the cell cycle. Again, the fluorescent signal correlates with the DNA content and allows rough quantification of DNA in individual cells.

Though the relevance of bacteria in biogeochemical cycles is rarely visible, intracellular deposition of elemental sulfur is an exception. The highly refractive granules are deposited intercellularly in chemolithoautotrophic bacteria capable of oxidation of reduced sulfur compounds. The presence of sulfur globules is a reliable indicator of this metabolic feature. Though small sulfur globules are sometimes difficult to discriminate from other inclusions (Fig. 1.2 d), large and numerous sulfur granules have a characteristic appearance under the light microscope (Fig. 1.2 e). Large protein inclusion bodies commonly occur in recombinant bacteria upon gene overexpression. The inclusions are sometimes big enough to influence the cell shape, as depicted in Fig. 1.2 f.

The cell cycle in eukaryotes may have been the first complex dynamic event to have been elucidated by light microscopy. In recent years, events in the bacterial cell cycle have also been documented. This application illustrates that macromolecular assemblies inside prokaryotic cells may very well be detectable under the (fluorescent) light microscope. In Fig. 1.2 i, the protein FtsZ – an important component of the division apparatus in prokaryotes – is located by immunofluorescence. The protein plays a key role during septum formation and forms a ring-like polymer at the site of cell division.

1.2.1.2 Eukaryotic cell structures at the light microscopic level

Fungal and plant cells

Since eukaryotic cells are normally larger than prokaryotic cells and compartmented by membraneous organelles, discrimination of their large intracellular structures is easier than in bacterial cells. Nevertheless, not all compartments are easy to detect in any cell.

Of all the types of eukaryotic cells, plant cells, due to their size and their rigid, highly refractive walls, are the most appropriate objects for visualization by light microscopy. In Fig. 1.3 a, meristematic cells from the root tip of *Vicia faba* provide an overview of the structures that may be differentiated with light microscopes. The cell wall, vacuoles, and the nucleus are the most conspicuous structures. It is difficult to differentiate smaller compartments: proplastids, possibly mitochondria, and cisternae of the endoplasmic reticulum (see also Fig. 1.3 b)

Because of their apparent simplicity, yeast cells are often used to serve as model eukaryotes. Cellular structures other than the wall, the vacuole, and the nucleus are very difficult to distinguish without application of fluorescent dyes (Fig 1.3 c). Many yeasts also produce capsular material, usually phosphomannans, β-linked mannans, heteropolysaccharides containing pentoses, glucuronic acid residues, and D-galactose. The cell wall itself consists of several layers that cannot be resolved by light

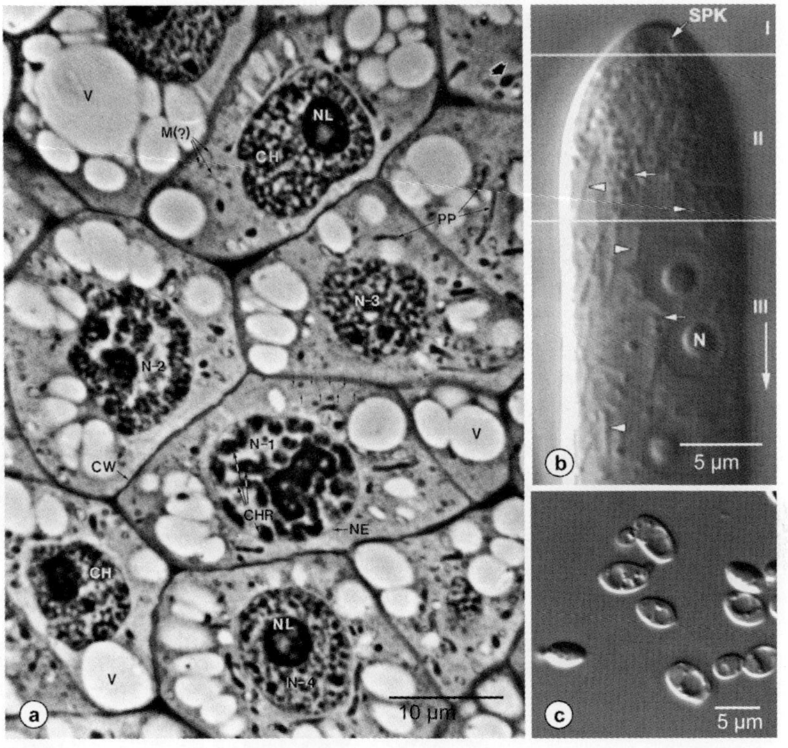

Fig. 1.3 Fungal and meristematic plant cells (overview):
(a) Thin section from the root-tip meristem of *Vicia faba*, showing the visible structures in plant cells (CW = cell wall, N1-N4 = nuclei, NL = nucleoli, NE = nucleus envelope, CH = chromatin, CHR = chromosomes, PP = proplastids, M? = mitochondria?, V = vacuole, small arrows depict thin membranous or filamentous structures at the resolution limit) (Gunning and Steer, 1996).
(b) Tip of an *Allomyces macrogyna* hypha. Region I of the hypha contains the Spitzenkörper (SPK), responsible for cell wall synthesis. Region II is characterized by numerous vesicles (arrows) and mitochondria (arrowhead). Region III contains nuclei (N), mitochondria and vesicles are less prominent (McDaniel and Robertson, 2000).
(c) Differential interference contrast image of yeast (*Saccharomyces*) cells. Large vacuoles fill 1/3 to 1/2 of the cell (the IN-VSEE Project, Arizona State University)

microscopy. Nevertheless, characteristic bud scars are visible, especially after fluorescence staining, as depicted in Fig. 1.4 a.

In most fully developed plant cells, the walls appear thick and multi-layered. Combinations of staining solutions also allow rough chemical differentiation of wall layers. Pectinous middle lamellae, cellulose walls with various degrees of lignification, may be discriminated (Fig. 1.4 c, d).

The large vacuoles appear bright and homogeneous. While vacuoles in meristematic plant cells are small and numerous (see Fig. 1.3 a), a single vacuole fills more than 50 % of the cellular volume in a mature cell. The tonoplast separates the vacuolar lumen from the cytoplasmic exterior. Vacuoles act as storage containers for carbohydrates, protein, lipids, and secondary metabolites, as well as compartments of osmoregula-

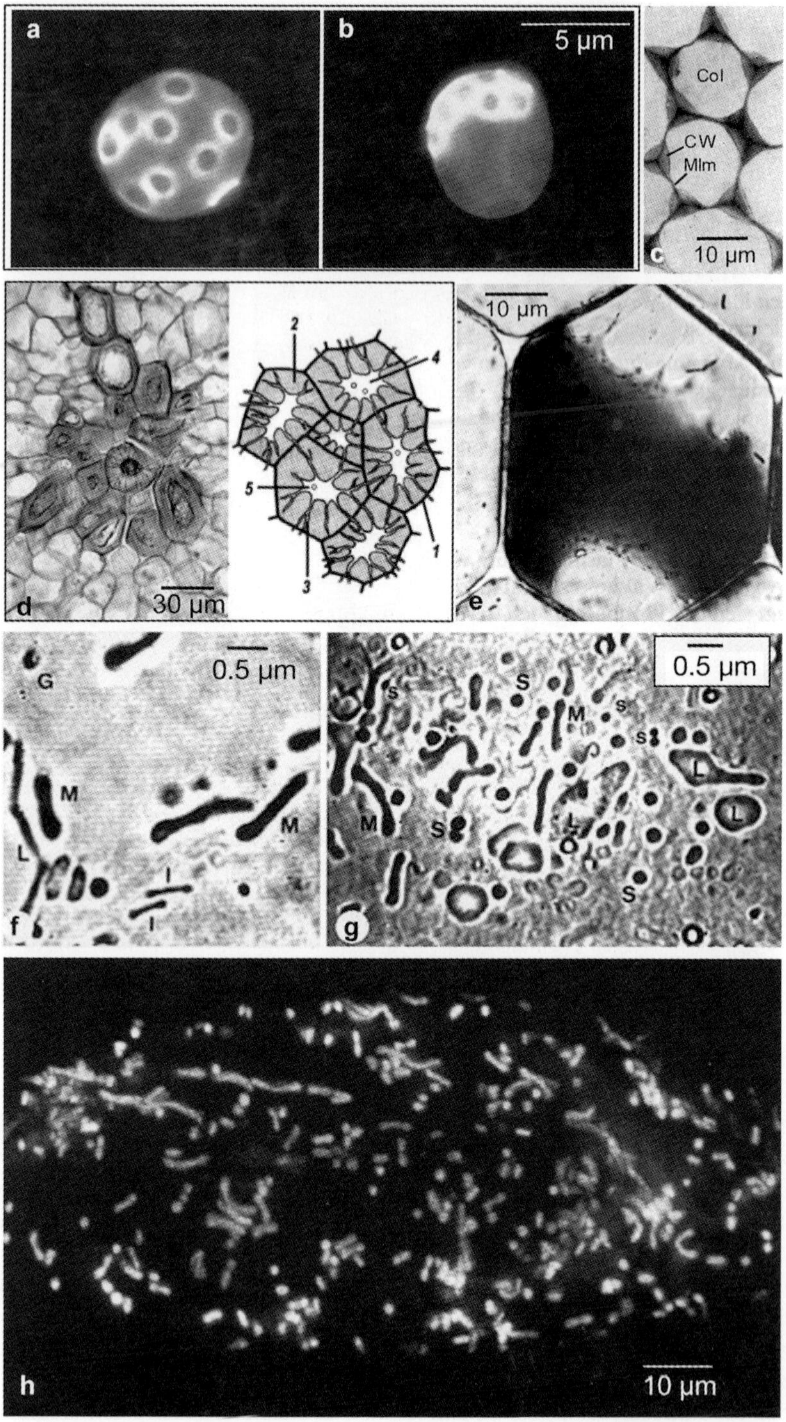
a
b
5 µm
Col
CW
Mlm
c
10 µm
2
4
5
1
3
d
30 µm
10 µm
e
G
0.5 µm
M
L
M
I
I
f
0.5 µm
S
M
S
S
L
L
M
S
L
S
g
h
10 µm

◀ **Fig. 1.4** Compartments in fungal and plant cells – cell walls, vacuole, mitochondria:
(a, b) Fluorescence images of yeast cells stained with Calcofluor White. Bud scars appear bright. Figure a depicts a mutant cell with random distribution of bud scars on the cell surface, (b) the wild-type cell (Chant et al., 1991).
(c) Angular collenchyma in *Cucurbita pepo*. The middle lamella appears as a dark line after staining with hemalaun (Mlm = middle lamella, Col = lumen of the cell, CW = cell wall).
(d) Isodiametric sclereids (stone cells) with very thick lignified cell walls, and numerous (branched) pits in peach (*Pyrus communis*), stained with phloroglucinol (stained red due to the presence of lignin). The sclereids are embedded in the fruit ground tissue (1 = primary wall, 2 = secondary wall, 3 = branched pit, 4 = cell lumen, 5 = cross-sectioned pit) (John Cheeseman, Univ. Illinois; diagram modified from Weier et al., 1974).
(e) Plasmolysis of an epidermal cell from *Rhoeo spathacea*. The vacuole, filled with colored sap, occupies nearly the total volume of the cytoplasm (from Nultsch and Grahle, 1983).
(f, g) Components of the cytoplasm in an epidermal cell of *Allium cepa*. The cells of the inner epidermis of the onion were examined with a UV microscope at a wavelength of 310 nm. The higher resolution of this microscope compared to conventional light microscopes allows a clearer image of small organelles (f1: M = mitochondrion, I = small, mitochondria-like inclusions, L = leucoplast, G = Golgi apparatus (?). f2: Legend as in f1, S = Large sphaerosomes, ss = small sphaerosomes (membrane-surrounded vesicles) and particles with unknown function (from Gunning and Steer, 1996).
(h) *Nicotiana* cell with mitochondria, viewed by fluorescence microscopy after staining with Rhodamine-123 (from Gunning and Steer, 1996)

tion and intracellular protein degradation. Vacuoles sometimes contain various visible inclusions such as polymetaphosphate (volutin) granules. In higher plants, the vacuoles are often filled with water-soluble pigments (Fig. 1.4 e).

Although mitochondria are visible as faint structures under the standard light microscope, high resolution or staining with fluorescence markers provide clearer images of mitochondrial shape and distribution. Large mitochondria of the *Allium cepa* epidermis are visualized by use of short-wavelength ultraviolet light. (Fig. 1.4 f, g). Specific staining is also possible with fluorescent agents. The vital stain Rhodamine-123 is predominantly accumulated in mitochondria. The fluorescent image (Fig. 1.4 h) illustrates the shape and distribution of the organelles in cultured *Nicotiana* (tobacco) cells.

Plastids, especially pigmented plastids (chloro- and chromoplasts), are the most characteristic compartments of plant cells. Under appropriate conditions, thylakoidal grana structures (i.e., thylakoid membrane stacks) are visible as small disks inside chloroplasts (Fig. 1.5 a). Autofluorescence of the photosynthetic pigments allows detection of plastids by fluorescence microscopy without application of any additional stain (Fig. 1.5 b). Like mitochondria, plastids contain DNA that may be visualized under the fluorescence microscope by staining with DAPI (4',6-diaminidino-2-phenylindole) (Fig. 1.5 c). The differentiation of plastids may not only lead to photosynthetically active chloroplasts. Chromoplasts contain carotenoids and xanthophylls, responsible for some orange-reddish fruit colors. Mature chromoplasts are regularly spindle-shaped (Fig. 1.5 d). The loss of pigments (by mutation, for example) in certain tissue types may lead to colorless plastids (leucoplasts). Sometimes, plastids with storage function are also termed leucoplasts. Among these, the starch-containing amyloplasts are most important and are widespread in plant storage tissue. The development of starch grains in plastids is especially easy to visualize after iodine (I/KI) staining of

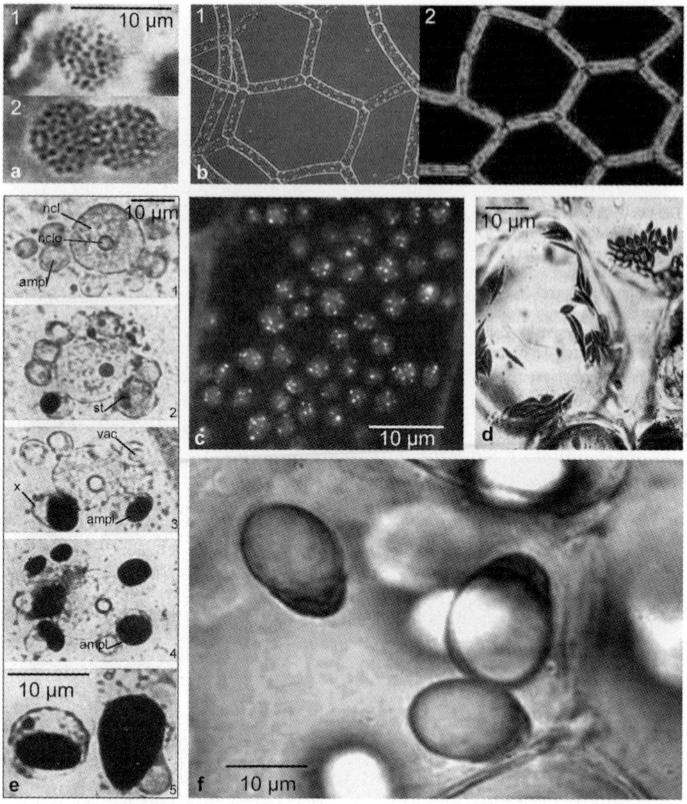

Fig. 1.5 Compartments in plant cells – plastids:
(a) The lens-shaped chloroplasts in cells from *Spinacia oleracea* are depicted in top view. Grana stacks of the thylakoid membrane are visible as small, dark, disk-shaped dots in the lumen of the chloroplast. The bright stroma is located among the grana. 1 = single chloroplast; 2 = dividing chloroplast (from Braune et al., 1983)
(b) *Hydrodictyon reticulatum*. Bright-field (1) and fluorescence (2) images. Use of chlorophyll autofluorescence is a simple method for selective imaging of chloroplasts in plant cells (Peter V. Sengbusch – b-online).
(c) Fluorescence microscopic image of chloroplasts of *Elodea*. Plastid DNA (nucleoids) are stained with the DNA-specific fluorescence marker *DAPI (4',6-diaminidino-2-phenylindole)*. Each fluorescent dot is equivalent to one nucleoid (from Kleinig and Meier, 1999).
(d) Spindle-shaped chromoplasts in the mesocarp cells of *Rosa rugosa* (from Nultsch and Grahle, 1983).
(e) *Solanum tuberosum*. Development of starch granules in the amyloplasts (1–4). Starch appears dark after treatment of the sample with iodine solution. Nucleus (ncl), nucleolus (nclo), amyloplast (ampl), starch (st), vacuole (vac), (x) shows an amyloplast with the stroma displaced by a growing starch granule. 5: Two amyloplasts with starch granules. Magnification in 2–4 is the same as in 1 (from Braune et al., 1983).
(f) Massive starch accumulation in the chloroplasts of *Pellionia* (Urticaceae). The crescent-shaped "cap" is still green and photosynthetically active. Though small starch granules regularly occur in chloroplasts, voluminous granules in photosynthetically active plastids are exceptional (Peter V. Sengbusch – b-online)

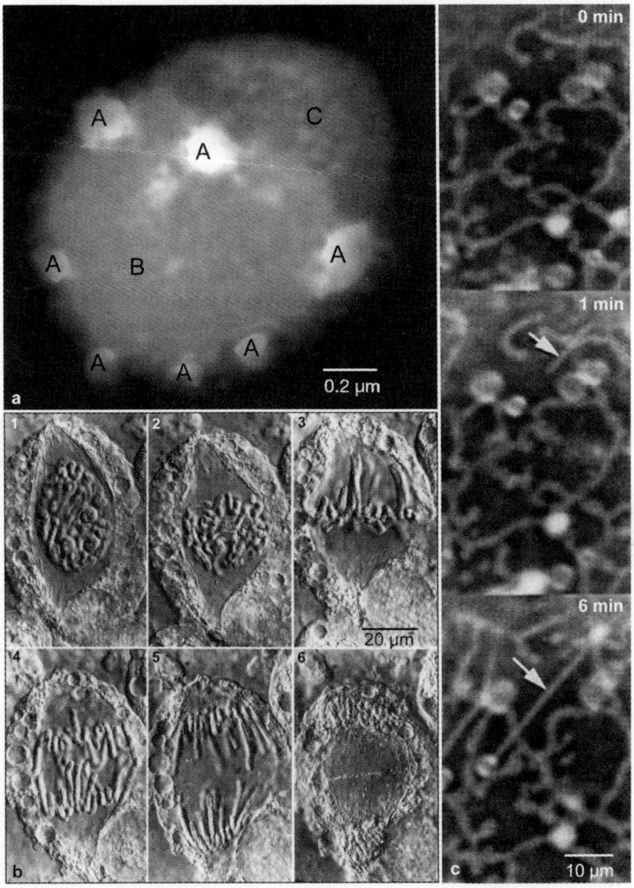

Fig. 1.6 Cellular compartments: nucleus and endoplasmic reticulum.
(a) Organization of sub-compartments of the yeast nucleus as visualized by indirect immunofluorescence (from Wente et al., 1997).
Three functional "compartments" have been differentially marked within the nucleus. One contains the majority of the chromatin (B), stained red with ethidium bromide; the nucleolus (C) is stained blue with a nucleolus-specific antibody and the fluorescent marker CY5; telomers (A), are stained green with another specific antibody (FITC as fluorescent marker).
(b) Mitosis phases in an endosperm cell of *Haemanthus katharinae*. Differential interference contrast (from Kleinig and Mayer, 1999).
1. Prophase: Polar caps and nucleus envelope are clearly visible; nucleoli and condensed chromatin are in the nucleus.
2. Prometaphase: Nucleus envelope is fragmented, centromers move to the equatorial plain.
3. Metaphase: Spindle apparatus is completed, chromosomes form the metaphase plate.
4. Anaphase: Kinetochores of the chromatides move polewards.
5. Late anaphase.
6. Telophase: forming of the daughter nuclei, beginning of cell division.
(c) Confocal scanning laser microscopy images of details in the peripheral area of a living tobacco BY-2 cell. Endoplasmic reticulum and other membrane systems were stained with rhodamine-123. The images were taken after 1 min and 6 min after the initial image (0 min). Arrows indicate a changing structure (from Gunning and Steer, 1996)

starch, as depicted in Fig. 1.5 e. Finally, the starch grain fills most of the volume of the mature amyloplast. Photosynthetically active chloroplasts also regularly contain starch. Exceptionally, large starch grains also occur in this plastid type (Fig. 1.5 f).

The large nucleus and (condensed) chromatin are easily detectable in most eukaryotes. Most fundamental discoveries of structural features during the cell cycle have been achieved by (bright-field) light microscopy. Though classical staining methods for bright-field microscopy (see Fig. 1.3 a) are fully sufficient to resolve the nuclear structure (nucleolus, stages of chromatin condensation), the use of fluorescent probes reveals additional features. The organization of an inter-phase yeast nucleus is depicted in Fig. 1.6 a. Fluorescent dyes specifically stain the chromatin (the majority of the volume) and the nucleolus, where ribosomal DNA is transcribed and ribosomes are assembled, as well as the repressor activator protein 1, an abundant nuclear protein that binds to the repetitive sequences found at all yeast telomers. Bright spots mark the positions of the telomeric sequences. Inside an inter-phase nucleus, nucleoli are visible as more highly condensed entities. The use of interference contrast and differential interference contrast microscopes allows the condensed chromatin to be visualized without additional staining. Thus, some dynamic events (such as chromosome segregation) may be directly observed without damaging the sample (Fig. 1.6 b).

The Golgi apparatus, the endoplasmic reticulum, and cytoskeletal elements are also visible by light microscopy when fluorescent stains are applied. The fine structures of these compartments remain unresolved, but the dynamics of these compartments (membrane flow, transport) are visible. Vital staining of the endoplasmic reticulum in the cell periphery of cultured plant cells (the cortical endoplasmic reticulum) reveals the extensive movement in a network of fine vesicles and tubules (Fig. 1.6 c).

Localization of Golgi stacks and dynamics of the Golgi apparatus are depicted in Fig. 1.7. A suitable marker system for this compartment is an antibody specific for a carbohydrate residue of a glycoprotein family predominantly localized in the Golgi apparatus of plants. The fluorescently labeled antibody reveals the position of Golgi stacks in intact cells. In cells treated with the fungal metabolite brefeldin A, the Golgi

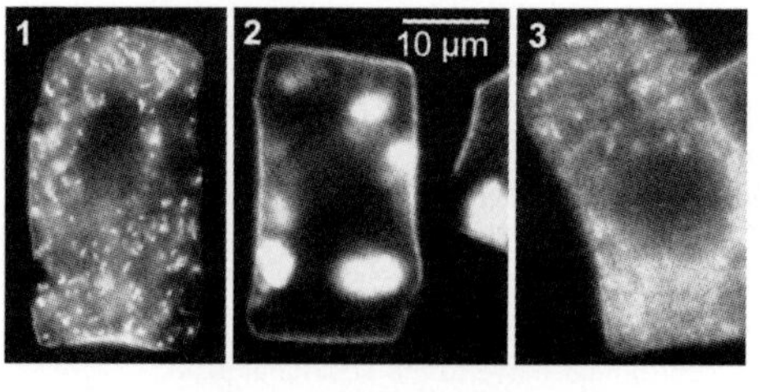

Fig. 1.7 Cellular compartments: Golgi apparatus.
Fluorescence microscopic images of osmotically intact cells from the root tip of *Zea mays*. The Golgi apparatus is labeled by immunofluorescence (from Gunning and Steer, 1996).
1. Normal state. The dark area in the center of the cell is the unlabelled nucleus.
2. Cell treated with the antibody brefeldin A, which inhibits a variety of membrane transformations. Cisterna stacks are dispersed because the transport from the ER to the Golgi apparatus is interrupted. Golgi membranes form big clusters.
3. Cell treated with brefeldin A as in 2, after 2 h recovery

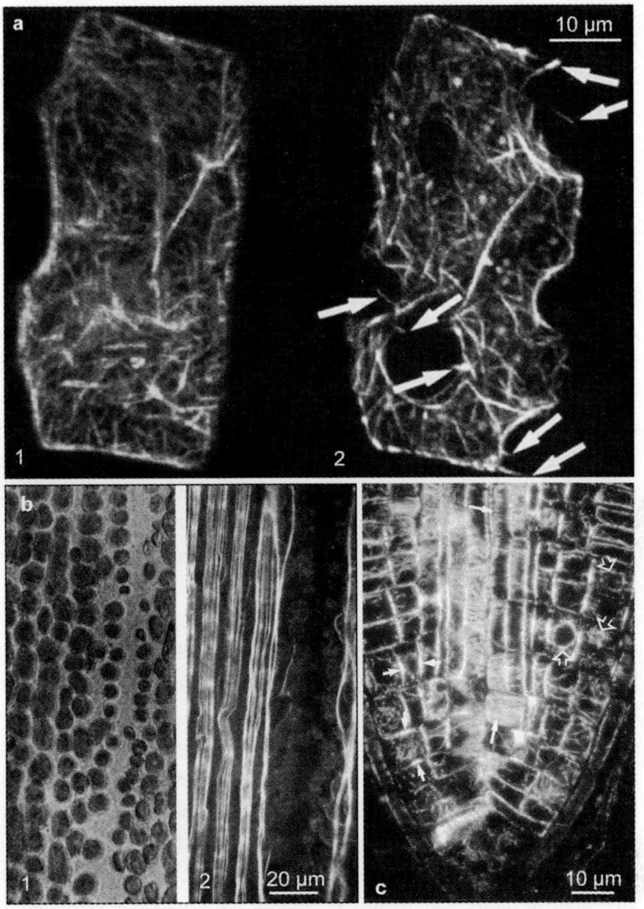

Fig. 1.8 Cytoskeletal elements in plant cells:
(a) Phalloidin, marked with rhodamine, was injected into this epidermal cell of *Tradescantia* for detection of actin filaments. The actin cytoskeleton is visible. After gentle plasmolysis the cytoplasmic membrane is detached from the cell wall. Arrows indicate actin filaments that extend from the protoplast to the cell wall (from Gunning and Steer, 1996).
(b) Detail of an internodial cell of *Chara* in bright-field (left) and fluorescence (right) image. Actin filaments associated with the rows of chloroplasts are labeled with a fluorescence-marked anti-actin-antibody. The actin-free zone separates the two oppositely directed cytoplasma streams (from Kleinig and Meier, 1999)
(c) Median longitudinal section from the apical root meristem of *Azolla*. Microtubules were visualized by immunofluorescence microscopy. In elongating cells, cortical microtubules perpendicular to the long axis of dividing cells can be seen marking the division plane (long arrows). Short arrows indicate preprophase ribbons, which appear in cells shortly before division. Some cells are forming mitosis spindles (wide arrows) (from Gunning and Steer, 1996)

apparatus becomes disassembled. After termination of brefeldin treatment, the stacks become redistributed.

Although plant cells are stabilized by rigid cell walls, they have complex cytoskeletons, especially involved in intracellular transport of subcellular compartments, chromosome segregation, and cell wall formation. Staining with phalloidin-rhodamine, a fluorescent complex that binds to actin, reveals the presence of the actin cytoskeletal network. Figure 1.8 a shows the presence of actin in an epidermis cell of *Tradescantia* before and after plasmolysis. In plasmolyzed cells, some thin filaments clearly protrude from the cytoplasmic membrane to the cell wall. Actin filaments are responsible for the transport of chloroplasts. The large internodial cells of the green alga *Chara* contains streaming rows of chloroplasts, obviously attached to actin filaments. The filaments are localized with anti-actin antibodies, coupled to actin, and visualized by fluorescence microscopy. Microtubules indirectly determine the cellular size by determination of the sites of synthesis of cellulose fibrils. The microtubular network is depicted in a root tip of the aquatic fern *Azolla*, stained with fluorescently labeled anti-tubulin antibodies. The distribution of microtubules in cells depends on the stage of the cell cycle (Fig. 1.8 c).

Animal cells

Animal cells exhibit the same degree of complexity as cells from higher plants. Numerous staining techniques for cytological differentiation in biomedical research have been developed for staining of compartments. Most of these techniques are only relevant for medical diagnostics, and so are not presented in detail here. Again, fluorescent techniques reveal localization and dynamics of cytoskeleton, mitochondria, ER, and Golgi apparatus. At the light microscopic level, the most characteristic features of animal cells are the lack of a thick, possibly multi-layered, wall and the absence of plastids and large vacuoles. The outermost layer of a typical animal cell exposed to the environment is formed by the glycocalyx, oligosaccharide chains that usually contain sialinic acid and produce a negative surface charge. Extracellular glycoproteins and polysaccharides have been found to be associated with the oligosaccharides.

The role of the intracellular stabilizing structure is therefore more pronounced. In the absence of an outer wall, the cellular shape is directly determined by the cytoskeletal elements. These are detectable by use of stains or specific antibodies. Although the structures of all single cytoskeletal fibers are below the resolution limit of light microscopes, the dynamics of these systems have been extensively investigated by (time-resolved) microscopic fluorescence techniques. Thick fiber bundles are also visible in phase or interference contrast. Besides their involvement in intracellular transport, the cytoskeletal elements also play key roles in cell movement.

Intermediary filaments, which are absent in plant cells, are involved in maintenance of the overall cellular shape. Localization of these filaments is again routinely performed with the aid of specific antibodies, as shown for bundles of cytokeratin and vimetin filaments in cultured animal cells (Fig. 1.9 a, b).

Actin filaments (F-actin) are composed of actin monomers, representing the most abundant cellular protein. The filaments are responsible for cellular movements, including gliding, contraction, and cytokinesis. The association of actin filaments with the protein myosin is responsible for muscle contraction. Phase contrast images of certain cell types allow visualization of these elements. The distribution of "stress fibers", as shown in Fig. 1.9 c, correlates with the immunofluorescent image of actin, as depicted in Fig. 1.9 d. The fibers are composed of numerous actin filaments.

Microtubules also act as scaffolds to determine cell shape, and provide "tracks" for cell organelles and vesicles to move on. Microtubules also form the spindle fibers for separating chromosomes during cytokinesis. Localization of actin, myosin, and tubulin (as microtubule monomer) is depicted in Fig. 1.10 a. Though in principle quite similar, cytokinesis in animal cells differs from that in plant cells in some aspects.

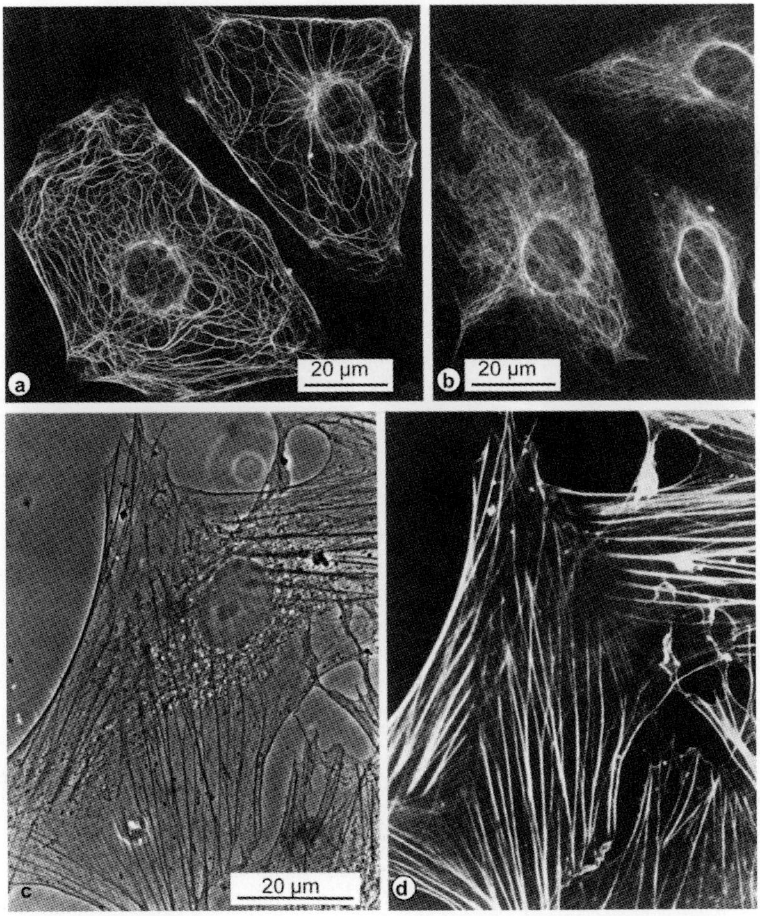

Fig. 1.9 Specialized features in animal cells: cytoskeletal elements.
(a, b) Intermediate filaments of kidney cells from the rat kangaroo visualized by indirect immunofluorescence (a, cytokeratin filaments; b, vimentin filaments) (from Kleinig and Meier, 1999).
(c, d) Actin filaments (stress fibers) in rat cells. The same detail seen in phase contrast (b) and by indirect immunofluorescence (c) (from Kleinig and Meier, 1999)

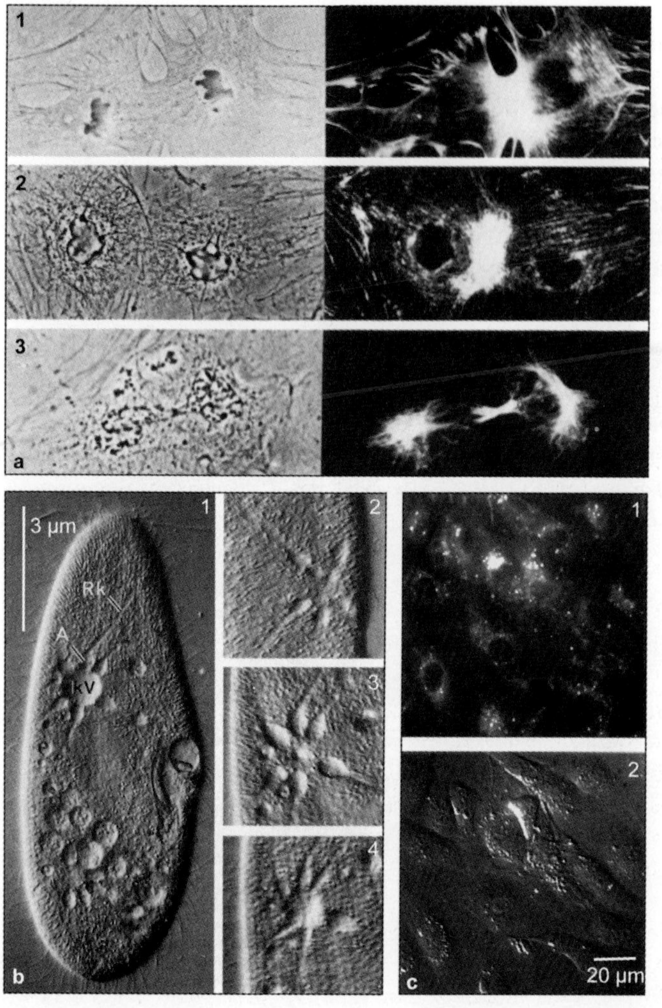

Fig. 1.10 Specialized features of animal cells: cytokinesis and contractile vacuoles in ciliates:

(a) Cytokinesis of animal tissue culture cells (PtK2-cells). Detection of actin (1), myosin (2), and tubulin (3) at the plane of division by indirect immunofluorescence (right; phase contrast images are shown at the left half of the figure) (from Kleinig and Sitte, 1986).

(b) Contractile vacuole of the ciliate *Paramecium*. 1: Cell with contractile vacuole (kV), ampullae (A), and radial canals (Rk). 2–4: Stages of the filling of the vacuole. Interference contrast (from Kleinig and Meier, 1999).

(c) Punctate fluorescence of a marker after endocytosis by cultured cells. (1) Cells observed in epifluorescence. (2) Same field of cells as in A, observed in interference contrast. Dye is distributed in a punctate pattern throughout the cytosol (from Niles and Malik, 1999)

Centrioles, organizing structures for microtubuli, are present only in animal cells and lower plants. This feature cannot be visualized directly by light microscopy but gives rise to some differences in the appearance of cytokinetic events in animal and plant cells. In cilia and flagella, microtubuli are the key components for the tubulin-dynein molecular motor system.

Animal cells do not contain large vacuoles. The compartments responsible for degradation of cellular compounds are lysosomes, usually much smaller than plant vacuoles. Of highest complexity are contractile vacuoles from several protists. The vacuoles have an osmoregulatory function: they export water to the cellular exterior. A system consisting of ampoules and radiary channels transfers water to the vacuole, from which it is transported to the cellular exterior by fusion of the vacuole with the cytoplasmic membrane. The sequence is depicted in Fig. 1.10b 2–4. The contractile vacuole may be viewed as a modified Golgi apparatus. Other exocytotic and endocytotic processes are also more abundant in animal cells than in plant cells. Specially adapted fluorescent techniques allow resolution of endo- and exocytotic processes at cell surfaces. Incorporated fluorescent particles are visible under fluorescence microscopes (Fig. 1.10c).

Plant tissue

Most routine applications for light microscopic methods still involve visualization of tissue structures and detection of targets in tissue. Because of the presence of rigid cell walls and – in comparison with tissue of higher animals – a low degree of organ complexity, plant tissue structures are easy to differentiate by bright-field microscopy.

Besides meristems (tissue characterized by cell division, an exceptional feature in permanent tissues), five types of tissues ("permanent tissue") are distinguished: dermal tissues (epidermis, cork, bark), ground tissue (parenchyma), secretion tissue, collenchyma/sclerenchyma tissues (collenchyma, sclereids, fibers), and vascular tissues (vessel elements, xylem, phloem).

Epidermal cells are tightly packed and have no intercellular spaces, thus providing a means to minimize loss of water and the invasion of biotic or abiotic agents from outside the plant (Fig. 1.11 a). The epidermis consists of ground epidermal cells (or pavement cells), trichomes (unicellular or multicellular plant "hairs", Fig. 1.11 b), and stomata (Fig 1.11 c). The outer cell walls of epidermal cells in particular may be very thick and may occupy a large proportion (20–30 %) of a cell volume. Evaporation of water is minimized by a cuticule (consisting of cutin), as depicted in Fig. 1.11 a 2. Cork and bark represent other types of dermal tissue. They are thick, rigid, and multi-layered. Cork derives from subepidermal meristem. Multiple layers of cork, sequentially deriving from new meristems in woody plants, build up bark.

Parenchyma represents the bulk mass of tissue in the plant, it embeds vascular tissue, and is surrounded by dermal tissue. Cells have a low degree of differentiation, usually thin primary walls, but may also show a certain degree of lignification (Fig. 1.12 a). Cells of the parenchyma vary in structure (isodiametric, prosenchymatic, star-like) and function (assimilation, storage of assimilates and water, aerenchyma).

Collenchyma is a living tissue composed of elongated cells with thick, non-lignified primary walls. Such cells are most closely aligned physiologically with parenchyma

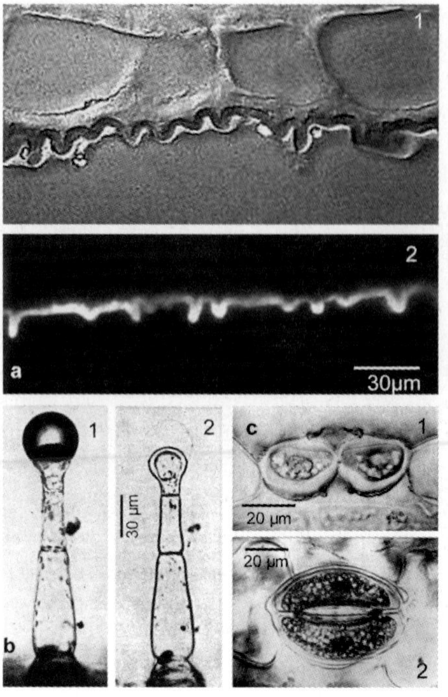

Fig. 1.11 Plant epidermis:
(a) 1: Cross-section of the East African Apocynaceae *Tabernaemontana pachysiphon* show that the cuticle of the lower leaf surface covers the foldings of the epidermal cell walls like a tablecloth. Section stained with the lipophilic marker Sudan IV (Peter v. Sengbusch, b-online, Univ. Hamburg). 2: The fluorescent dye Acridine Orange exclusively stains the cuticule (Peter v. Sengbusch, b-online, Univ. Hamburg).
(b) Glandular trichomes with unicellular head and unbranched stalk. Capitate trichomes (glands at the "head" of the trichome) from *Cucurbita pepo*. 1: a native hair with a drop of excrete on its top; 2: the same hair after ethanol treatment; the drop has been dissolved. A thin cuticular layer on the top of the hair is visible.
(c) Stomata from a *Lilium candidum* leaf in side view (1) and top view (2). Stomata are responsible for gas transfer and typical derivative of the leaf epidermis. They are in most cases located at the lower leaf surface

cells. Primary wall thickenings are deposited unevenly, and in a variety of patterns (Fig. 1.4 c, 1.12 b).

Various types of secretion tissue are important for storage, secretion, and excretion of plant products (Fig. 1.12 c). Secretory products may be released from the cell into canals (as in resin ducts), subcuticular cavities (at the tips of glandular trichomes, Fig. 1.12 c1), or onto the surface of the organ (as in nectaries, hydathodes, and salt glands). They may also accumulate in a vacuole of a cell in the form of crystals (as in crystal-containing idioblasts; Fig. 1.12 c2) or amorphous inclusions (as in tannin and oil cells). A special type of internal secretory system is composed of laticifers that produce, for example, latex.

All higher plants possess vascular tissue for transport of water from the root into the stem and leaves (xylem) and for the flux of assimilates down to all plant organs (phloem). Xylem and phloem become differentiated from a specialized meristem, the procambium. In most cases vascular bundles in stems and roots of higher plants are collateral: their xylem and phloem are arranged parallel to each other. In monocotyledons (Fig. 1.13 a), all cells of the procambium differentiate into vascular tissues, whereas in gymnosperms and dicotyledons a layer of dividing cells, the vascular cambium, remains between xylem and phloem. This produces secondary vascular tissues that contribute to the growth of plant organ in thickness (secondary growth, Fig. 1.13 b). Water-conducting cells of both primary and secondary xylem are tracheary elements. They are of two types – tracheids and vessel members – both of which are dead at maturity and have lignified cell walls of various sculptures. Tracheids are long, narrow cells extending in the direction of water movement.

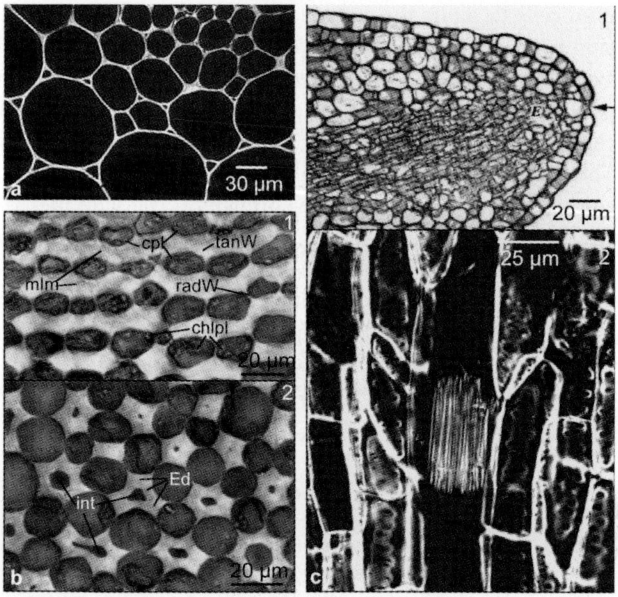

Fig. 1.12 Plant tissue: ground tissue, collenchyma, and simple secretory structures:
(a) Parenchymatic cells and relatively large intercellular spaces in a cross-section through the stem of *Geum urbanum*. Staining with coriphosphin (Peter v. Sengbusch, b-online, Univ. Hamburg).
(b) Collenchyma is a typical supporting tissue outside vascular bundles in herbaceous plants. 1) *Sambucus nigra*. Lamellar collenchyma (tanW = thickened tangential walls, radW = thin radial walls, mlm = middle lamella, chlpl = chloroplasts, cpl = cytoplasm. 2) *Petasites hybridis*. Lacunar collenchyma. The edges of the cells (ed) are thickened by wall material and in most cases enclose a large intercellular space (int) (from Braune et al., 1983)
(c) Secretory structures. 1) Hydathodes secrete (exude) water, which appears as droplets on the leaf surface (guttation). Two types of hydathodes are recognized: active (usually in the form of glandular trichomes; see fig. 1.11 c), which are not connected to tracheary elements, and passive (usually located at leaf margins or tips) as shown here for cabbage (*Brassica oleracea*). Here, water is released from the xylem, and then filters through intercellular spaces of the epithem (E), formed by modified mesophyll cells. At the tip of the hydathode region, a stoma (arrow) is seen. This is permanently open and allows the flow of water to the surface of the leaf. 2) Single secretory cell (idioblast) from the stem of *Dracaena* sp. The cell contains long needles (raphides), which lie in parallel bundles. Such crystals show birefringence with polarized light (John Cheeseman, Univ. Illinois, Urbana)

Together with xylem, the second complex conducting tissue of vascular plants is the phloem. Its main function is the long-distance transport of sugars and other photosynthates from the source (mature leaves), or reserves (germinating seedlings) towards the sinks, such as roots, developing reproductive structures (flowers, fruits, and seeds), meristems, and young leaves. Like xylem, phloem consists of several types of cells, including conducting and parenchyma cells, phloem (or bast) fibers (Fig. 1.13 b), sclereids, and, in some plants, secretory ducts and laticifers.

Gymnosperms have sieve cells as their sieve elements, while sieve tubes are characteristic of flowering plants. Each sieve tube element is accompanied by one or more, usually dense, cytoplasmic cells known as companion cells (Fig. 1.13 c). The companion cells lie along the sieve element and are in fact the product of longitudinal division from a common mother cell. Companion cells play an important part in loading sieve tubes with photosynthates in source areas and unloading in sink areas. They also provide ribosome-less sieve elements with essential proteins and ATP. Companion cells and sieve elements together constitute a functional unit in food conductance. They possess specific intercellular communications in their common walls, in which several plasmodesmata on the companion cell side merge at the site of the middle lamella into the sieve pore on the sieve side. In gymnosperms, typical companion cells are absent in the phloem (Fig. 1.13 d).

Root tissue structures exhibit some considerable differences: assimilating parenchyma is usually missing. The root vascular tissue is located in the center, surrounded by a rhizoderm or exodermis, a parencymatic root cortex, an endodermis, and the meristematic pericycle (Fig. 1.13 e).

Fig. 1.13 Plant tissue: vascular tissue. ▶

(a) Semi-thin cross-section of a vascular bundle from *Zea mays*, stained with coriphosphin. This "closed" bundle is typical for monocotyledons: The mature bundle does not contain meristematic cells (cambium; i.e., secondary growth is impossible). Phloem upper half of the bundle and xylem are located side by side, which is typical for stems of higher plants. The phloem contains larger sieve elements and smaller companion cells large xylem elements in the lower half of the bundle. Lignified cells surrounding the bundle appear darker than the cells of the bundle and the parenchyma, due to the dark dye of these cells (John Cheeseman, Univ. Illinois, Urbana). The vascular bundle measures about 250 nm in diameter.
(b) A semi-thin cross-section from a stem of *Tilia cordata*. Detail of the secondary phloem (secP, the cambium (Cb) and secondary xylem. Besides sieve elements and companion cells, the phloem contains bast fibers with thick cell walls. The secondary xylem (secX) consists of various vessel elements and parenchymatic cells (Braune et al., 1983). Permanently dividing cambium cells allow secondary growth: formation of multiple radial layers of xylem and phloem (from Braune et al., 1983). The image shows a 230×150 μm section.
(c) *Cucurbita pepo*. Typical appearance of sieve tubes and companion cells in cross-section. Each large sieve tube (st) is accompanied by a smaller, dark-stained companion cell (cc; pP = phloem parenchyme, cb = cambium) (from Braune et al., 1983). The image shows a 250×300 μm section.
(d) A semi-thin cross-section of a pine (*Pinus strobus*) stem, showing the region around the cambium. Since several cell types (companion cells in the phloem, some types of vessel elements in the xylem) are absent, the tissue pattern appears quite uniform. The dark stained cells at the bottom represent secondary xylem and, as is typical of gymnosperms, are primarily composed of tracheids (John Cheeseman, Univ. Illinois, Urbana). The vascular cambium (with thin radial walls of cambial cells) is the second or third layer of cells outside of the secondary xylem. Mature secondary phloem cells can be seen towards the top and are recognized by their dark deposits representing callose (stained blue due to the Aniline Blue staining) (John Cheeseman, Univ. Illinois, Urbana). The image shows a 200×250 μm section.
(e) Semi-thin cross-section of a *Ranunculus* sp. root, showing a fully differentiated vascular cylinder. As is typical of roots, both the xylem (X) and the phloem (P) follow a centripetal order of differentiation (the oldest elements are farthest from the center of the vascular cylinder). The vascular cambium (Cb) is located between the xylem and the phloem. Note complete secondary cell walls (sW, stained due to their contents of lignin and suberin) in the endodermis cells lying opposite the phloem, and thin cellulosic endodermal walls (eW), readily permeable for water and solutes, located opposite water-conducting xylem "arms". The thin-walled endodermal cells are often designated passage cells since they are more likely to transport materials through pits via xylem arms than the thick-walled endodermis cells. Also seen are the one-layered pericycle (pc) (situated inward and adjacent to the endodermis) and cortical cells with amyloplasts (stained dark due to the presence of starch) (John Cheeseman, Univ. Illinois, Urbana). The image shows a 400×600 μm section

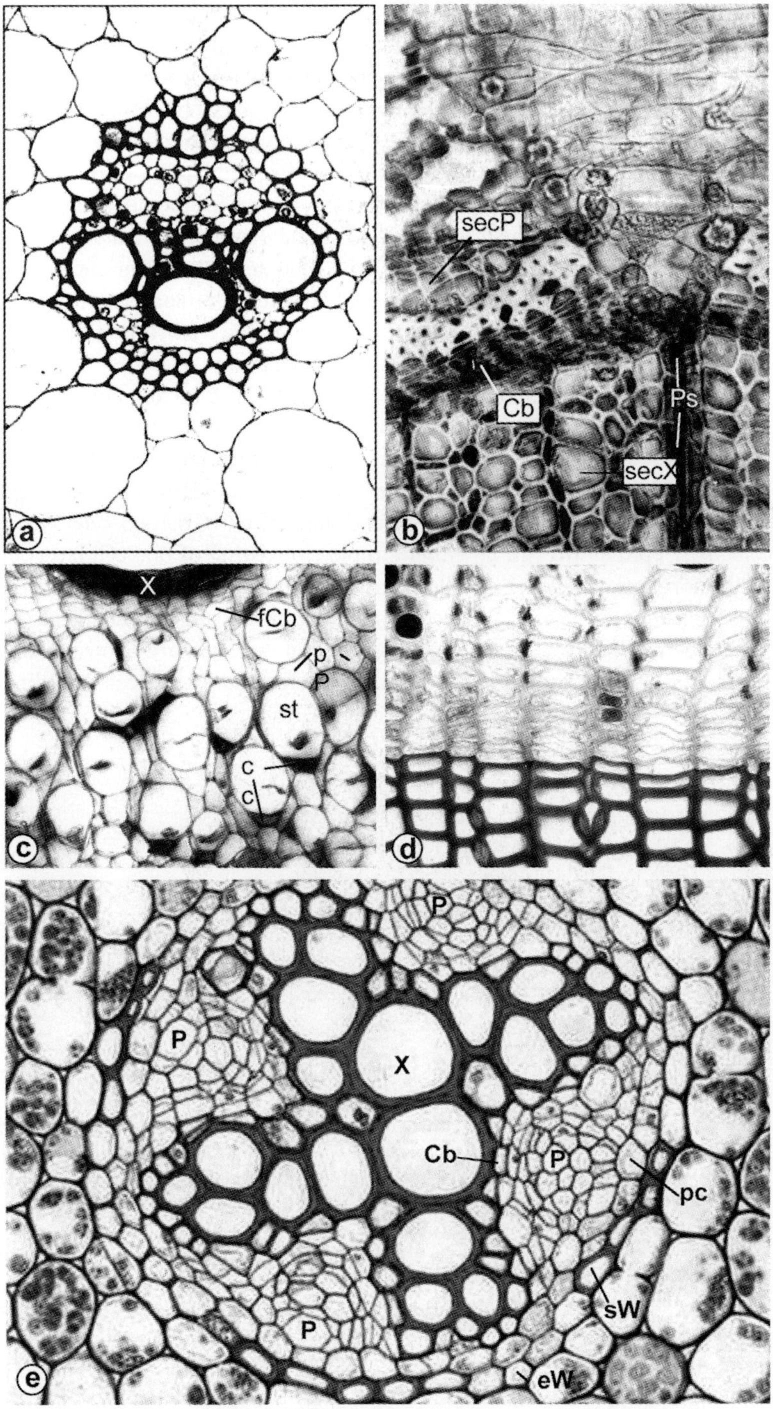
a
secP
Cb
Ps
secX
b
X
fCb
p
P
st
c
c
c
d
P
P
X
Cb
P
pc
sW
P
eW
e

Animal tissue

Because of their high complexity, identification and differentiation of cell tissues in various animal organs needs a high level of experience. About two hundred cell types are grouped into functional types (gametes, epithelial cells, sensory cells, nervous cells, muscle cells, blood cells, cells forming extracellular matrix). The cells are usually organized as tissue, forming numerous complex organs. Most cytological preparation techniques and stains have been specially developed for the differentiation of animal (especially human) tissue structures for medical diagnostics. Some are also of special relevance for (non-medical) biotechnology. In spite of the experience needed for histological differentiation, the characteristic structures of some tissues allow a rough orientation over a specimen, as in the orientation of a thin section through an arthropod gut depicted in Fig. 1.14.

Epithelial tissue lines body cavities (see Fig. 1.15). Functions of epithelial cells include transport of materials into, out of, or around the body, protection of the internal environment against the external environment, and secretion of products. The shapes of epithelial cells allows differentiation into columnar (elongated, see Fig. 1.15 a), cuboidal (cube-shaped, see Fig. 1.15 b) and squamous (flattened cells, see Fig. 1.15 c) cells. Any epithelium can be either simple (consisting of one cell layer) or stratified (with more than one layer of cells). Simple epithelia are in general adapted to absorption and secretion, while stratified epithelia protect against mechanical stress or chemical agents. In Fig. 1.15, only simple types of epithelia are shown. Glands may be formed by a single layer of epithelial cells, such as the goblet cells that line the intes-

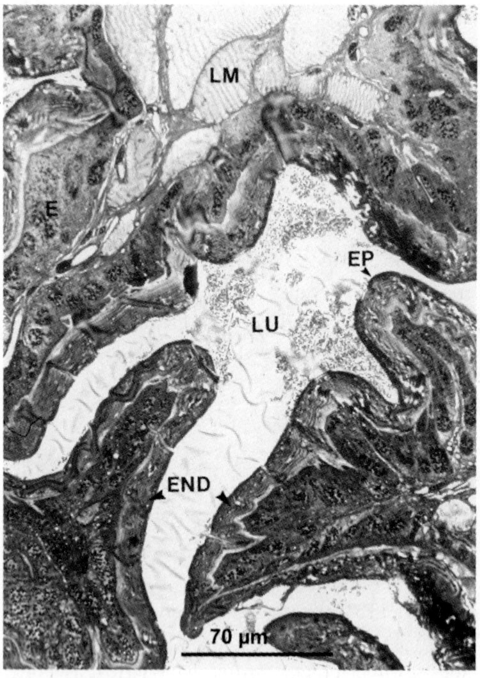

Fig. 1.14 A complex specimen with various animal tissue types.
Though some structures appear to be unique for the specimen, other tissue types may be discriminated by their location and appearance. Semi-thin section through the colon (part of the hindgut) of a locust. A layer of epithelium cells, covered by non-cellular epi- (EP) and endocuticula (END) enclose the lumen (LU) of the colon. Muscle tissue (LM) is visible in the upper part of the image (Ursel Fürstenau, Univ. Göttingen)

Fig. 1.15 Simple epithelia:
(a) Section of the secretory epithelium in the human small intestine (jejunum). The epithelium is formed by one layer of elongate, cylindrical cells (Gwen V. Childs, Univ. Texas, Medical Branch). The image shows a 130×200 μm section.
(b) Epithelia of human kidney tubuli. The sectioned cells appear flattened or as small squares (cuboidal epitelium) (Gwen V. Childs, Univ. Texas, Medical Branch). The lumen of a tubulus measures approximately 50 μm in diameter.
(c) Detail of the human kidney. G, Glomerulus; C, Capsula glomeruli. The capsula is formed by extremely flattened cells, appearing as a thick line. Even the nuclei are difficult to detect (squamous epithelium) (Milton Wolf and Mark Scarbrough, Univ. Kansas). The glomerulus capsule measures about 500 μm in diameter

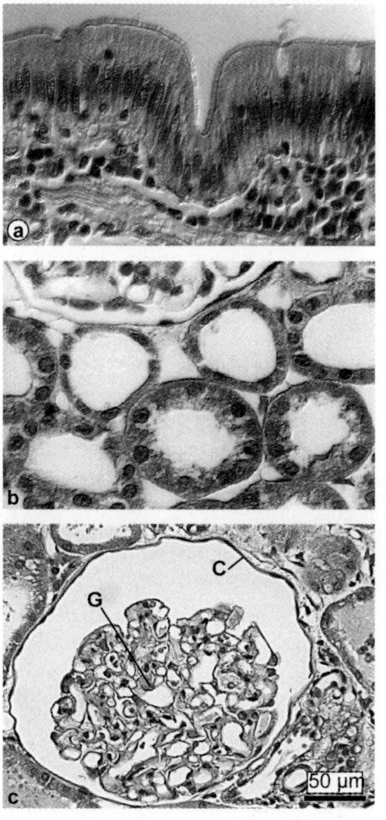

tine (see Fig. 1.15 a). Under a light microscope, no intercellular space is visible. The tight packing of cells produces the appearance of a layer rich in nuclei, clearly separated from other tissue.

The cells in connective tissue are separated from one another by a non-cellular matrix. The matrix may be solid (as in bone), soft (as in loose connective tissue), or liquid (as in blood). The occurrence of fibrous material, embedded in the intracellular matrix, is common. It is possible to differentiate according to the density of the fibrous matrix, into loose connective tissue (LCT, loose, non-ordered network of fibers), and fibrous connective tissue (FCT; densely packed bundles of fibers). LCT occurs beneath epithelium in skin and many internal organs, such as lungs, arteries, and the urinary bladder. This tissue type also forms a protective layer over muscle, nerves, and blood vessels. FCT occurs in tendons, which connect muscle to bone. Ligaments are also composed of FCT, and connect bone to bone at a joint. The typical appearance of collagen fibers is depicted in Fig. 1.16 a. Cartilage is a specialized connective tissue in which specialized cells, chondrocytes, produce a firm, resilient matrix of ground substances and fibers that resists compression with modest rigidity and flexibility. Three kinds may be distinguished by the composition of the matrix: hyaline (depicted in Fig. 1.16 c) and elastic cartilage as well as fibrocartilage. Bone is a hard connective tissue with cells, osteocytes, in a large quantity of matrix material, and serves for support, attachment, leverage, protection, and mineral storage. In order to provide great strength and rigidity with some elasticity, the matrix is composed of densely packed collagen fibrils infiltrated with bone mineral as fine crystals of calcium salts resembling hydroxyapatite crystals. Mineral constitutes about 65 % of the dry weight of bone (Fig. 1.16 b). The matrix is strong but dense, and so nutritive fluids cannot diffuse freely through it. Osteocytes therefore have many long processes extending through canaliculi (tiny channels) and making contact with one another and, indirectly, with blood vessels. The cell body lies in a cavity, a lacuna, in the matrix. Tooth components represent specially adapted bones, predominantly consisting of dentine (cell-free, hard material made up of collagen fibrils, impregnated with crystals of calcium salt) and cementum (thin layer of bone-like material, with osteocyte-like cells).

Striated muscle tissue is easily to detect in a light microscopic image: muscle cells form homogeneous fibrillar bundles with characteristic cross striations, as depicted in Fig. 1.16 f. Cardiac muscle fibers are a type of striated muscle found only in the heart. Skeletal (striated) muscle fibers function in conjunction with the skeletal system for voluntary muscle movements. The bands are areas of actin and myosin deposition in the cells. In general, smooth muscle tissue (Fig. 1.16 d) does not show transverse striations as present in cardiac muscles and, in highest density, in skeletal muscles. These fibers are components of structures in the digestive system, reproductive tract, and blood vessels.

Peripheral nervous tissue is also easily to detect through its characteristic structure. A peripheral nerve is composed of several nerve fibers surrounded by layers of con-

---

Fig. 1.16 Appearance of most important animal or human tissue types: ▶

(a) Typical appearance of extracellular fibrous material (collagen fibers). Here, numerous bundles of the extracellular matrix molecules are strongly stained in the coronal dental pulp (human). Under polarized light these fiber bundles appear birefringent and have a red, orange, yellow, or greenish color. They are therefore probably composed of various types of collagen molecules. Picrosirius Red staining (from Hillmann and Geurtsen, 1997). The image shows a 180×200 μm section.

(b) Cross-section of a human bone. b1: A system of longitudinal and transverse channels (dark stained) provides the bone with blood vessels. The osteocytes are arranged in concentric layers, embedded in the bone matrix. The image shows a 420×700 μm section. b2: Tiny radial channels (canaliculi) allow interconnection of the cells by filamentous processes. These channels are visible after staining of the specimen with thiamine picric acid (Gwen V. Childs, Univ. Texas, Medical Branch). The image shows a 130×180 μm section.

(c) Hyaline cartilage with developing chondrocytes in the upper part of the image (perichondrium). Nests of mature chondrocytes are deeper in the cartilage, secreting the surrounding matrix (Milton Wolf and Mark Scarbrough, Univ. Kansas). The image shows a 500×325 μm section.

(d) Layer of smooth muscle cells from an artery wall (from Kühnel, 1995). The image shows a 120×160 μm section.

(e) The cross-section through peripheral nerve fibers shows numerous darker axons surrounded with myelin sheaths. The image is a small part of a peripheral nerve. Although a whole nerve is composed of several sheathed nerve fiber bundles, blood capillaries and connective tissue, the cross-section of a myelinated axon has a typical appearance. Differentiation of unmyelinated axons is more difficult (Milton Wolf and Marc Scarbrough, Univ. Kansas). The image shows a 100×100 μm section.

(f) Longitudinal section through striated muscle. The characteristic pattern is clearly visible. The muscle fibers can be up to a few centimeters long, the diameter varies between 10 and 100 μm in different muscle tissues (Milton Wolf and Marc Scarbrough, Univ. Kansas). The image shows a 110×135 μm section

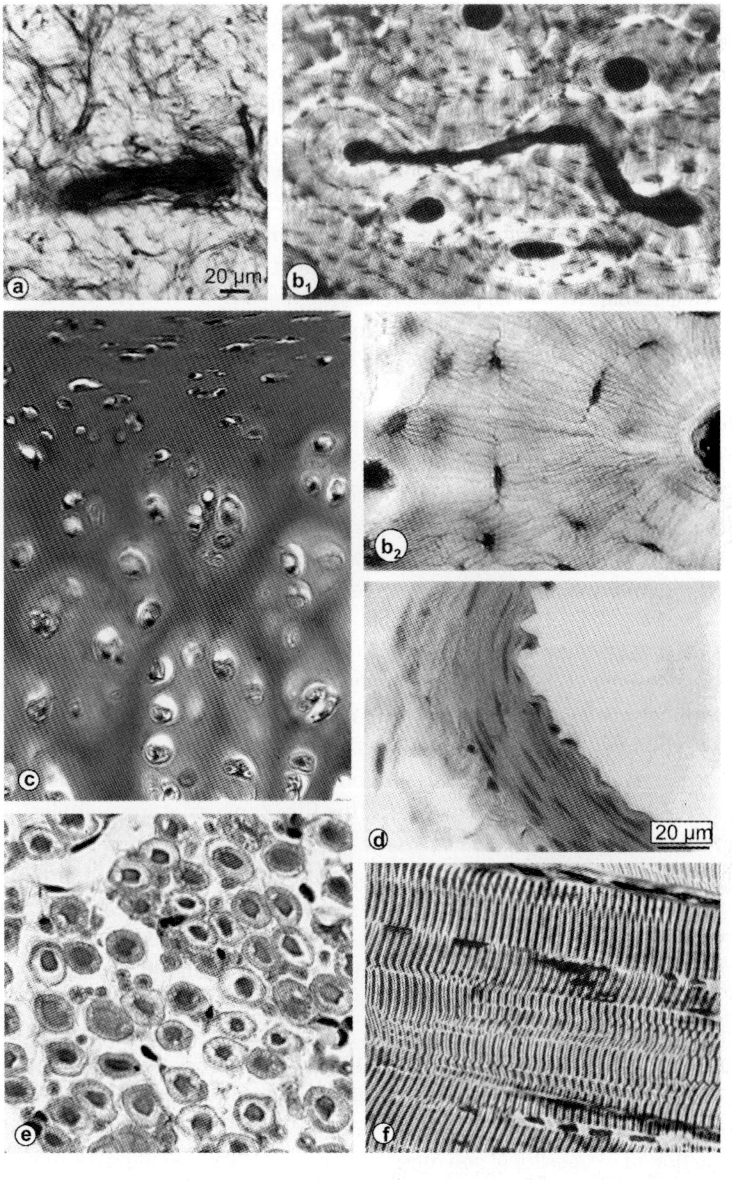

nective tissue outside the nerve. In cross-section, the lumen of the central axon inside the fiber and the surrounding myelin sheath may be clearly differentiated and has, in cross-section, a typical appearance (Fig. 1.16 e). Longitudinal sections show bundles of elongated, wavy cells.

1.2.2
Monitoring of Interactions of Organisms

Microorganisms are present in nearly all (both natural and technical) processes in which enough liquid water is available for at least a short time period. They can be indispensable, as in fermentation processes and in wastewater treatment, or they can be disruptive and hazardous agents in food production or in the pharmaceutical industry. In most processes, growth of microorganisms regularly occurs on a surface. This surface may be non-living matter, or it may be other unicellular organisms and living tissue.

Regulation of most processes should therefore also include biofilm monitoring.

1.2.2.1 Biofilms and biofouling

Numerous applications for imaging and monitoring of biofilms in environmental and biotechnological processes are now available. Where thick specimens have to be monitored, confocal microscopy is superior to standard microscopic techniques (see Section 1.3). Nevertheless, monitoring and surveillance of biofilm growth with standard bright-field and fluorescence microscopes is possible.

Biofilm development

Development of bacterial biofilms is a process that depends on the properties of the liquid phase in which the organisms are suspended, the surface for colonization, and the microorganisms. In principle, the stages of biofilm development are similar in various systems. On surfaces in contact with liquids, a conditioning film is build up on the surface within seconds. This film consists of macromolecules that become bound to the surface by electrostatic and van der Waals interactions. The properties of this film, only some nm in thickness, influence the subsequent initial step of biofilm formation. Monitoring of biofilms regularly requires application of (epi)fluorescence techniques, because thick, non-translucent surfaces often have to be monitored. The initial stages of biofilm formation show the attachment of microorganisms on the surface as a function of time. Quantification of the data gives information on the initial stages of biofilm formation (i.e., the phases of reversible and irreversible adsorption of organisms). With equivalent fluorescent techniques it is possible to monitor the initial stages of biofilm growth: the formation of microcolonies on the surface (Fig. 1.17). Maturing of the biofilm during subsequent stages is accompanied by production of extracellular polymeric substances (EPS; see Fig. 1.18 a). For monitoring of biofilms in these stages of development, confocal or deconvolution microscopy is helpful in order to obtain a highly resolved image of a thin slice of the sample or a reconstructed three-dimensional impression. Like tissue samples, however, biofilms may also be mechanically sectioned after embedding in an appropriate matrix (e.g., gelatin, paraffin, or resins) to visualize the structure in the axis vertical to the surface (Fig. 1.18 c, Fig. 1.19). Application of stains with affinity for exopolysaccharides reveals the extended masses of extracellular polymers in mature biofilms. Together with their enclosed organisms and other particles, the EPS acts as rigid glue, providing the struc-

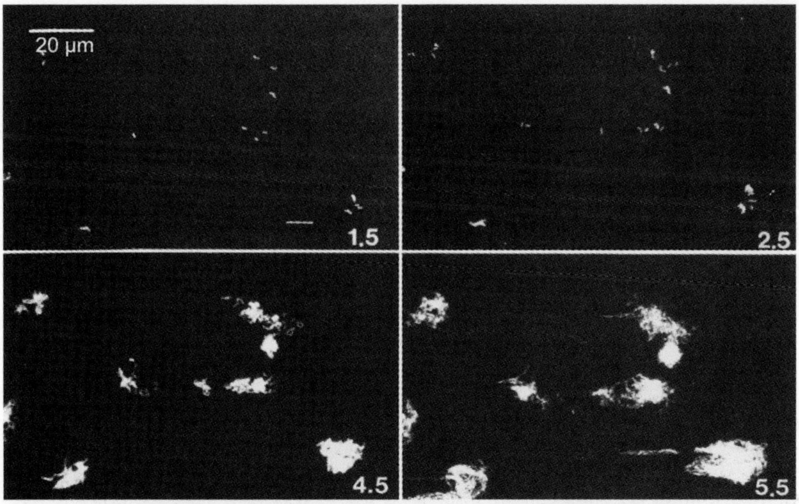

Fig. 1.17 Initial stages of biofilm formation. Dark-field photomicrographs showing growth and microcolony formation during surface colonization by *Vibrio harveyii*. The times (h) after inoculation with the flow cell are indicated. Initial microcolonies enlarge by growth of their resident cells but also by aggregation of planktonic cells. (Korber et al., 1995)

tural integrity of the film. Most biofilms do not cover the surface with a smooth layer of uniform thickness, but leave ducts and lacunae in the surface as long as the shearing forces of a streaming liquid on the biofilm surface allow it. These ducts may be of importance for the supply of deeper biofilm layers with nutrients, but also result from pieces of the biofilm that become detached from the surface (Fig. 1.18 b). These pieces, regularly released from mature biofilms, may cause severe problems, since they float off and clog up tubes or filter systems even in sectors in which no biofilm has developed. Clogging of membrane filters with a resident biofilm has been demonstrated by various microscopic techniques (see Fig. 1.18 c).

Biofilms do not develop only in systems in which liquid water is continuously present. Surfaces exposed to air are also regularly covered with a biofilm consisting of microorganisms. Depending on the environmental conditions, the dominating organisms may not only be bacteria; fungi, green alga, and perhaps higher organisms (lichens, mosses) can also cover surfaces exposed to the atmosphere. These films are abundant in natural habitats (such as on rock surfaces), but are also present on artificial surfaces (glass, steel, stone buildings, etc.). These films may be non-damaging and with no relevance, but may also be involved in biocorrosion, i.e., successive decomposition of the substratum. Instead of days or weeks, the films need several years or decades to reach maturity (i.e., until no further increase in biomass can be observed). Since the corrosive action of the films often results in a very tight association between cells and substratum, quantification of biomass (or biovolume development) is better performed in situ: when the organisms are still attached to the surfaces. Figure 1.19 depicts an endolithic biofilm. In this widespread sub-aerial biofilm type, the action of microorganisms successively pene-

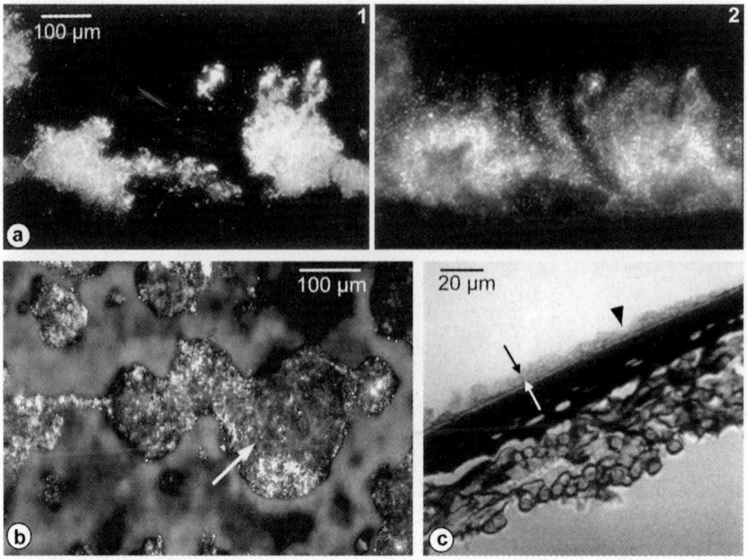

Fig. 1.18 Differentiation of cells and extracellular material in biofilms.
(a) Cellular DNA and EPS staining of biofilm cross-sections. The biofilm contains *Pseudomonas aeruginosa* and *Klebsiella pneumoniae* grown on a glucose minimal medium in continuous culture on solid supports (at the bottom of both images). Staining with ethidium bromide specific for nucleic acids (used as markers for the microbial cells) (1). Staining with Calcofluor indicates the distribution of EPS (2). The fluorescence of the visualized independently by use of appropriate filters. These images demonstrate that cell-free areas of the biofilm are filled with EPS. (Ricardo Murga, Dirk deBeer, Rohini Srinivasan, and Phil Stewart, Center for Biofilm Engineering, Montana State University, Bozeman)
(b) Face-on view of a biofilm consisting of discrete aggregates of microbial cells in a slime matrix (arrow), separated by interstitial voids or water channels. The biofilm consists of *Pseudomonas aeruginosa*, *Pseudomonas fluorescens*, and *Klebsiella pneumoniae* grown in a flow cell. The biofilm does not form a homogeneous surface layer. Channels allow quick exchange of oxygen and nutrients and thus higher growth rates than closed biofilm layers. The differential staining of the components was performed with fluorescein for the bulk liquid and the nucleic acid stain propidium iodide for the cells. (Paul Stoodley, Center for Biofilm Engineering, Montana State University, Bozeman; de Beer et al., 1994)
(c) Vertical thin section of a polyamide-membrane, covered with biofilm, visualized by bright-field microscopy. The biofilm (marked by two arrows) grew on a reverse-osmosis membrane during seven days of operation and reduces the efficiency of the system. The film is not of uniform thickness, but contains voids and indentations (arrowhead). (Flemming, 1995)

trates a soluble stone matrix (such as limestone). The organisms, usually fungi, green algae, and cyanobacteria, live between a few micrometers and one centimeter below the surface of the substratum (marked by arrows in Fig. 1.19 a). Quantification of the development in this biofilm type was performed by estimation of the surface areas of sectioned cells after visualization by fluorescence microscopy and calculation of the respective volume in a defined rock area. The development of the biofilm based upon these data is shown in a growth curve (Fig. 1.20). In this example, the biofilm needs several decades until no increase in biovolume is observed.

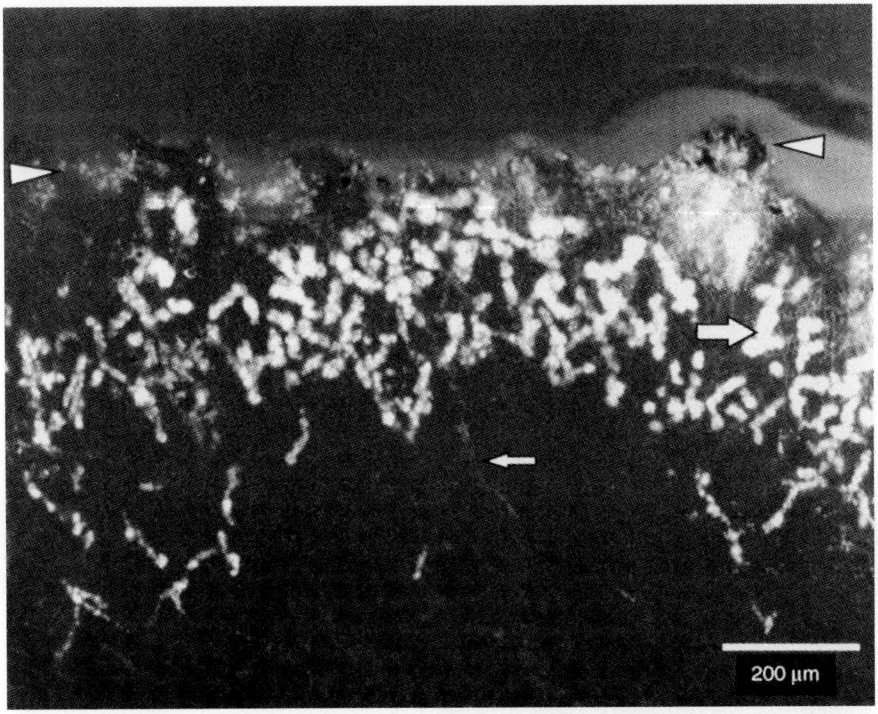

Fig. 1.19 Fluorescence images of thin sections through a rock surface with a rock-dwelling (endolithic) microbial biofilm. The thin section was prepared with a saw microtome vertical to the rock surface (arrowheads). Green algae (and/or cyanobacteria) show bright autofluorescence in the upper layer of the biofilm (large arrow). Fungal filaments exhibit weak fluorescence (small arrow). (Christine Flies, Univ. Göttingen)

Compositions of biofilm populations

In situ labeling techniques allow the identification of single cells or specific surface antigens. Oligonucleotide probes attached to fluorescent markers are regularly used for detection of organisms. These probes bind to specific regions of ribosomal RNA and may be selected in such a way that organisms may be identified as members of large phylogenetic groups as well as down to the species level. Whereas species-specific oligonucleotide probes are useful for identification of a specific, already characterized species in a biofilm, markers with group-specificity are used to determine overall community composition. This allows monitoring of a microbial community during biofilm development or after modification of environmental conditions (such as after application of an antibiotic agent).

Markers specific for the eubacterial domain are helpful for estimation of the total amount of bacteria in a biofilm sample, especially when a small number of bacteria are scattered in a sample (Fig. 1.21). This method is normally more accurate than cultivation methods for determination of bacterial cell numbers in a sample, because organisms from environmental samples are often uncultivable on (conventional) media.

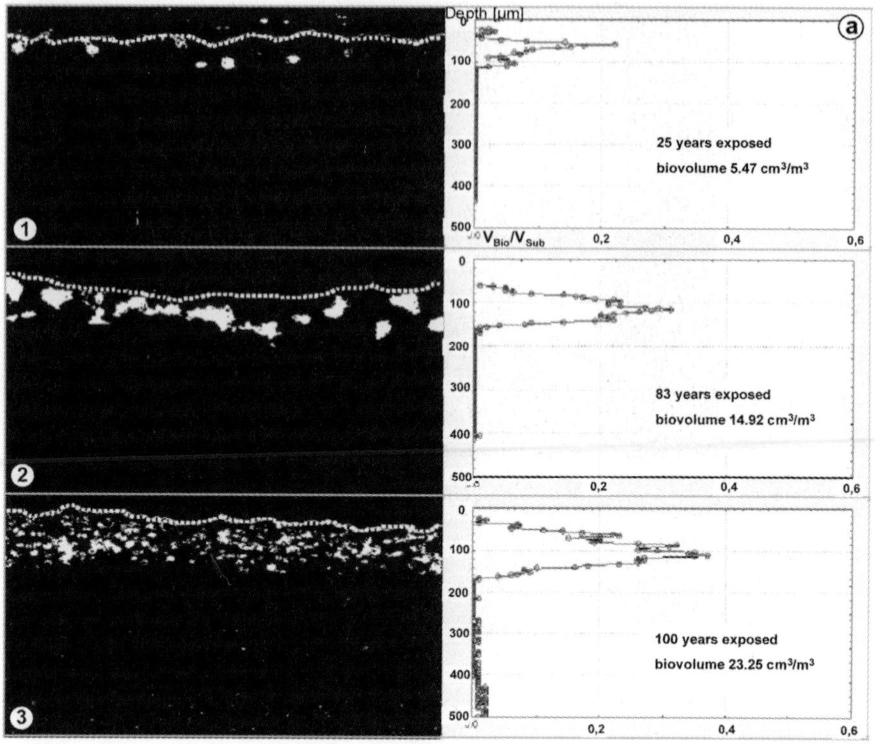

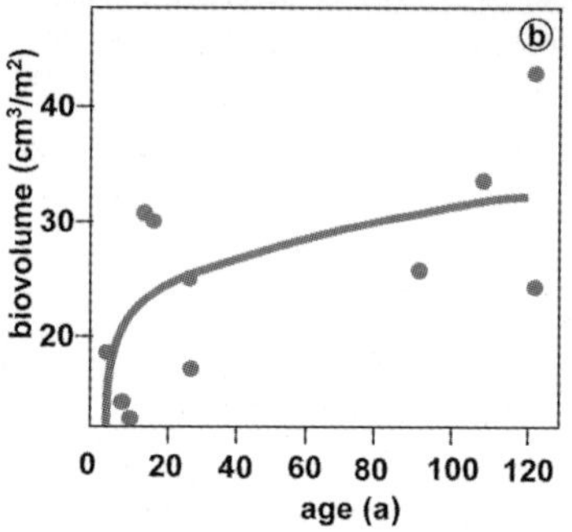

Fig. 1.20 Development of an endolithic biofilm.
(a) Development of a biofilm as depicted in 1.19. Samples were taken from rocks of a glacier foreland with documented glacier retraction (i.e., calculable exposure times of the rocks to the atmosphere). The biovolume occupied by the fluorescent cells was calculated from thin sections of binarized images and used for the respective depth profiles (right part of the figure). V_{Bio} = Volume of biomass; V_{Sub} = Volume of substratum.
(b) Biovolume development after analysis of fluorescent images as indicated in (a), documenting the increase of biovolume over several decades. (Pohl, 2000)

This is documented in Fig. 1.22, in which comparative analysis of activated sludge has been performed among the groups of proteobacteria. This group includes most Gram-negative bacteria and is subdivided into five major branches, as depicted in Fig. 1.22 a.

Figure 1.22 b shows a comparison of community structure analysis in an activated sludge sample, as determined by in situ hybridization, dot blot hybridization of extracted nucleic acids, and classification of colonies after cultivation (Wagner et al., 1993). To demonstrate the basic principles of the methods, differentiation between the large subclasses of proteobacteria was performed (with a eubacterial marker for definition of the 100 % value). Depending on the method, the percentage of

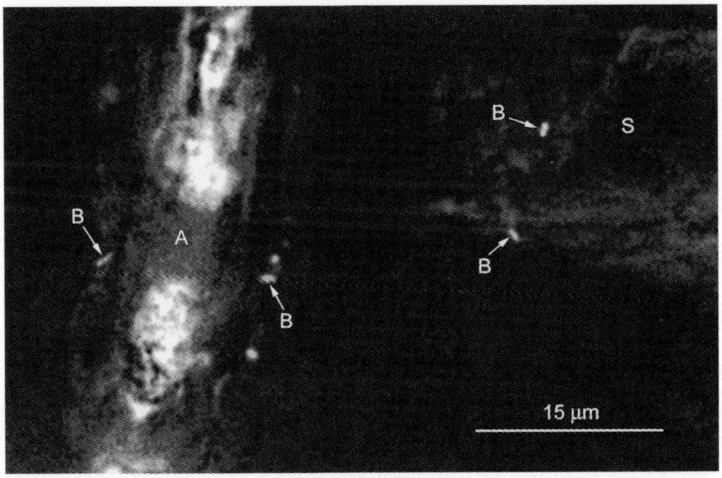

Fig. 1.21 Fluorescence in situ hybridization of eubacteria labeled in situ in an algal mat. Although a bright-field image did not reveal the presence of bacteria, use of fluorescence imaging with the aid of an eubacterial marker revealed the presence of some bacteria (B) in a biofilm sample (see also Figs. 1.29 e, f; 1.30) attached to algal filaments (A) and mineral grains (S). (Rudolph Reimer, Univ. Göttingen)

the subclasses contributing to the sample varies. In situ hybridization reveals that organisms of the γ-subgroup are less common in this sample than the α- and β-subgroups, whereas members of the γ-subgroup are easier to cultivate on media. Most of them belong to the enterobacteria, aeromonads, and *Acinetobacter*. Dot-blot hybridization reflects the ribosome content of a specific subclass and is a measure of the contribution of a group to the total metabolic activity. Here, members of the β-subgroup dominate. Only a few members of this group are cultivable. This example illustrates a crucial problem for characterization of microbial communities: cultivation of isolated organisms is a prerequisite for classical identification and also for designing specific targets for identification in a biofilm matrix. In order to provide probes specific for uncultivable organisms, total community DNA has to be analyzed. After characterization of the specific ribosomal RNA genes, oligonucleotide probes may be prepared for identification of the respective uncultivable organism in the biofilm. The method also allows novel taxonomic subgroups to be localized; Fig. 1.23 shows a sludge floc with several as yet unknown bacteria of the γ-subgroup. The organisms were isolated from a deteriorated biological phosphorus removal reactor and may be competitors to the phosphate-accumulating bacteria, thus disrupting biological phosphate removal from wastewater.

For monitoring of biological wastewater treatment, organisms with specific metabolic features usually have to be detected. Knowledge of the presence and distribution of large phylogenetic groups in sludge flocs is therefore of limited value. However, identification of bacteria at lower phylogenetic levels is also possible, thus allowing indirect estimation of the metabolic features in this group. The example in

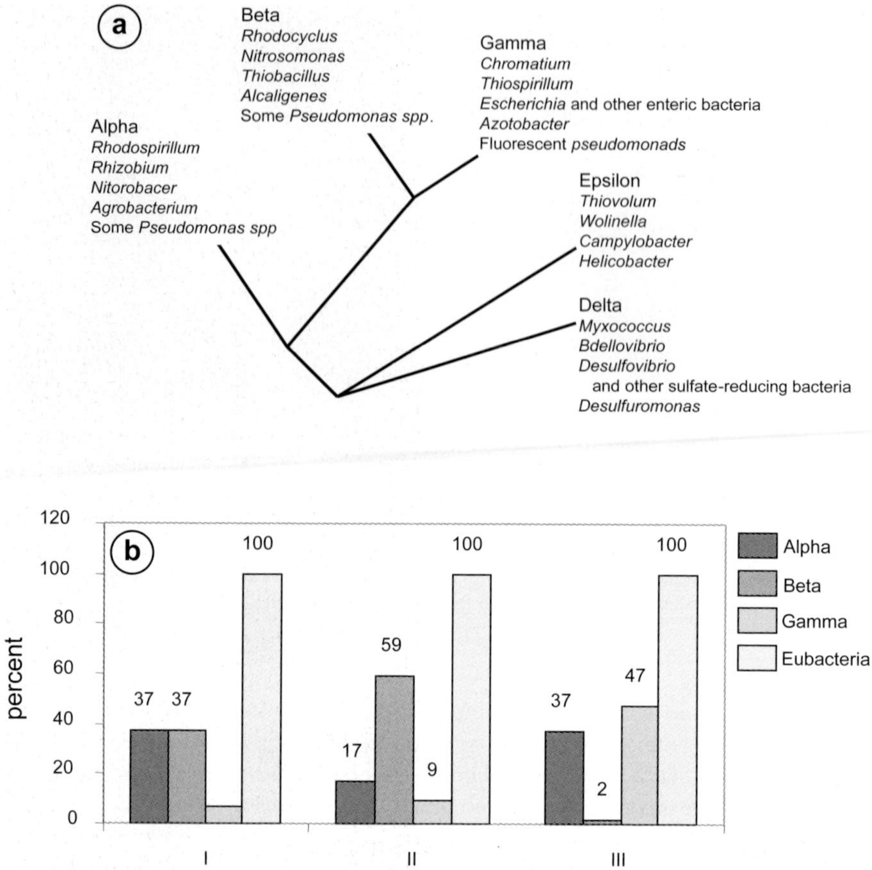

Fig. 1.22 (a) Simplified phylogenetic tree showing the large subgroups of Proteobacteria. Standard plating media are often adapted to enteric bacteria and related organisms. Members of the γ-subgroup are thus often overestimated after plating of activated sludge samples.
(b) Comparison of community structure analyses in an activated sludge sample as determined by in situ hybridization (I), hybridization of extracted nucleic acids reflecting the amount of ribosomes from the respective group (II), and classification of colonies after plating (III). The results are standardized over the values obtained with a marker specific for all proteobacteria and other eubacteria. (Wagner et al, 1993)

Fig. 1.24 shows the application of probes specific for members of *Desulfobulbus* spp. that have been used to locate sulfate-reducing bacteria in activated sludge.

The predominance of the sulfate reducers in the upper (i. e., exposed to the medium and aerobic) part of the biofilm could be quantificated after in situ hybridization with specific probes in the specimen shown in Fig 1.24. (Okabe et al., 1999). In a similar way, members of the genus *Acinetobacter* may be specifically labeled in order to estimate their contribution to the decomposition of organophosphorous compounds (Fig. 1.25). Often, up to 70 % of cultivable cells in activated sludge have been identified

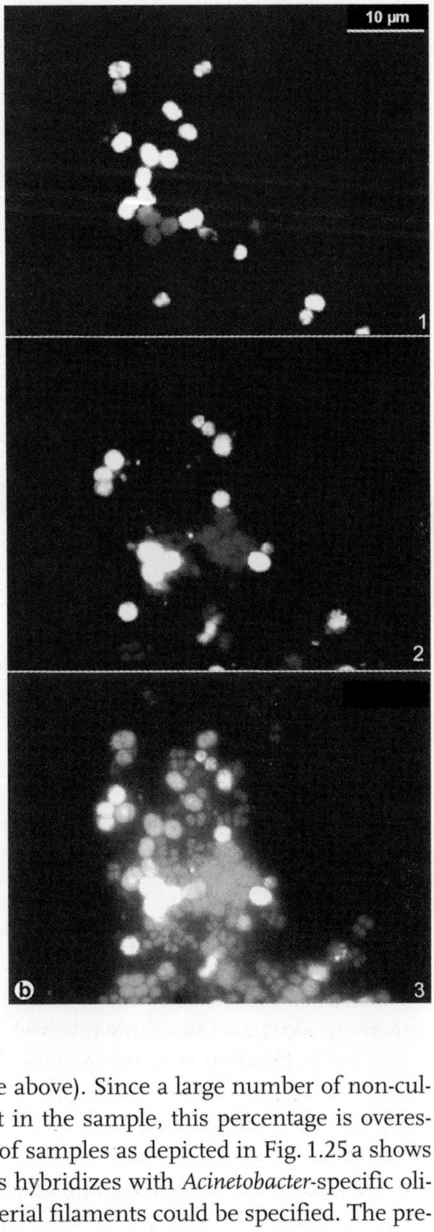

Fig. 1.23 In situ hybridization with probes directed against unculturable organisms. The probes have been obtained after analysis of total community DNA. Parts (1) and (2) depict the results for the same image sector after application of probes specific for two different members of a subgroup of the γ-subclass of the Proteobacteria. In (3), the fluorescence image of both markers and a marker specific for all Proteobacteria is depicted. (Nielsen et al., 1999)

as members of the genus *Acinetobacter* (see above). Since a large number of non-cultivable cells from other genera are present in the sample, this percentage is overestimated after plating. In situ hybridization of samples as depicted in Fig. 1.25 a shows that only a minor fraction of the organisms hybridizes with *Acinetobacter*-specific oligonucleotide probes. In a similar way, bacterial filaments could be specified. The presence of a large amount of bacterial filaments in activated sludge can contribute to poor settlement of the activated sludge flocs in sedimentation tanks, resulting in sludge bulking. Organisms of the genera *Haliscomenobacter, Leptothrix,* and *Sphaerotilus* are frequently detected in these samples and are differentiated here by in situ hybridization (Fig. 1.25 b, c).

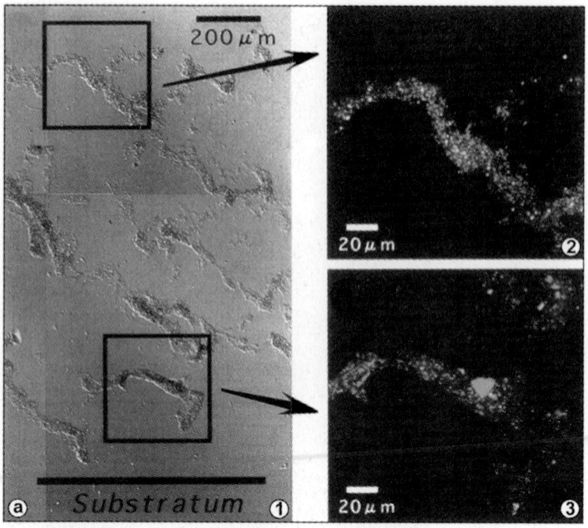

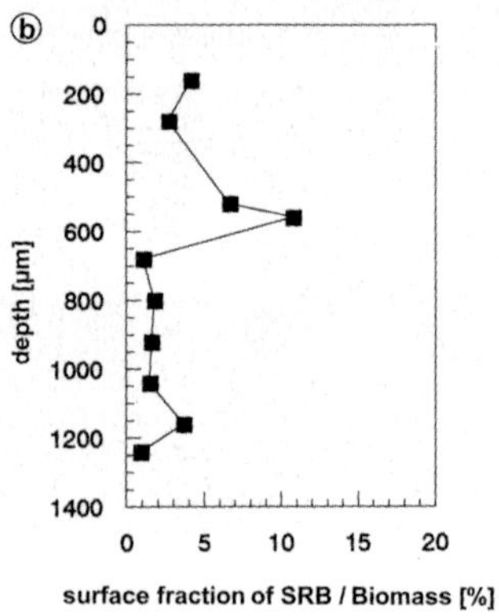

Fig. 1.24 Detection of *Desulfobulbus* in sludge flocs:
(a) In situ hybridization of vertical sections (20 µm thick) of an aerobic wastewater biofilm with fluorescently labeled SRB660 probe (specific for *Desulfobulbus* spp.). (1) Vertical section of the entire biofilm (differential interference contrast). The biofilm thickness is about 1,500 µm. (2, 3) Confocal laser scanning microscopic images of the boxed fields.
(b) Depth profile of *Desulfobulbus* distribution after quantification of the results as depicted above (e.g., by counting fluorescent ad non-fluorescent organisms). The reliability of the result strongly depends on the specificity of the marker and background fluorescence of the sample. (Okabe et al., 1999)

Localization of enzyme activity is a method used for direct monitoring of metabolic activity. In Fig. 1.26, organisms involved in the process of biological phosphorus removal are co-localized with the enzyme of interest, phosphatase. Phosphatases produce fluorescent, water-insoluble precipitates with the artificial substrate ELF-PO_4. Sites of high phosphatase activity are detectable in an activated sludge floc (Fig. 1.26 b). Combination with specific bacterial markers show that 17–20 % of the *Cytophaga-Flavobacterium* group displays phosphatase activity.

Damaging and beneficial effects of biofilms

Numerous organisms in biofilms cause corrosion of their substratum. Any colonized material – plastic, metal, stone, glass, and of course biopolymers – may be altered, dissolved, or decomposed upon microbial activity. Nevertheless, not all colonizers damage their substratum, and biofilms may also sometimes have a protective effect. The impacts of microorganisms on surfaces regularly require sub-light (e.g., scan-

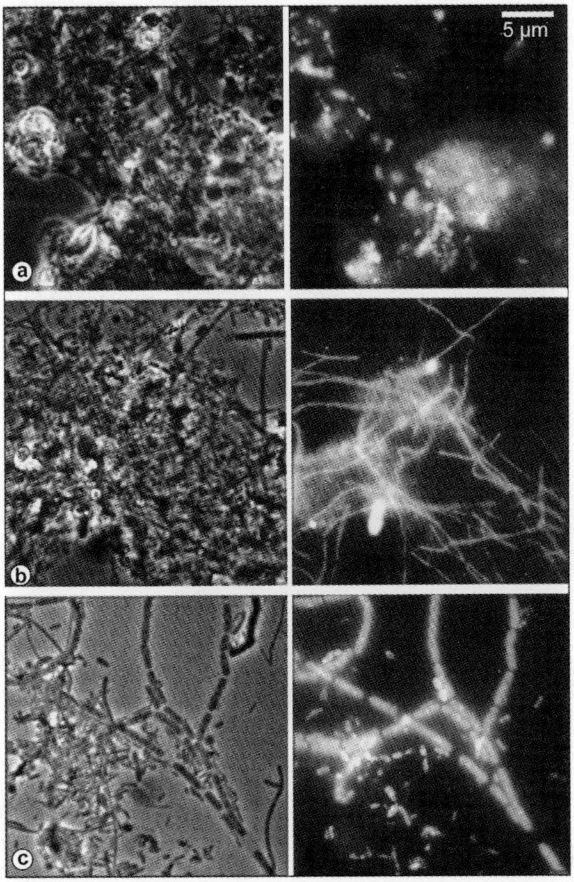

Fig. 1.25 In situ identification of bacteria of the genus *Acinetobacter* and threadlike bacteria in activated sludge (left: image in phase contrast; right: epifluorescence):
(a) In situ hybridization with an *Acinetobacter*-specific probe.
(b) In situ hybridization with a *Haliscomenobacter hydrossis* specific probe.
(c) Simultaneous in situ-hybridization with two probes specific for the detection of *Leptothrix* spp. and *Sphaerotilus* spp.
(Wagner and Amman, 1996)

ning electron microscopy) microscopic resolution in order to monitor the alterations at an early stage. Obvious alterations of the substratum at the light microscopic level are depicted in Fig. 1.27 a–c. The mechanisms resulting in damage to a substratum differ with respect to the attacking organism and the chemical nature of the substratum itself. The micropit on limestone is caused by the action of chelating organic acids (such as oxalic acid), produced by various endolithic algae and fungi, whereas the alterations of the complex biopolymers wood (Fig. 1.27 b, c) and wool, normally caused by various fungi, but also by bacteria, may be traced back to the activity of enzymes. These (bio)polymer-degrading enzymes are extracellular and are often attached to the surface of the organisms; monomers may thus be taken up immediately. The detection of extracellular chitinase has been performed similarly to that of phosphatase activity (see above), by monitoring of a fluorescent cleavage product of chitinase. As depicted here, the activity is connected with the cells that grow on the surface of a chitin-coated carrier (Fig. 1.27 d, e).

Metal corrosion caused by microbial biofilms may still be the most expensive damage caused by microbial biofilms. The formation of soluble salts from metal surfaces

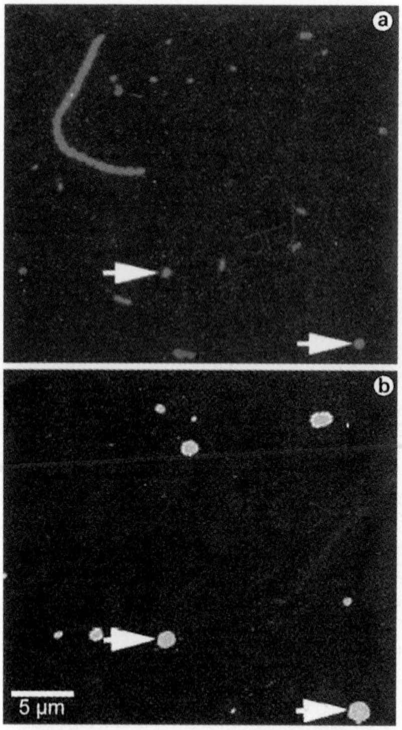

Fig. 1.26 Detection of phosphatase activities in activated sludge flocs:
(a) Fluorescence image depicting cells after treatment with the CF319a oligonucleotide probe (arrows), specific for the *Cytophaga-Flavobacterium* group.
(b) Fluorescence image of same field of view as 1 above, using a filter combination that resolves crystals of ELF, indicating areas of phosphatase activity (arrows).
(van Ommen Kloeke and Geesey, 1999)

has therefore been extensively investigated. In metal corrosion, biogenic effects interfere with non-biogenic effects. Not all biofilms appear to be detrimental; if the organisms do not produce corrosive agents the film coating may even be protective. Figure 1.28 a shows an uncoated metal surface exposed to growth medium, compared with a metal surface that has been coated with a biofilm during the time of exposition. Corrosion as visible in the light microscope and as determined by mass loss is reduced after biofilm coating.

In both natural ecosystems and agriculturally used land, some beneficial aspect of biofilms may be their contributions to pedogenesis (biogenic rock weathering, see Fig. 1.29 a; an unwanted process if the stone is used as building material) and soil stabilization. This is important in areas where higher vegetation is missing due to extreme climatic conditions. These crust-like biofilms (also termed cryptobiotic soil crusts, 1.29 b) cover areas on sandy soils, especially in arid and semiarid deserts and rangelands. The short periods of growth (such as after rare, but often heavy rainfall) and the high complexity of the crusts need long developing times, generally years instead of days or weeks. The influence of soil crusts on soil hydrology and soil nitrogen content differs with respect to soil type, climatic conditions, presence of higher vegetation, etc., but is of relevance for the ways in which these areas may be of agricultural use (usually rangeland in extensive grazing management). Besides green algae and/or cyanobacteria (Fig. 1.29 c, d), cryptogams (lichens, small mosses) may also

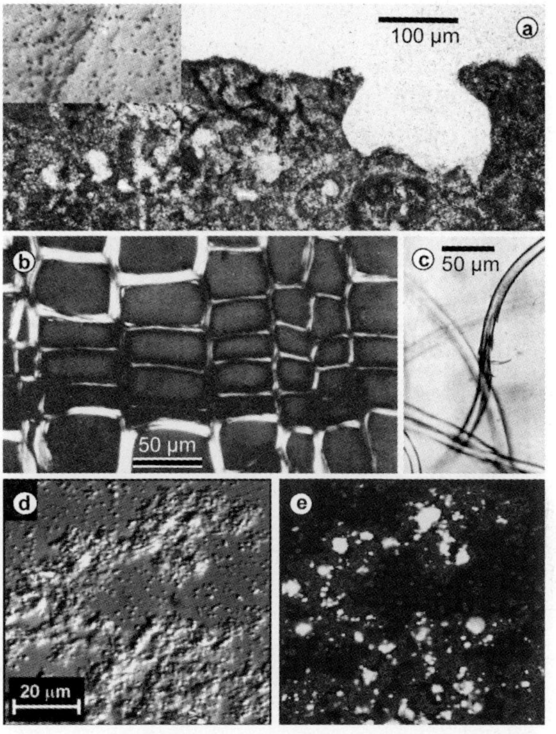

Fig. 1.27 Damaging activities of biofilms:
(a) Micropit in limestone, caused by the action of microfungi. The pit has been formed by a fungal fruiting body (now missing). The inset shows an approximately 1×2 cm sector of the limestone surface densely covered with these pits (Christine Flies, Univ. Göttingen).
(b) Scots pine cross-section under polarized light, showing loss of birefringence under polarized light (dark cell walls), caused by chemical alterations of the cellulose cell wall by microbial attack (Greaves and Levy, 1968).
(c) Wool fiber from felt, damaged by microbial attack (Sharpley and King, 1972).
(d) Bacterial cells grown on an artificial chitin surface (Baty et al., 2000).
(e) Detection of chitinase activity by monitoring of a fluorescent cleavage product of chitinase (Baty et al., 2000)

be present in mature soil crusts. These crusts are several mm up to 1 cm thick, some of them layered (Fig. 1.29 e, f). Besides fertilization of soil (an effect that may be of minor importance in many cases because of the low productivity of these crusts), they play an important role in soil stabilization through agglutination of soil particles, as depicted in Fig. 1.29 c and 1.30. Light microscopic observation of relatively undisturbed specimens of films on mineral surfaces requires specialized embedding and thin sectioning because of the presence of mineral components. The specimens in Fig. 1.30 (and also the rock-inhabiting biofilm in Fig. 1.19) were prepared by grinding of thin sheets of embedded material.

1.2.2.2 **Plant-microbe interactions: symbiosis**

Evaluation of beneficial, neutral, and also pathogenic interactions between microorganisms and higher plants is of outstanding importance in the agricultural sciences. The organisms may live in more or less close association with the host, as endo- or ectosymbionts. A variety of microbial species are also found in the rhizosphere, in more or less close association with roots. Rhizobium-legume and mycorrhiza symbiotic associations in particular have been investigated in detail at the structural and functional levels, because of their importance for nutrient supply to their host (crop) plants. Various species of *Pseudomonas, Klebsiella, Enterobacter*, and *Azospirillum*, however, are also found in association with plant roots. Combinations of bacterial hybridization

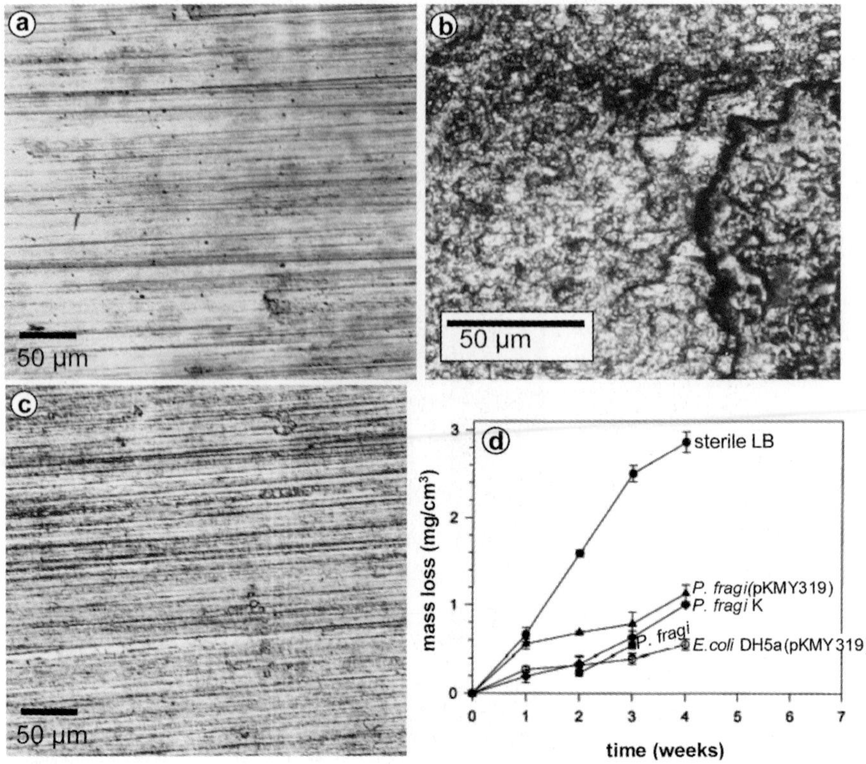

Fig. 1.28 Effect of a biofilm on a carbon steel surface (images taken with a confocal laser scanning microscope):
(a) Untreated sample.
(b) Sterile sample after exposure to a complex growth medium (Luria-Bertani medium) for two weeks, with a heavily corroded surface.
(c) Sample exposed as in (b), but the medium was inoculated with *Pseudomonas fragi*. After removal of the film, scratch marks as on the fresh surface are still visible.
(d) Determination of mass loss as a measure of corrosion of steel samples covered with biofilms from various bacterial strains (as indicated) compared with corrosive mass loss in sterile medium.
(Jayaraman et al., 1997)

markers, labeled with different fluorescent dyes, can reveal the attachment of the organisms on the surfaces of root hairs (Fig. 1.31 a, b). These associations are studied because of their (possible) importance in reducing the need for fertilizer application and in increasing crop yields. Nitrogen-fixing *Klebsiella* and *Enterobacter* species adhere in vitro to roots of several gramineae, including cereals (Haahtela et al., 1985). The free living nitrogen-fixing bacteria *Azospirillum brasilense* adhere to root hairs of tomato, pepper, cotton, several cereals, and sugarcane. The adhesion of some *Pseudomonas* strains also seems to promote plant growth.

Adhering symbionts are only in loose interaction with their hosts. Invasion of the plant tissue needs numerous specific steps of the symbiont as well as of the host cell. Establishment of endosymbiosis between a microorganism and a plant host is accom-

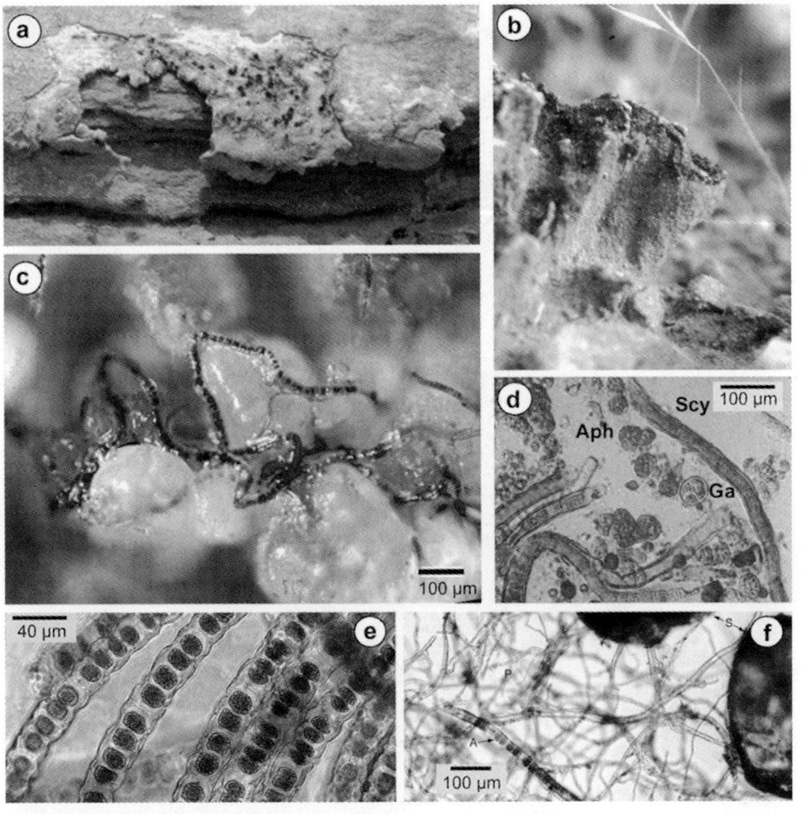

Fig. 1.29 The role of biofilms in pedogenesis and soil stabilization:
(a) Sedimentary rock colonized by a rock-inhabiting lichen biofilm.
(b) Sandy soil covered with thin dark cryptobiotic soil crust.
(c) Filaments of a green alga agglutinating sand particles (here the conjugate algae *Zygogonium ericetorum*).
(d) Endolithic cyanobacteria from a crushed alpine limestone rock (Aph: *Aphanocapsa* spec., Scy: *Scytonema* spec., Ga: *Gloeocapsa alpina*) (Pohl, 2000).
(e, f) Two layers of a cryptobiotic soil crust (3–5 mm in thickness), gelatin sections. Gelatin embedding of the sample prior to sectioning holds the organisms in place. (e) Upper algal layer, dominated by *Zygogonium ericetorum*, (f) lower layer, containing fungal filaments.
(Rudolph Reimer, Univ. Göttingen)

panied through various cytological alterations of the invading microorganism and the host. The structural and mechanistic similarities between symbiotic and pathogenic interactions are obvious. Both cases involve various stages of specific recognition (either from one of the partners or from both), entering into physical contact, and a series of physical reactions upon penetration of the pathogen or symbiont by the host. Figure 1.32 shows some basic cytological features occurring during interaction between the plant and endosymbiont (either pathogenic or beneficial). Though endosymbionts penetrate the plant cell wall, they remain separated from the cytoplasm of

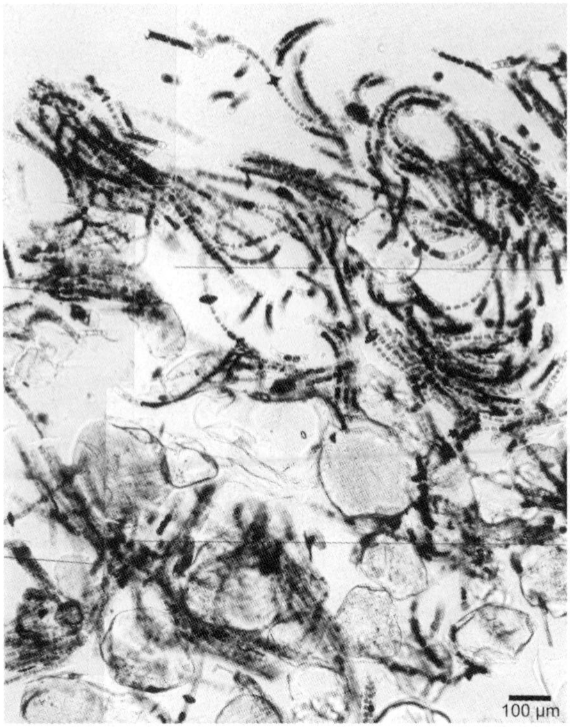

Fig. 1.30 Thin section of a cryptobiotic soil crust. The overview shows massive colonization by filamentous algae in the upper layer. Lower layers contain sand particles. The sample has been embedded in acrylic resin prior to grinding with a saw microtome (Rudolph Reimer, Univ. Göttingen)

the host cell by a membrane originally derived from the plant's cytoplasmic membrane. The symbiont is located in a separate compartment, also termed a "symbiosome". Matrix material is often deposited between this membrane and the symbiont. Parallels between the plant symbiosome and the phagosomes of animal cells are obvious; in animal cells, too, parasites live inside a (modified) phagosome compartment. The tools for investigation of these interactions are similar, and the structural alterations of both symbiotic partners may be monitored down to some important cytological events at the light microscopic level.

Fluorescence techniques have become an indispensable tool for investigation of *Rhizobium*-legume interaction. Figure 1.33 illustrates two elementary stages of the Rhizobium-legume interaction. Figure 1.33 a shows an alfalfa root hair (emerging from a rhizodermis cell) that has been infected with a mixed population of *Sinorhizobioum meliloti*. The bacteria have been marked with different fluorescent markers. They are situated inside a tubular infection thread that grows down the inside of the root hair and the body of the rhizodermis cell. After exiting the cell, the bacteria enter the intercellular space and induce growth of new infection threads in the underlying cell layers. Nodulation factors of the bacteria induce differentiation of cells from the root cortex into nodule primordia that become penetrated by the inward-growing infection thread. Branching of the thread at this stage of infection ensures that a sufficient number of nodule cells become colonized. When bacteria exit the infection

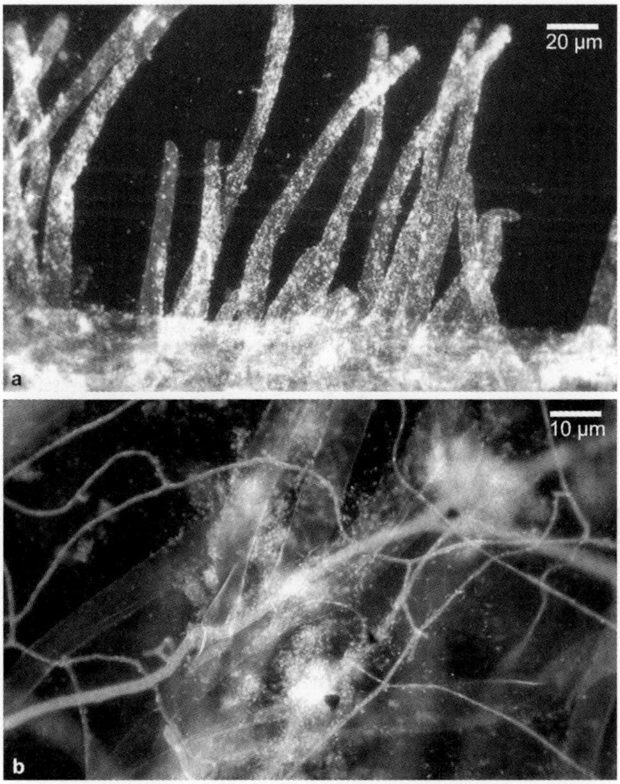

Fig. 1.31 Colonization of root surfaces by microorganisms:
(a) Root hairs and the root surface (bottom of micrograph) of red clover with bacteria extensively coating the root hairs. The specimen was treated with dilute Acridine Orange to obtain bright fluorescence of bacterial cells against less fluorescent root hairs.
(b) Fungal filaments across the rhizosphere of spring wheat, together with mainly red fluorescent bacteria. Root hairs are visible in the background. (Gunter Trolldenier, Univ. Hannover; Trolldenier, 1996)

thread and thereby enter the cytoplasm of nodule cells, they differentiate into irregularly shaped, large cells (bacteroids), as depicted in Fig. 1.33 c and d. They are located in the cytoplasm of the host cell, surrounded (either singularly or in groups) by an additional membrane envelope derived from the host cytoplasmic membrane.

Even more widespread than plant-bacterial interactions are those between plants and fungi. After attaching to the surface, pathogenic and symbiotic fungi eventually penetrate and colonize plant tissues. Associations between fungi and plant roots (or rhizoids) have been observed in all groups of plants, including mosses. This mycorrhiza is build up between plants and chytridiomycetes, zygomycetes (arbuscular mycorrhiza), and true fungi (ascomycetes, basidiomycetes). The fungus serves the tree by supplying water-dissolved minerals, whereas the plant supplies mycorrhizal fungi with carbohydrates and vitamins. Interactions with various structural characteristics

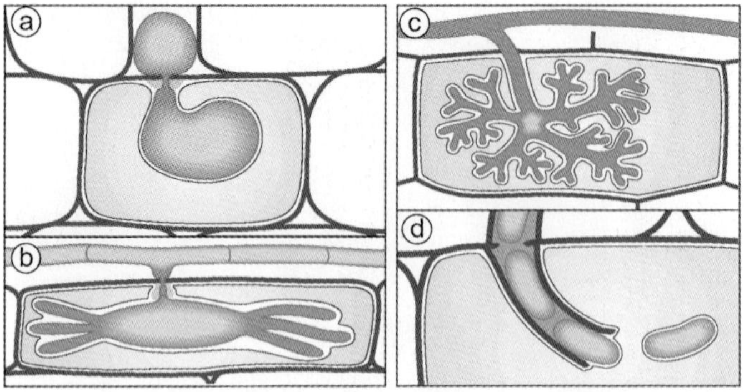

Fig. 1.32 Examples of plant cell invading microorganisms. In all cases, the invading microbe is separated from the cytoplasm of the plant by a thin perimicrobial (symbiosome) membrane.
(a) Haustorial complex of the oomycete *Peronospora* within a leaf parenchymatic cell.
(b) The haustorial complex of *Erysiyphe graminis* within a leaf epidermal cell.
(c) The arbuscule of *Glomus* sp. within a root cortical cell.
(d) An infection thread and bacteroid within a root nodule cell.
(Parniske, 2000)

are used to distinguish mycorrhizal interactions. Ectomycorrhiza is characteristic of many tree and shrub species from boreal (cold) and temperate climates. The fungus forms a coat around the root tip, and hyphae penetrate between cells of the root cortex to form a branched structure, the Hartig net (Fig. 1.34 a–d, f), but do not penetrate cells. In endomycorrhiza, the fungal hyphae penetrate between and into the cells of the host, where they form branched haustoria (Fig. 1.34 e, f). The infection structures of the fungus in the tissues are highly variable, transition forms existing between ecto- and endomycorrhiza. Vesicular-arbuscular mycorrhiza is the most common endomycorrhiza type, characterized by the highly branched haustorium (arbuscule). Only a limited number of arbuscular mycorrhiza forming species (all of them are zygomycetes) have been described; most belong to the genus *Glomus*. The development of symbiosis is initiated when a fungal hypha touches a rhizodermis cell, where it forms an appressorium (an attachment structure). Invasion of the root either between two rhizodermal cells or by penetration of a cell is facilitated by lytic enzymes (cellulases, proteases) excreted by the fungus. The development of the fungus is influenced by the plant, and a single fungal species may show significantly different morphological growth patterns depending on the plant host. After the growing hypha has reached the inner root cortex, the hypha penetrates the wall – but not the cytoplasmic membrane – of a cell and differentiates into an arbuscule. The fungal cell wall becomes progressively thinner, which allows intensive contact between host and symbiont cells. The life span of an arbuscule is limited to a few days, whereas the host cell remains undamaged and may host a new arbuscule.

Other rhizosphere organisms – especially bacteria – also influence the process of mycorrhiza formation and the mycorrhiza itself. The mantle surface and the interhy-

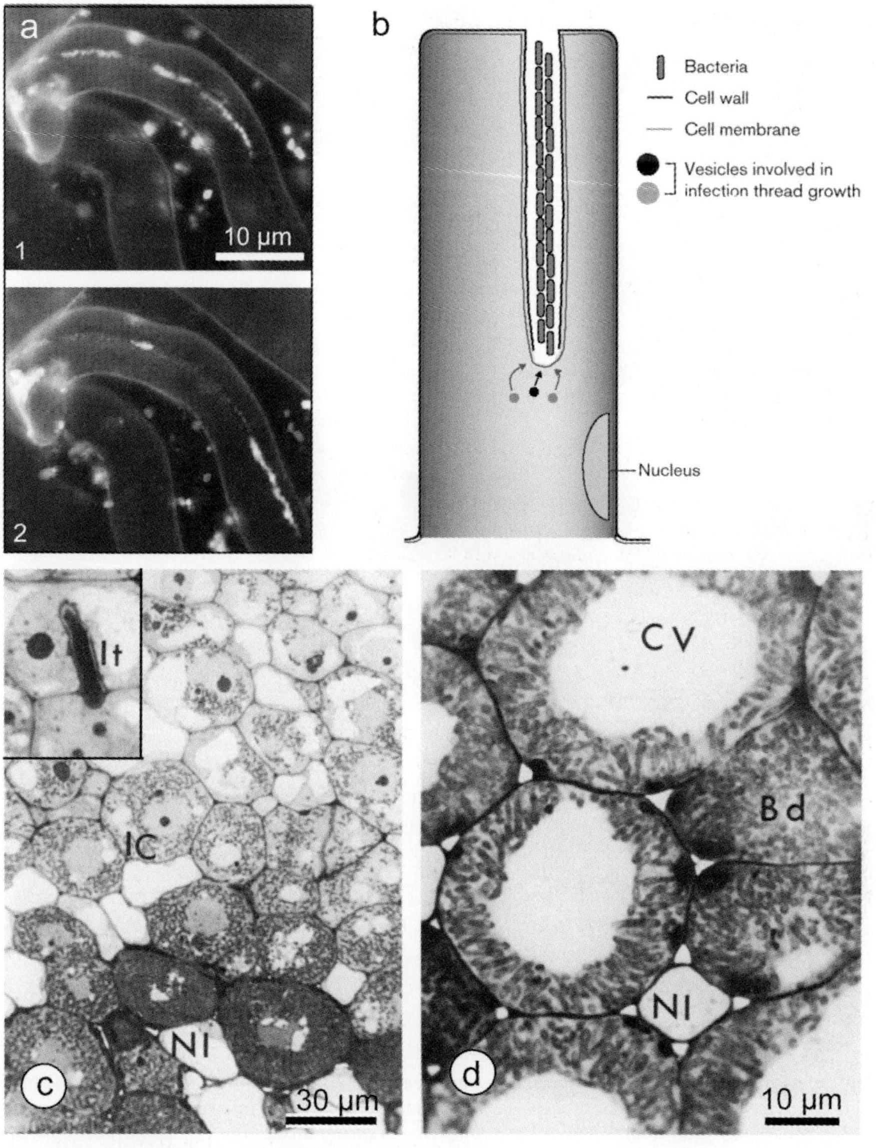

Fig. 1.33 Important stages of *Rhizobium*-legume interaction:
(a) Alfalfa root hair infected with a mixed population of differentially labelled *Sinorhizobium meliloti*. One strain has expressed the red fluorescent protein DsRed, the other a variant of Green Fluorescent protein. The different color channels are depicted in (1) and (2). The experiment shows that the bacterial population inside infection threads is derived from the clonal expansion of a few invading cells (Gage and Margolin, 2000).
(b) Schematic diagram of infection thread formation in alfalfa. The bacteria remain topologically outside the root until they bud from the thread as depicted in Fig. 1.32 (d). (Gage and Margolin, 2000)
(c) Thin section of cells from a root nodule containing non-infected (NI) and infected cells (IC). The inset shows an infection thread (It) (Pabel and Yang, 1981).
(d) Enlarged view of fully developed bacteroids (Bd) as depicted in the lower part of (c) in overview. NI: non-infected cell, CV: central vacuole (Pabel and Yang, 1981)

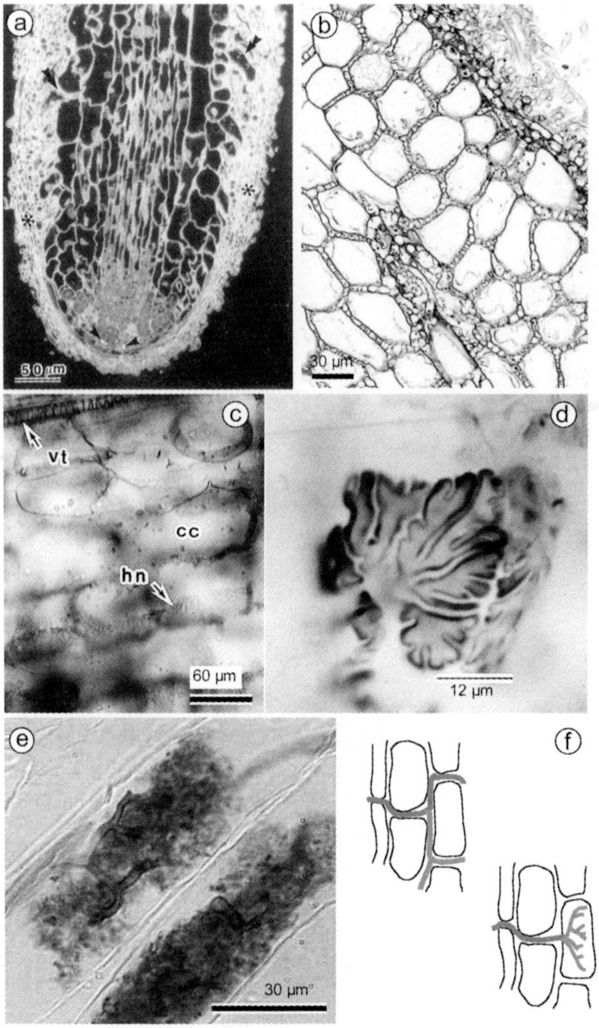

Fig. 1.34 Cytology of Ectomycorrhiza:
(a) Ectomycorrhiza of *Eucalyptus pilularis-Hydnangium carneum* embedded in LR White resin and stained with acriflavine-HCl. Section of root apex showing a thick fungal mantle (marked with asterisks), some Hartig net hyphae (double arrowheads), and small starch grains in the root cap (arrowheads) (Peterson, 1991).
(b) Bright-field image of a thin section showing root cortical cells surrounded by small fungal Hartig net hyphae.
(c) Root cortical cell of *Pinus densiflora* surrounded by Hartig net hyphae (hn) in overview. cc: cortical cell; vt: vascular tissue (Gill et al., 1999).
(d) Hartig net palmetti surrounding a root cortical cell of *Pinus densiflora* (Gill et al., 1999).
(e) The arbscule of an endomycorrhiza (Jim Deacon, Univ. Edinburgh).
(f) Schematic view of ecto- and endomycorrhiza. Ectomycorrhiza (upper image) surrounds root cortical cells, endomycorrhiza builds up intracellular (but not intracytoplasmic) arbuscules. Both strategies allow an intensive exchange of water, minerals and nutrients

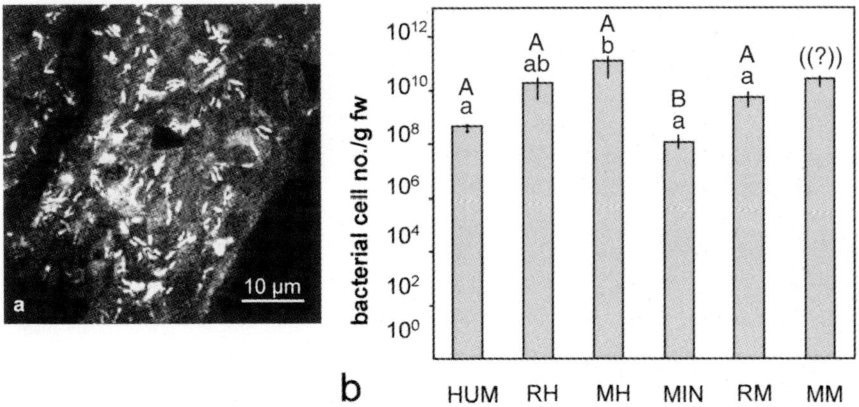

Fig. 1.35 Bacterial colonization of Ectomycorrhiza:
(a) Fluorescence image of bacteria colonizing the hyphal mantle and the ectomycorrhizal mantle on beech roots. The bacteria (predominantly belonging to the γ-subgroup of Proteobacteria) are visible as small elongated bright spots on the mantle surface (Mogge et al., 2000).
(b) Results of an experiment documenting the colonization of mycorrhiza by bacteria. In a greenhouse experiment, Scots pine was grown on two soil layers taken from a natural forest stand: a humus horizon (HUM) and a mineral horizon (MIN). The cell number was determined after fluorescence microscopy of Acridine Orange stained samples. The samples show highest cell densities in the mycorrhizal roots (MH, MM for the two soil types), the youngest non-mycorrhizal roots (RH, RM) and bulk soil (HUM, MIN) (Heinonsalo et al., 2001)

phal spaces of mycorrhiza are densely colonized by bacteria (Fig. 1.35 a). As in environmental biofilms, only a small fraction of these bacteria are cultivable under known conditions, but diversity and distribution patterns of the organisms may be worked out in a way similar to that described above for biofilm samples: in situ hybridization with group-specific probes or probes derived from the community DNA extracted from the original sample allow the composition of the rhizosphere community to be quantified. Here, group-specific probes are of special use when the amount of an important group, such as nitrogen-fixing bacteria, is to be evaluated and compared with the data for cultivable organisms.

Pathogenic plant-microbe interactions

In spite of the structural similarities between the beneficial symbiotic and pathogenic associations between host and microorganism (Fig. 1.32), the onset of symbiotic associations is facilitated by the host, whereas pathogenic microorganisms induce a variety of defense mechanisms in the plant cell, triggered by elicitors. Elicitors are pathogen- or plant-derived signal molecules belonging to a wide variety of chemical types, and induce the host's response mechanisms. Plants produce numerous agents in response to bacterial, fungal, nematode, and insect attack or mechanical injury, and these may inhibit the invading agent. Plants may also be pretreated with elicitors in order to enhance the inhibiting reaction. The structural changes occurring in response to elicitor-mediated response are visible at tissue and cellular levels. A very conspicuous reaction is depicted in Fig. 1.36 a, showing heavy deposition of amorphous ma-

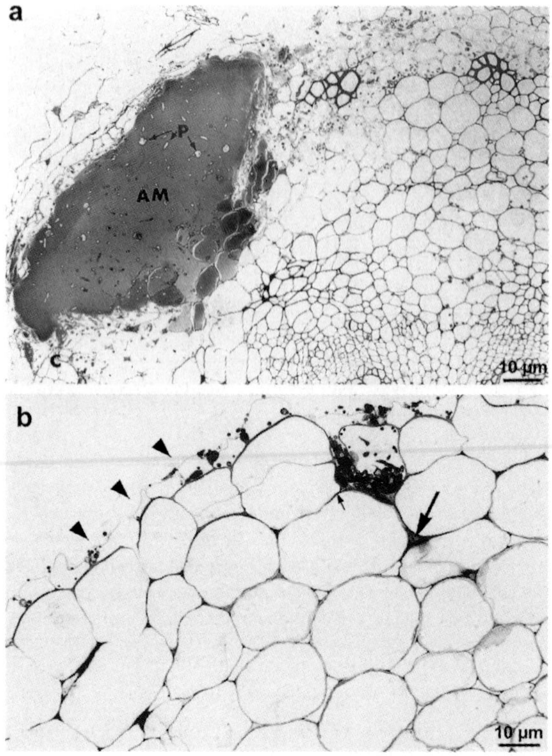

Fig. 1.36 Sensitized cucumber and tomato plants:
(a) Cross-section of a cucumber root after treatment with BTH (acibenzolar-S-methyl) and 96 h after infection with *Pythium ultimum*. Staining with Toluidine Blue. In the root cortex, heavy deposition of amorphous material is visible.
(b) Cross-section of a tomato root, treated as described in (a), but infected with *Fusarium oxysporum*. Sections of fungal filaments are mainly located at the rhizodermis (arrowheads), Invasion of rhizodermal cells and intercellulars (as seen in the upper right part of the image) obviously induces local deposition of dark-stained material in cell walls and intercellular spaces (arrows) of the root.
(Benhamou and Nicole, 1999)

terial after infection with the pathogenic fungus *Pythium ultimum*. The effect has been enhanced by sensitization of the plant with BTH (benzo(1,2,3)thiadiazole-7-carbothioic acid S-methyl ester, acibenzolar-S-methyl). The compound activates a number of resistance-associated genes. Here, the deposition of material containing large amounts of phenolic compounds has been amplified in BTH-treated cells as compared with untreated cells. In a similar way, a stimulated tomato root infected with *Fusarium oxysporum* deposits material in intracellular spaces (Fig. 1.36 b).

The clearly visible alterations in the tissue structure may be traced back to activation of diverse secondary metabolism enzymes, resulting in the production of antimicrobial agents, lignin, or phenolics. Localization of the metabolically active sites in the plant may be achieved by detection variously of the involved antimicrobial agent itself, the enzyme(s) involved in its synthesis, or the respective mRNA species. Treatment of cultured parsley cells with a peptide elicitor of fungal origin causes rapid and large changes in the levels of various unsaturated fatty acids. In situ RNA/RNA hybridization in parsley leaf buds demonstrates the presence of ω-3 fatty acid desaturase mRNA accumulated in sharply defined areas around the infection sites. The unsaturated fatty acids are possibly important precursors of specific cellular signals such as jasmonate or of polymers associated with local suberization of cells or reinforcement of cell walls (Kirsch et al., 1997; see Fig. 1.37).

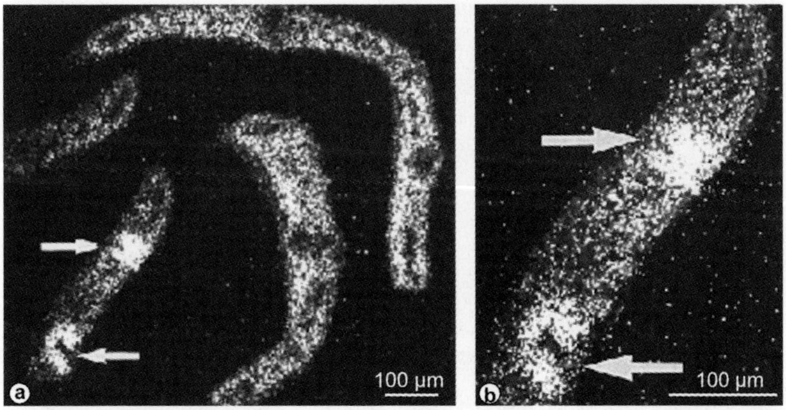

Fig. 1.37 In situ localization of ω-3 fatty acid desaturase mRNA in infected and uninfected areas of parsley leaf buds.
(a) Cross-section hybridized with ω-3 fatty acid desaturase mRNA 6 h after inoculation with spores of Phytophtora sojae.
(b) Larger magnification of a selected area from (a).
(Kirsch et al., 1997)

In this respect, the reaction of the plant cytoskeleton is of some interest. Cytological studies have documented the movement of the cytoplasm, nucleus, and other organelles to the infection site (Hardham and Mitchell, 1998). This feature mainly reflects the high transport activity of Golgi vesicles containing wall polymer (precursors) to the infection site of the cell (Fig. 1.38).

As already mentioned, surface features of both partners mediate specific recognition between plant and microorganism. Whereas pili are the significant factors for surface attachment of the bacterial cell, the interaction between the pathogenic fungus and the host is structurally far more complex. Important stages of fungal infection may be analyzed by use of specific antibodies directed against structures of the parasite or the host that are relevant for the infection process. Thus, adhesion as the first step of infection by a fungal pathogen may be monitored by localization of the adhesive agent. Application of a monoclonal antibody raised against the *Phytophtora cinnamomi* spore surface components allowed the zoospore adhesive agent to be followed. This is stored in small vesicles in the cortical cytoplasm at the site of the zoospore that contacts the host plant. The adhesive is then secreted onto the host surface (Fig. 1.39). Several lytic enzymes of the fungus are involved in the next steps of infection (penetration of the cuticula, degradation of middle lamellae of the host parenchymatic cells). Fungal infection structures (haustoria) are directly visible by light microscopy and – due to their surface composition – are detectable by specific marker systems. Monoclonal antibodies specific for wall components in haustoria but not other fungal or plant cell wall components were used to localize haustoria of *Melampsora lini* (flax rust fungus) in flax (Fig. 1.39 e). Other specific structures of the fungus (cnida, germ tubes, intracellular hypha) may also be localized in a similar way. These localization techniques require a well characterized target for the respective structure (such as an oligo-

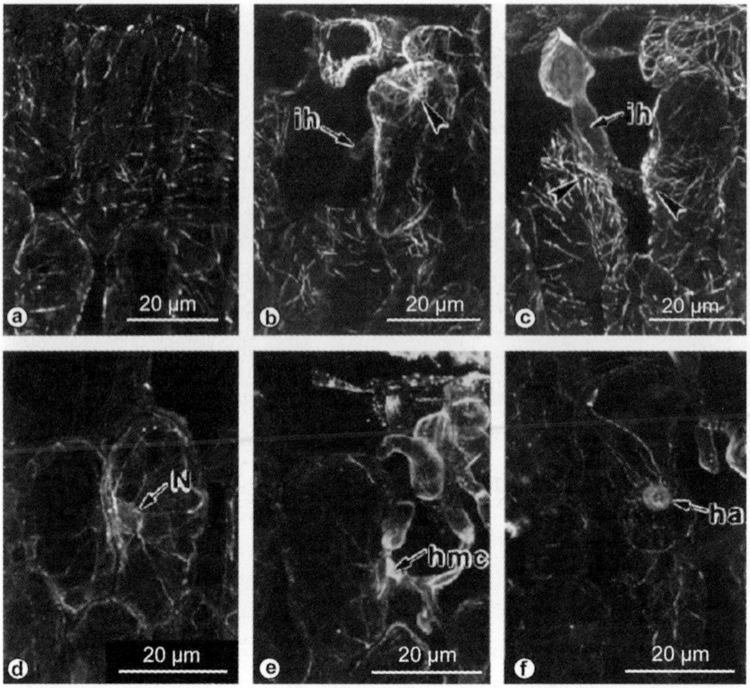

Fig. 1.38 Fluorescence images (taken with a confocal microscope) of microtubules and actin microfilaments immunofluorescently labeled with anti-β-tubulin and anti-actin, respectively, in sections of flax leaves
(a) Uninfected leaf showing transversely oriented cortical microtubules.
(b, c) Leaves infected with *Melampsora lini*. Microtubules in leaf parenchyma cells are focused on regions in contact with an infection hypha (arrowheads). ih: infection hyphae.
(d) Uninfected leaf showing actin microfilaments centered on the nucleus (N) in a leaf parenchyma cell.
(e, f) In leaves infected with *Melampsora lini*, actin microfilaments are focused on a region in contact with a mother cell of a haustorium (hmc) and around the haustorium itself (ha). (Kobayashi et al., 1994)

saccharide residue or a protein) and a detection system, such as a monoclonal antibody or a lectin (Hardham and Mitchell, 1998; Murdoch et al., 1998).

1.2.2.3 **Animal-microbe interactions**

Like plant surfaces, all epithelia of animals are colonizable (and indeed colonized) by microorganisms. Most of these interactions are non-hazardous or beneficial (especially in the gut), but some may of course be pathogenic. In addition to intensive studies concerning interactions of pathogenic microbes in tissue, non-medical biotechnological approaches include analysis of gut biota (rumen microorganisms in cattle and "probiotic" strains in the human intestine, for example). The most intensively colonized tissue surfaces are in the intestinal systems of animals and humans. Their microbial populations are indispensable for digestion of nutrients and supply of the

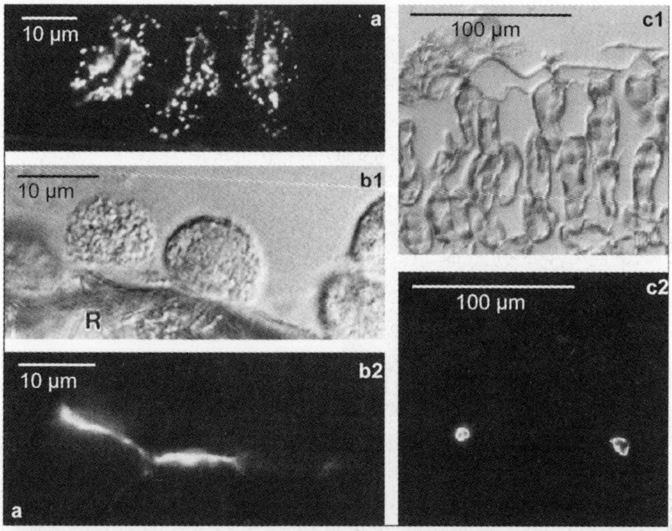

Fig. 1.39 First steps of the infection by the phytopathogenic fungi:
(a, b) Adhesion of spores of *Phytophtora cinnamomi*. Immunolabeling of putative adhesive material in zoospores (a) and cysts (b1, b2) with a fluorescently labeled monoclonal antibody (Vsv-1). The labeled vesicles are located in the cortical cytoplasm on the ventral surface of the zoospores. When the zoospores encyst at the surface of a root (R), they orient so that their ventral surface faces the root and the Vsv-1 antigen is secreted from the ventral vesicles onto the root surface. b1 and b2 show the same detail with interference contrast (1) and fluorescence (2).
(c) A section of an infected flax leaf labeled with a monoclonal antibody which reacts with a component in the walls of *Melampsora lini* haustoria but not with any other fungal or plant cell walls. Both images depict the same field of view, with differential interference contrast (1) and fluorescence (2).
(Hardham and Mitchell, 1998)

host with degraded macromolecules, vitamins, etc. Often, microorganisms are not directly monitored in situ, but in fecal samples instead. For differentiation of organisms in samples, fluorescent markers are again useful. DAPI stains (for bacteria) and Calcoflour White (for fungal hyphae), used in combination with image processing, are applicable for quantification of the total organism counts in gut segments, as documented for the analysis of interactions between microorganisms and earthworms (Schönholzer et al., 1999; see Fig. 1.40 a). Green fluorescent protein (GFP) again proved to be a useful marker system for monitoring the survival of specifically introduced strains in a gut system. A GFP-gene-containing plasmid, introduced into several lactic acid bacteria, allows control over the survival of the organisms in complex gut biota. After passage through the gut and exposition to air, GFP fluorescence allows the viable organisms to be counted. These techniques are helpful for documentation of the fact that the (in this case probiotic) strains of lactic acid bacteria are present in the gut only for a limited time until they are obviously displaced by the resident gut flora (Fig. 1.40 b).

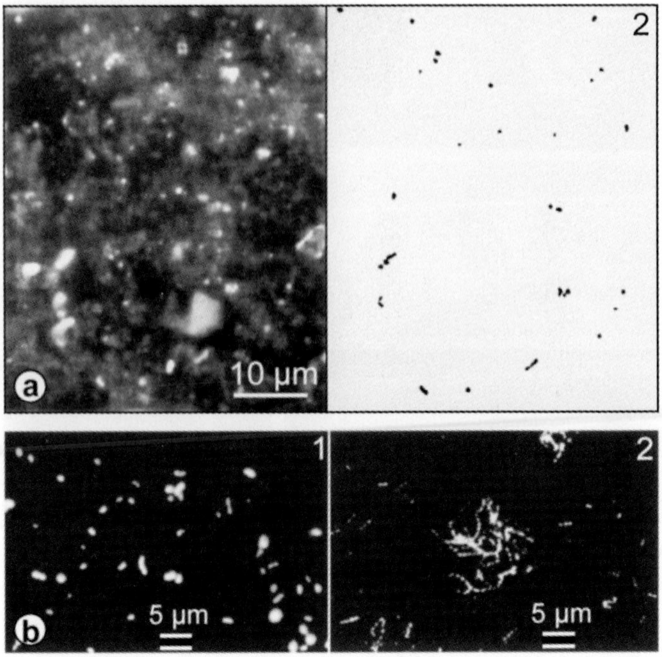

Fig. 1.40 Detection and quantification of bacteria in fecal samples:
(a) Two stages from image processing of a foregut sample from *Lumbricus terrestris* after in situ hybridization with a eubacterial marker. Left: color image; right: the same detail as in the color image, detection of bacteria after deletion of all non-bacterial objects (Schönholzer et al., 1999).
(b) Fluorescence image of two bacterial strains expressing GFP. 1: *Lactococcus lactis*; 2: *Enterococcus faecalis*. The strains are used for population dynamic studies in model gut biota. (Scott et al., 1998)

Fig. 1.41 Structure and dynamics of microbial populations in the intestinal tract: ▶
(a, b) Radial (a) and longitudinal (b) oxygen and hydrogen profiles in gut of the termite *Reticulitermes flavipes*. The profiles illustrate the high variability of the environmental conditions in the gut, mainly caused by the surface-to-volume ratio of the gut and, subsequently, by the microbial community. (Brune and Friedrich, 2000)
(c) Abundance of cells detected by fluorescence in in situ hybridization in various fractions (wall, non-attached contents of the lumen, flagellates) of the P3b/P4 region of the termite gut. The markers used for quantification are depicted in "Flagellates". In "Wall" and "Contents", a fraction of the cells marked with EUB338, specific for eubacteria, could be detected with the more specific markers. Sulfate-reducing bacteria (SRB382Db) adhere to the wall, but are absent in the other fractions, whereas the "contents" fraction mainly contains members of the α-subclass of proteobacteria (ALF1b). Other markers: BET42a: β-subclass of proteobacteria; HGC69a Gram-positive bacteria with high DNA G+C content; CF319a *Cytophaga-Flavobacterium* group. (Berchtold et al., 1999).
(d) Dynamics of the microbial composition of a human volunteer. Two probes, termed Bif164 and Lowgc2P, were used to detect members of the genus *Bifidobacterium* and of a group of Gram-positive bacteria related to the *Clostridium leptum* group. Typically, the total number of bacteria and also of the bacterial subgroups fluctuates markedly. Fluctuations of the *Bifidobacterium* cell number are very pronounced. (Franks et al., 1998)

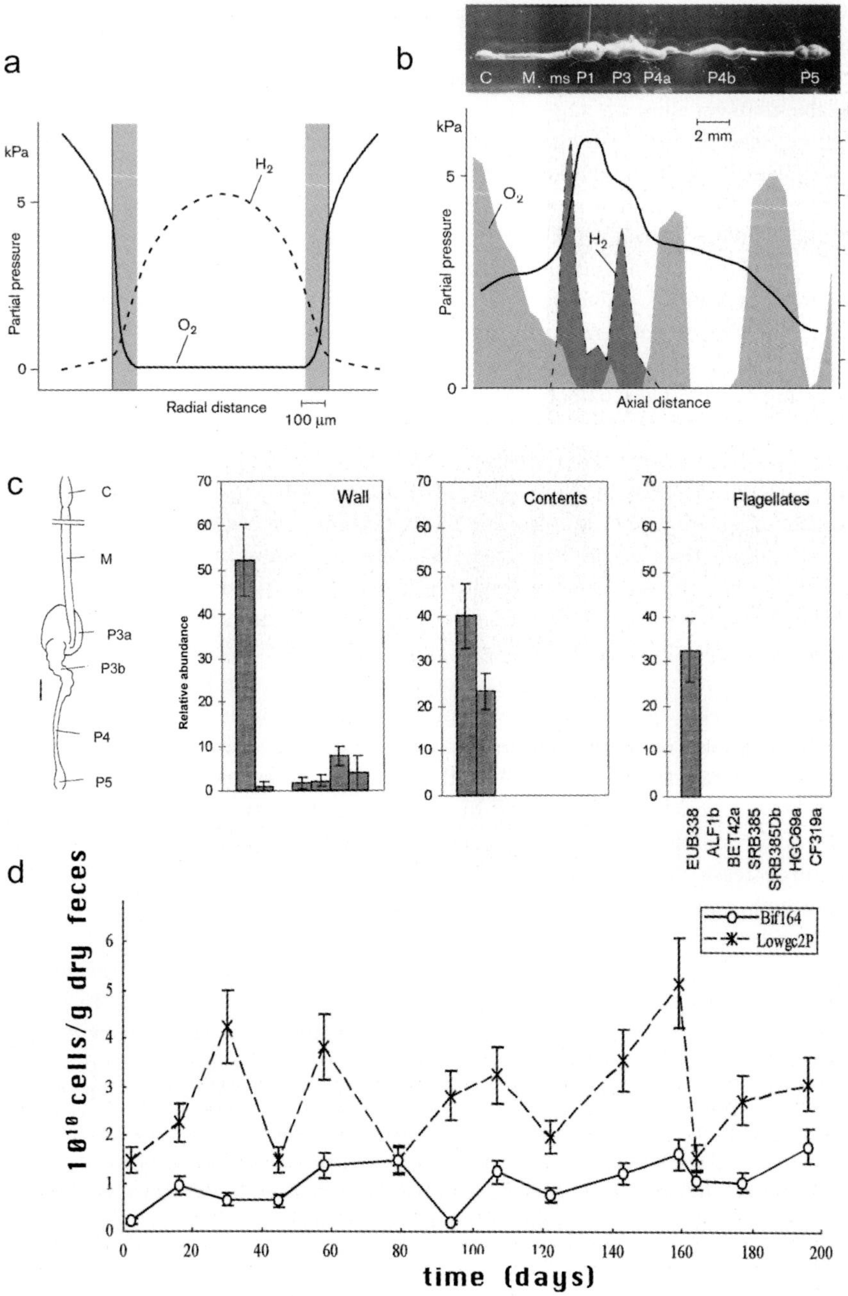
a
kPa
5
0
Partial pressure
H2
O2
Radial distance
100 µm
b
C M ms P1 P3 P4a P4b P5
kPa
pH
2 mm
Partial pressure
O2
H2
Axial distance
c
C
M
P3a
P3b
P4
P5
Relative abundance
Wall
Contents
Flagellates
EUB338
ALF1b
BET42a
SRB385
SRB385Db
HGC69a
CF319a
d
10^10 cells/g dry feces
Bif164
Lowgc2P
time (days)

When specific oligonucleotide probes are available, detection in animals or their excretions is possible. RNA hybridization probes were used for detection and quantification of eubacteria in human fecal probes. At present, appropriate combinations of available markers allow detection of about two thirds of the fecal microflora at least at the genus level. In healthy humans, *Bacteroides* and *Clostidium coccoides / Eubacterium rectale* (detected with one marker) constituted half of the fecal flora. Relative and absolute amounts of cells may be quantified by automated counting of every fluorescent cell under the fluorescence microscope. Most of these investigations have been performed on feces. Correlation of the composition of microbial populations with the conditions of the microenvironment, however, requires in situ monitoring. In this respect, the termite gut has become an attractive model system (Fig. 1.41). Since termites contain a variety of organisms involved in wood degradation, their gut is an interesting reservoir for biopolymer-degrading microorganisms. Figure 1.41 a and b reflects the heterogeneity of the gut environment with oxygen, hydrogen, and pH profiles. The composition of the gut biota varies with respect to the various compartments and the cross-section of the gut, as indicated in Fig. 1.41 a–c. The non-colonizable organisms are only accessible by application of specific probes. The first in situ estimations revealed the heterogeneity of the population in regions P3b and P4 (part of the paunch and colon) in the three large "compartments" of the gut region: associated with the wall, in the contents of the gut lumen, and as symbionts of the gut-inhabiting termites (Fig. 1.41 c).

Variations of human intestine biota over time are reflected in Fig 1.41 d. Here, the varying amounts of two representative groups – again distinguished by use of specific fluorescent probes on human fecal samples – are depicted. These variations are individually different and not associated with pathogenic processes. However, the interaction between microorganisms and human tissue is probably the most important field in infection biology. Since the reliability of the therapeutic effects of antibiotics tends to decrease, mechanisms of attachment to a tissue surface – usually the first step of infection – and all subsequent steps of interaction have to be understood at the structural level. In principle, the tools for investigation are similar to the approaches in plant pathology described above. Some mechanisms of attachment, invasion, and respective cellular responses are similar in principle at the cellular level. In higher animals, of course, the response of the immune system on microbial attack is a unique feature. Intracellular parasites in animal cells are mainly bacteria and protozoa. The two groups have evolved different mechanisms of entering a host cell. At the structural level, the outcome appears to be similar: the organisms persist (and multiply) mainly inside vacuolar compartments derived from invagination of the cytoplasmic membrane (phagosome). The phagosome may have been fused with a lysosome to form a phagolysosome (but modified in such a way that the invaders do not become digested). At least some bacterial pathogens may actively lyse the vacuole and persist free in the cytosol of the host cell. Figure 1.42 summarizes the various modes of interaction between pathogenic microorganisms (mainly bacteria) and a eukaryotic cell, such as an epidermal or a phagocytic cell. The interaction with the host may be restricted to adherence at the cell surface, but may also include entry into the cell, where the organism may persist in phagosomes but may also enter the cytoplasm after pha-

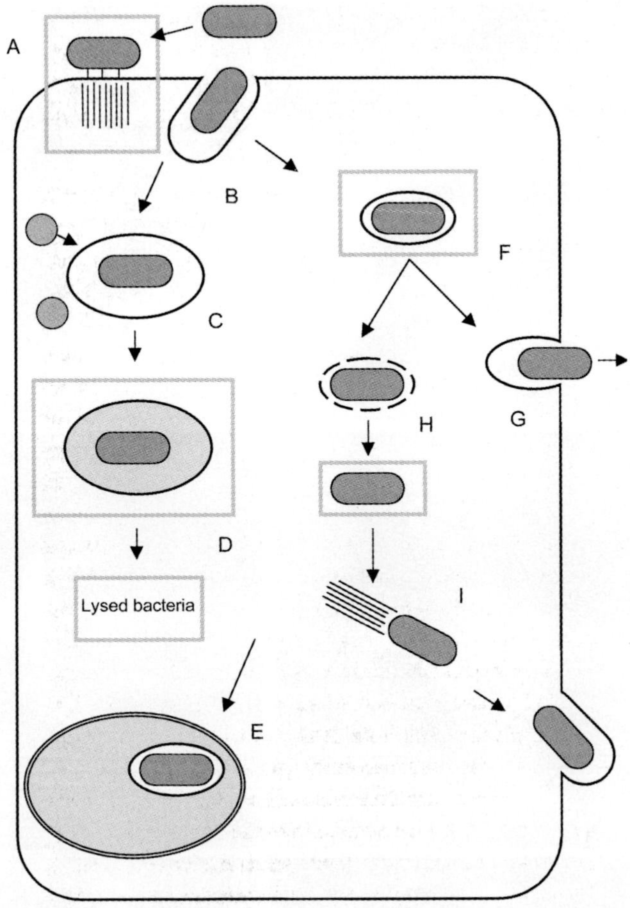

Fig. 1.42 Pathogenic microorganisms in various host cell compartments. Though numerous bacteria persist adhered at the cell surface (such as *Staphylococcus*, A), adherence is also the first prerequisite for invasion. Entry of the cell occurs through phagocytosis (B). The vesicle may become a phagolysosome after fusion with lysosomes (C, D). This process leads to death of non-adapted bacteria, but several pathogens may survive and persist in phagolysosomes. Alternatively, the vacuole does not develop into a phagolysosome (F). The bacteria may persist in this compact vacuolar compartment, leave the cell by exocytosis, or may enter the cytoplasm after lysis of the vacuolar membrane (H). Free bacteria actively move around in the cytoplasm by directed polymerization of an actin tail and may also re-enter neighboring tissue cells (I). Some records also account for the invasion of the nucleus by bacteria (E)

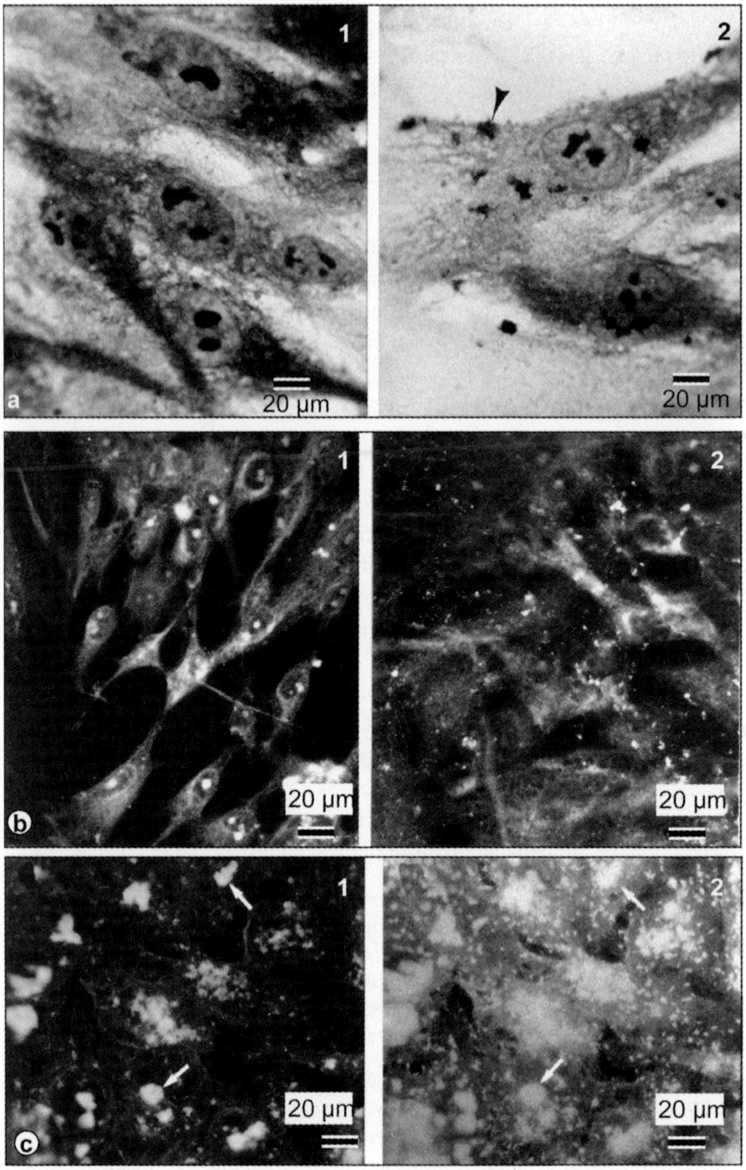

Fig. 1.43 Adherence of *Mycoplasma penetrans* to host cells:
(a,b) Adherence of *Mycoplasma penetrans* to cultured human cells 0 h (1) and 2 h (2) after infection. (a) Giemsa staining; (b) Fluorescence staining with FITC. Arrowhead indicates a cluster of *Mycoplasma* cells.
(c1) Anti-tubulin MAb and anti-mouse FITC conjugate were used to stain tubulin (arrows point at aggregated tubulin).
(c2) Same field as panel a1, except that mycoplasmas were labeled with anti-*M. penetrans* antibody and anti-rabbit rhodamine conjugate (arrows). The sites of large accumulations of tubulin correspond to the locations of *Mycoplasma* clusters (arrows). (Giron et al., 1996)

gosome lysis. Various means of exiting the host cell have been observed. The presence of rickettsia (and eukaryotic parasites) in the nuclei of host cells has been reported. The events summarized in Fig. 1.42 may all be followed by light microscopy and may be deduced from experiments with appropriate markers specific for lysosomes, phagosomes, and bacterial cells. Though structural details are not visible under the light microscope, important cytological events concerning cytoskeletal reorganization and phagosome and lysosome localization may be detected.

Adherence on the host cell and the first step of entry are depicted here for *Mycoplasma penetrans*. Members of the *Mycoplasma* group are ubiquitous parasites on the membranes of animal (as well as plant) cells, and have therefore been objects of research on interaction between bacterial parasites and eukaryotic host cells since the end of the nineteenth century. Their action on host cells is often not severely damaging, since they do not produce specific toxins. Nevertheless, members of the group may cause severe chronic infections of the respiratory tracts of animals and humans. Various species are often found as contaminants in animal cell cultures, while some members of the group have also been observed inside infected cells. They are thus also viewed as causative agents of chronic and systemic infections.

Mycoplasmas may be visualized by Giemsa staining as well as by use of fluorescent dyes (Fig. 1.43). As already depicted for the adherence of microorganisms to plant cells, adherence of *Mycoplasma* to cultured human cells also triggers a signal that promotes cytoskeletal changes. The application of antitubulin antibodies revealed that the sites of large accumulation of tubulin correspond to the location of *Mycoplasma* clusters (Giron et al., 1996). Pathogen-induced alterations of the cytoskeleton are common events during adhesion and, especially, penetration of pathogens. Since membrane invagination and phagocytosis are directed by the cytoskeleton, the pathogen-induced alterations are a prerequisite for pathogen-directed entry of a normally non-phagocytic host cell. Otherwise, intracellular pathogens may depend passively on the uptake mechanisms of phagocytic cells.

Figure 1.44 shows examples of two steps of the invasion of host cells by *Salmonella*. By application of bacterial and lysosomal markers, the presence of bacteria and phagosomes before and after fusion with lysosomal vesicles (steps B–D in Fig. 1.42) may be deduced. Co-localization of both markers, together with bacterial cells, revealed that the bacteria are taken up by phagosomes, which become phagolysosomes some minutes after infection. Although non-pathogenic bacteria are lysed by the action of phagolysosomes, these pathogens may persist and use the phagoloysosomes as a vehicle.

The invasion of a single cell is, of course, only the first step in pathogenesis. Spreading in a tissue, an organ, or the whole host body will usually follow. An important entry port for pathogens is depicted in Fig. 1.45, which shows *E. coli* strains specifically attached to Peyer's patch lymphoid follicle epithelium. These specialized epithelia are part of the mucosal immune system and are important for development of immunological response against pathogenic organisms in the intestine. Peyer's patches are therefore a target site for oral vaccines. Sections obtained from the distal ileum of a rabbit infected with a rabbit enteropathogenic *E. coli* strain were stained with Toluidine Blue and hematoxylin/eosin, for (non-specific) visualization of tissue and bacteria, and this clearly illustrated strain-dependent variations in adherence to the tissue surface.

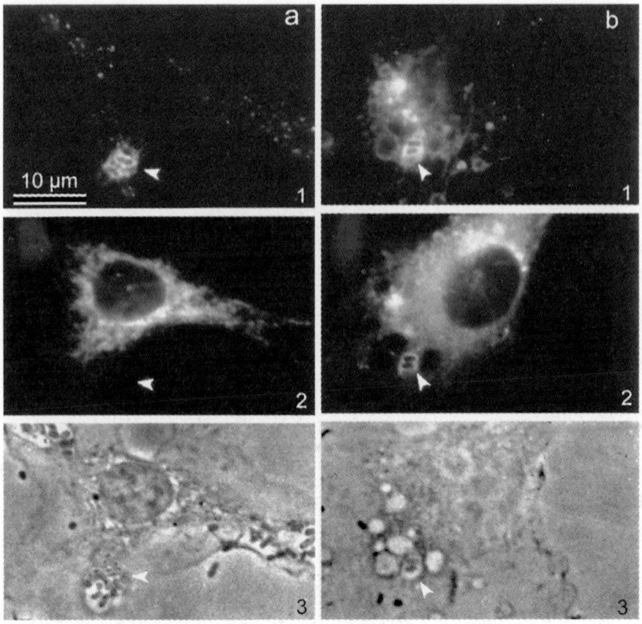

Fig. 1.44 Phagosome-lysosome fusion in macrophages containing *Salmonella typhimurium*

Macrophages were infected for 5 min with *Salmonella typhimurium* together with a phagosome marker for visualization of the phagosome interior in a fluorescence image. Lysosomal markers have been used for the visualization of lysosomes.

(a) 0 min after infection. 1: phagosomes (arrows) with bacteria displacing fluorescence of Texas Red labeled ovalbumin. 2: The sample area depicting fluorescence of a lysosomal marker. No lysosomal marker is visible in the phagosome (arrowhead). 3: bright-field phase contrast image of the area.

(b) 15 min after infection. 1: area equivalent to a1. 2: the lysosomal marker co-localizes with the phagosome marker (arrowhead) and the bacteria due to increased merging of phagosomes with lysosomes after longer incubation. 3: bright-field phase contrast image of the area. (Oh et al., 1996)

As in other biofilms, adherence is brought about by the expression of bacterial pili, which bind specifically to surfaces of tissue cells (von Moll et al., 1997). A similar experimental setup is presented in Fig 1.45 c, d with fluorescent-labeled cells of *Salmonella typhimurium* (Nickerson and Curtiss, 1997). The *Salmonella* strain depicted in Fig. 1.45 d is an *rpoS* mutant strain. This strain is unable to regulate a wide variety of genes in response to nutrient-limiting conditions and is less infective than a wild-type strain. The invasion of the epithelia and the underlying tissue is clearly visible. The cytological events visible by light microscopy are also summarized in Fig. 1.45 e, showing the *Salmonella*-type of invasion. M-cells are regularly used as entry ports by the bacteria, and these cells are key sites for antigen sampling. This feature is exploited by numerous pathogens in various ways. *Salmonella* enters M-cells and, via this cell

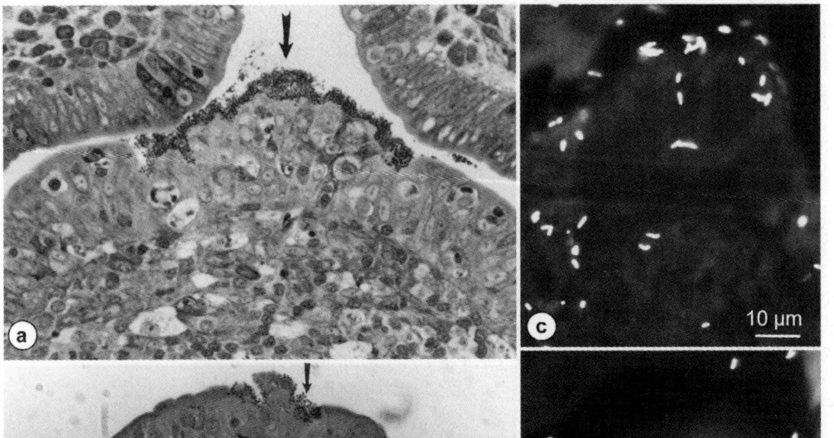

Fig. 1.45 Adherence of bacteria to lyphoid follicle tissue epithelia and intracellular movement:
(a, b) thin sections of rabbit Peyer's patch lymphoid follicle 24 h post-inoculation with two various *E. coli* strains exhibiting different degrees of adherence. Arrows depict adherence of bacteria to Peyer's patch lymphoid follicle epithelium (von Moll and Cantey, 1997).
(c, d) Fluorescence micrographs of a wild-type and a *rpoS* mutant strain of *Salmonella typhimurium* strains exhibiting different degrees of adherence and the respective colonization of lower tissue layers (Nickerson et al., 1997).
(e) Scheme illustrating the invasion of *Salmonella* by passage of M cells in Peyer's patches. Salmonella use these cells as entry ports and the associated macrophages as vehicles for dissemination and infection of the epithelial cells (Sansonetti and Phalipon, 1999).
(f) Actin tail of *Rickettsia rickettsii* in a mammalian cell. Actin of this cell line has been GFP tagged and is now visible as a fluorescent tail. At the tail end, the cell is visible as a less bright dot (labeled stained by immunofluorescence; bar represents 3 µm) (Heinzen et al., 1999)

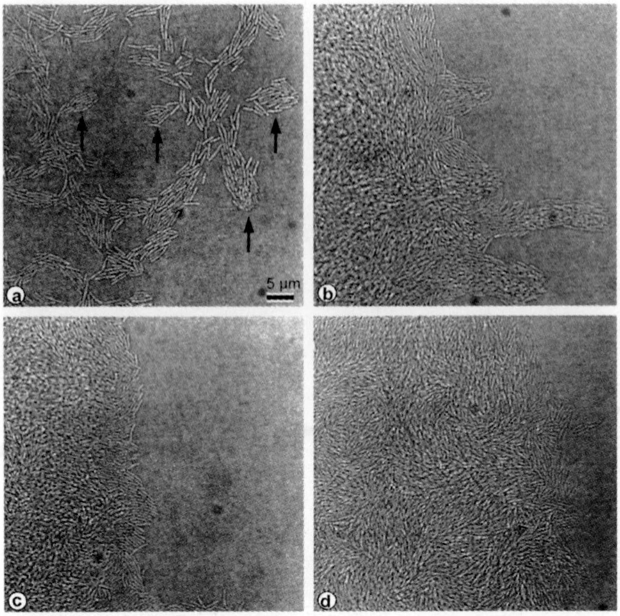

Fig. 1.46 One effect of quorum sensing in a *Pseudomonas aeruginosa* biofilm
Slide culture assay for twitching motility. The *P. aeruginosa* PAO1 wild-type and quorum-sensing mutants were inoculated onto slide cultures and incubated for 3 to 5 h at 37 °C. The areas in the photomicrographs represent the edges of cell growth. (a) Wild-type, (b–d) strains with various mutations in the regulation of quorum sensing. As a consequence, regulation of twitching motility is impaired, leading to the typical appearance of the edges. Twitching stains are infective and form small "rafts" (as indicated by arrows), whereas non-infective strains produce no rafts. All strains except the wild type form multiple cell layers (Glessner et al., 1999).

type, also into macrophages, as depicted in the schematic drawing. Macrophages are used as vehicles for spreading of the pathogen, either locally or inside the entire host.

An additional intracellular interaction step is used by parasites such as *Rickettsia* or *Listeria*. These are able to lyse phagosomes and are motile in the cytosol with the aid of the host's actin. The organisms use the propulsive force supplied by parasite-directed actin polymerization in the host cell. Many of the cellular events necessary for this movement have been defined. At the microscopic level, the most prominent structure is the actin tail, depicted in Fig 1.45 f.

The formation of many bacterial biofilms in interaction with living tissue appears to be regulated by quorum sensing systems. Essentially, the systems trigger cellular events that depend on cell density, such as the detachment of cells from a mature biofilm. The ubiquitous Gram-negative *Pseudomonas aeruginosa* in particular has become a model organism for investigation of regulatory events during biofilm formation. These events are governed in *Pseudomonas* by two quorum sensing systems, las and rhl, each of which has been shown to activate numerous genes in a hierarchical manner. Each system consists of a transcriptional activator and an autoinducer synthase. The autoinducers are N-acyl-homoserine lactones with variable acyl groups. These act as signal molecules in the quorum sensing system. Upon secretion, the freely diffusing molecules become accumulated in the biofilm proportionally to the cell density. A certain threshold concentration induces a number of alterations of the cells, connected to the structure of the biofilm community. Light microscopic methods reveal the differences in biofilm structure of wild-type and quorum sensing mutants (Fig. 1.46). The mutant strains are defective in a special type of motility ("twitching") essential for surface colonization, and so are unable to maintain a cellular monolayer. Since the quorum sensing system is relevant for interaction of organisms with both plant and animal tissue, these systems have great potential application. Disruption of the signaling events would be an interesting approach for modulation of biofilm formation and, hence, a possible way to interrupt infective processes.

References

1 Baty, A. M., Eastburn, C. C., Diwu, Z., Techkarnjanaruk, S., Goodman, A. E., Geesey, G. G., (**2000**). Differentiation of chitinase-active and non-chitinase-active subpopulations of a marine bacterium during chitin degradation. *Appl. Environ. Microbiol.* 66, 3566–3573.

2 Berchtold, M., Chatzinoas, A., Schönhuber, W., Brune, A., Amann, R., Hahn, D., König, H (**1999**). Differential enumeration and in situ localization of microorganisms in the hindgut of the lower termite *Mastotermes darwiniensis* by hybridization with rRNA-targated probes. *Arch. Microbiol.* 172, 407–416.

3 Benhamou, N., Nicole, M. (**1999**). Cell biology of plant immunization against microbial infection: the potential of produced resistance in controlling plant diseases. *Plant Physiol. Biochem.* 37, 703–719.

4 Braune, W., Leman, A., Taubert, H. (**1983**). Pflanzenanatomisches Praktikum I. 4th ed. Gustav Fischer, Stuttgart.

5 Brune, A., Friedrich, M. (**2000**). Microecology of the termite gut: structure and function on a microscale. *Curr. Opin. Microbiol.* 3, 263–269.

6 Carrió, M. M., Villaverde, A. (**2001**). Protein aggregation as bacterial inclusion bodies is reversible. *FEBS Lett.* 489, 29–33.

7 Chantt, J., Corrado, K., Pringle, J. R., Herskowitz, I. (**1991**). Yeast *BUD5*, encoding a putative GDP-GTP exchange factor, is necessary for bud site selection and interacts with bud formation gene *BEM1*. *Cell* 65, 1213–1224.

8 deBeer, D., P. Stoodley, F. Roe, and Z. Lewandowski (**1994**). Effects of biofilm structures on oxygen distribution and mass transport. *Biotech. Bioeng.* 43, 1131–1138.

9 de Pedro, M. A, Quintela, J. C, Höltje, J.-V., Schwarz, H. (**1997**). Murein Segregation in *Escherichia coli. J. Bacteriol.* 179, 2823–2834.

10 Den Blaauwen, T., Buddelmeijer, N., Aarsman, M. E., Hameete C. M., Nanninga N. (**1999**). Timing of FtsZ assembly in *Escherichia coli. J. Bacteriol.* 181, 5167–5175.

11 Flemming, H. (**1995**). Biofouling bei Membranprozessen. Springer, Berlin.

12 Franks, A. H., Harmsen, H. J. M., Raangs, G. C., Jansen, G. J., Schut, F., Welling, G. W. (**1998**). Variations of bacterial populations in human faeces measured by fluorescent in situ hybridization with group-specific 16S rRNA-targeted oligosaccharide probes. *Appl. Environ. Microbiol.* 64, 3336–3345.

13 Gage, D. J., Margolin, W. (**2000**). Hanging by a thread: invasion of legume plants by rhizobia. *Curr. Opin. Microbiol.* 3, 613–617.

14 Gill, W. M., Lapeyrie, F., Gomi, T., Suzuki, K. (**1999**). *Tricholoma matsutake* – an assessment of in situ and in vitro infection by observing cleared and stained whole roots. *Mycorrhiza* 9, 227–231.

15 Gillies, R. R., Dodds, T. C. (**1977**). Illustrierte Bakteriologie. Hans Huber, Bern.

16 Giron, J. A., Lange, M., Baseman, J. B. (**1996**). Adherence, fibronectin binding, and induction of cytoskeleton reorganization in

cultured human cells by *Mycoplasma penetrans*. *Infect. Immun.* 64, 197–208.
17 Glessner, A., Smith, R. S., Iglewski, B. H., Robinson, J. B. (**1999**). Roles of *Pseudomonas aeruginosa* las and rhl quorum-sensing systems in control of twitching motility. *J. Bacteriol.* 181, 1623–1629.
18 Gould, G. W., Hurst, A. (**1969**). The Bacterial Spore. Academic Press, London, New York.
19 Greaves, H., Levy, J. F. (**1968**). Microbial associations in the deterioration of wool under long-term exposure. In: *Biodeterioration of Materials*, Vol. 1, Academic Press, London (ed. J. J. Elphik), pp. 429–443.
20 Guerrero, R., Haselton, A., Sol, M., Wier, A., Margulis, L. (**1999**). *Titanospirillum velox*: a huge, speedy, sulfur-storing spirillum from Ebro Delta microbial mats. *Microbiology* 96, 11584–11588.
21 Gunning, B. E. S., Steer, M. W. (**1996**). Bildatlas zur Biologie der Pflanzenzelle. 4th ed. Gustav Fischer, Stuttgart.
22 Haahtela, K., Korhonen, T. K. (**1985**). In vitro adhesion of N_2-fixing enteric bacteria to roots of grasses and cereals. *Appl. Environ. Microbiol.* 49, 1186–1190.
23 Hardham, A. R., Mitchell, H. J. (**1998**). Use of molecular cytology to study the structure and biology of phytopathogenic and mycorrhizal fungi. *Fungal Genet. Biol.* 24, 252–284.
24 Hillmann, G., Geuertsen, W. (**1997**). Light-microscopical investigation of the distribution of extracellular matrix molecules and calcifications in human dental pulps of various ages. *Cell Tissue Res.* 289, 145–154.
25 Heinonsalo, J., Jörgensen, K. K., Sen, R. (**2001**). Microcosm-based analysis of Scots pine seedling growth, ectomycorrhizal fungal community structure and bacterial carbon utilization profiles in boreal forest humus and underlying illuvial mineral horizons. *FEMS Microbiol. Ecol.* 36, 73–84.
26 Heinzen, R. A., Grieshaber, S. S., Van Kirk, L. S., Devin, C. J. (**1999**). Dynamics of actin-based movement by *Rickettsia rickettsii* in Vero cells. *Infect. Immun.* 67, 4201–4207.
27 Jarchau, T. (**1985**). Untersuchungen zum Vorkommen und zur Häufigkeit von Proteinvorstufen bei der Bildung von 1- und F-Pili in *Escherichia coli*. Thesis, Univ. Göttingen.
28 Jayaraman, A., Earthman, J. C., Wood, T. K. (**1997**). Corrosion inhibition by aerobic biofilms on SAE 1018 steel. *Appl. Microbiol. Biotechnol.* 47, 62–68.
29 Kirsch, C., Takamiya-Wik, M., Reinold, S., Hahlbrock, K., Somssich, I. E. (**1997**). Rapid, transient, and highly localized induction of plastidial omega-3 fatty acid desaturase mRNA at fungal infection sites in *Petroselinum crispum*. *Proc. Natl. Acad. Sci. USA* 94, 2079–2084
30 Kleinig, H. (**1986**). Zellbiologie: ein Lehrbuch. Hans Kleinig, Peter Sitte. 2nd ed. Gustav Fischer, Stuttgart.
31 Kleinig, H., Maier, U. (**1999**). Zellbiologie. 4th ed. Gustav Fischer, Stuttgart.
32 Kobayashi, I., Kobayashi, Y., Hardham, A. R. (**1994**). Dynamic reorganization of microtubules and microfilaments in flax cells during the resistance response to flax rust infection. *Planta* 195, 237–247.
33 Korber, D. R., Lawrence, J. R., Lappin-Scott, H. M., Costerson, J. W. (**1995**). Growth of microorganisms on surfaces. In: *Microbial Biofilms. Plant and microbial biotechnology research series; 5* (eds. H. M. Lappin-Scott, J. W. Costerton). Cambridge University Press, Cambridge.
34 Kühnel, W. (**1995**). Taschenatlas der Zytologie, Histologie und mikroskopischen Anatomie. 9th ed., Thieme, Stuttgart.
35 Malandrin, L., Huber, H., Bernander, R. (**1999**). Nucleoid structure and partition in *Methanococcus jannaschii*: an archaeon with multiple copies of the chromosome. *Genetics* 152, 1315–1323.
36 Mayer, F. (**1986**). Cytology and morphogenesis of bacteria. Gebrüder Borntraeger, Berlin.
37 McDaniel, D., Roberson, R. W. (**2000**). Microtubles are required for motility and positioning of vesicles and mitochondria in hyphal tip cells of *Allomyces macrogynus*. *Fungal Genet. Biol.* 31, 233–244.
38 Mogge, B., Loferer, C., Agerer, R., Hutzler, P., Hartmann, A. (**2000**). Bacterial community structure and colonization patterns of *Fagus sylvatica* L. ectomycorrhizospheres as determined by fluorescence in situ hybridization and confocal laser scanning microscopy. *Mycorrhiza* 9, 271–278.
39 Murdoch, L. J., Kobayashi, I., Hardham, A. R. (**1998**). Production and characterisation of monoclonal antibodies to cell wall components of the flax rust fungus. *Eur. J. Plant Pathol.* 104, 331–346.
40 Nickerson, C. A., Curtiss, R. (**1997**). Role of sigma factor RpoS in initial stages of *Sal-*

monella typhimurium infection. *Infect. Immun.* 65, 1814–1823.

41 Nielsen, A. T., Liu, W., Filipe, C., Grady, L., Molin, S., Stahl, D. A. (**1999**). Identification of a novel group of bacteria in sludge from a deteriorated biological phosphorus removal reactor. *Appl. Environ. Microbiol.* 65, 1251–1258.

42 Niles, W. D., Malik, A. B. (**1999**). Endocytosis and exocytosis events regulate vesicle traffic in endothelial cells. *J. Membr. Biol.* 167, 85–101.

43 Nultsch, W., Grahle, A. (**1983**). Mikroskopisch-botanisches Praktikum für Anfänger. 7th ed., Thieme, Stuttgart.

44 Oh, Y-K., Alpuche-Aranda, C., Berthiaume, E., Jinks, T., Miller, S. I., Swanson, J. A. (**1996**). Rapid and complete fusion of macrophage lysosomes with phagosomes containing *Salmonella typhimurium*. *Infect. Immun.* 64, 3877–3883.

45 Okabe, S., Itoh, T., Satoh, H., Watanabe, Y. (**1999**). Analyses of the spatial distribution os sulfate-reducing bacteria and their activity in aerobic wastewater biofilms. *Appl. Environ. Microbiol.* 65, 5107–5116.

46 Pabel, J. J., Yang, A. F. (**1981**). Light and electron microscopic studies of the nodule structure of alfalfa. *Can. J. Microbiol.* 27, 36–46.

47 Parniske, M. (**2000**). Intracellular accommodation of microbes by plants: a common developmental program for symbiosis and disease? *Curr. Opin. Plant Biol.* 3, 320–328.

48 Peterson, R. L. (**1991**). Histochemistry of Ectomycorrhiza. In: *Methods in Microbiology, Vol. 23, Techniques for the study of Mycorrhiza* (eds. J. R. Norris, D. J. Read, A. K. Varma), pp. 107–120. Academic Press, London.

49 Pfennig, N., Trüper, H.G. (**1992**). The family *Chromatiaceae*. In: *The Prokaryotes*, 2nd ed. (eds: A. Balows, H. G. Trüper, M. Dworkin, W. Harder, K.-H. Schleifer), pp. 3200–3221. Springer, New York.

50 Pohl, W. (**2000**). Wechselwirkungen zwischen endolithischen Biofilmen und Karbonatgesteinen in alpinen Gebieten Mitteleuropas. Dissertation, Universität Göttingen.

51 Sansonetti, P. J., Phalipon, A. (**1999**). M cells as ports of entry for enteroinvasive pathogens: mechanisms of interaction, consequences for the disease process. *Semin. Immun.* 11, 193–203.

52 Schönholzer F., Hahn, D., Zeyer, J. (**1999**). Origins and fate of fungi and bacteria in the gut of *Lumbricus terrestris L.* studied by image analysis. *FEMS Microbiol. Ecol.* 28, 235–248.

53 Scott, K. P., Mercer, D. K., Glover, L. A., Flint, H. J. (**1998**). The green fluorescent protein as a visible marker for lactic acid bacteria in complex ecosystems. *FEMS Microbiol. Ecol.* 26, 219–230.

54 Sharpley, J. M., King, M. E. (**1972**). Laboratory analysis of problems in papermakers felts. In: *Biodeterioration of Materials*, Vol. 2 (eds. A. H. Walters, E. H. Hueck-van-der Plas), pp. 161–167. Applied Science Publ., London.

55 Trolldenier, G. (**1996**). Counting by Fluorescence Microscopy. In: *Methods in Soil Biology* (eds. F. Schinner, R. Ohlinger, E. Kandeler), pp. 13–17. Springer, Berlin.

56 van Ommen Kloeke, F., Geesey, G. G. (**1999**). Localization and identification of populations of phosphatase-active bacterial cells associated with activated sludge flocs. *Microb. Ecol.* 38, 201–214.

57 von Moll, L. K., Cantey, J. R. (**1997**). Peyer's patch adherence of enteropathogenic *Escherichia coli* strains in rabbits. *Infect. Immun.* 65, 3788–3793.

58 Wagner, M., Amann, R. (**1996**). Die Anwendung von in situ-Hybridisierungssonden zur Aufklärung von Struktur und Dynamik der mikrobiellen Biozönosen in der Abwasserreinigung. In: *Ökologie der Abwasserorganismen* (eds. H. Lemmer, T. Griebe, H.-C. Flemming), pp. 93–110. Springer, Berlin.

59 Wagner, M., Amann, R., Lemmer, H., Schleifer, K.-H. (**1993**). Probing activated sludge with oligonucleotides specific for proteobacteria: inadequacy of culture-dependent methods for describing microbial community structure. *Appl. Environ. Microbiol.* 59, 1520–1525.

60 Weier, T. E., Stocking, C. R. Barbour M. G. (**1974**). Botany: An Introduction to Plant Biology, 5th ed., John Wiley, New York.

61 Wente, S. R., Gasser, S. M., Caplan, A. J. (**1997**). The nucleus and nucleocytoplasmic transport in *Saccharomyces cerevisiae*. In: *The molecular and cellular biology of the yeast Saccharomyces* (eds. J. R. Pringle, J. R. Broach, E. W. Jones), pp. 471–546. Cold Spring Harbor Laboratory Press, Cold Spring Harbor.

1.3
Applications and Limitations of Confocal Laser Scanning Microscopes (CLSMs)

Though some images taken with a CLSM have already been presented in the previous section, the techniques described were devoted to "standard" light microscopy. Standard microscopes are affordable even for small laboratories and easily to handle in daily routine tasks. In spite of the fact that confocal microscopes have become affordable for larger departments or central facilities, they remain primarily a research tool rather than an instrument of daily routine.

The most important advantage of the CLSM is its ability to deliver thin optical sections through a thick specimen. The sections are not blurred by light below and above the focal plane. Through the assembling of a series of optical sections, three-dimensional information is obtained without the need for physical sectioning. Since confocal microscopy, in principle, produces an image composed of defined points of a specimen that are excited by a focused beam of light, the confocal image is a reflected light image. Although imaging may, in principle, be performed with backscattered light, most applications are performed with excited/emitted light of autofluorescent or fluorescent-stained samples. Again, the application of fluorescent dyes allows the localization of structures that remain invisible by bright-field imaging techniques. Although the intense focal spot rapidly bleaches photosensitive dyes, it is possible to select conditions that allow minimal total exposure time for each focused spot.

The optical section is the important image unit of the CLSM. The confocal fluorescent image may be combined with a (non-confocal) standard transmitted light image of the specimen. A series of optical sections along the z-axis (a z-series image stack) can be combined to produced extended focus images or used for reconstruction of side views (x/z sections) and three-dimensional representations of the specimen. In time-lapse studies, a series of optical sections can also be processed into a three- or four-dimensional data set (two or three spatial dimensions and time). For some time-lapse studies, single line scans, with very short recording times, may also provide sufficient information.

There are some limitations for opaque specimens. Laser light – the common light source for confocal microscopes – is sufficiently intense to excite a fluorophore deep in the specimen. However, the number of photons emitted is limited by the concentration of the dye (given a relatively unlimited amount of excitation light available from the laser). Deeper inside the tissue, most photons will be scattered and not reach the detector. In addition, differences in refractive indices within the tissue will distort the cone of light, corrupting confocality. Another problem is that high numerical aperture objectives generally have small working distances, or in other words the lower areas of thick tissues cannot be sectioned by the microscope.

1.3.1
Imaging of Cells and Cellular Components

CLSM is especially useful where visualization – or exclusion – of a specimen image in all three dimensions is important. The following examples therefore only describe some application examples in which CLSM is especially advantageous, such as where resolution of the z-axis is of special use. Optical sectioning allows the generation of sections down to 0.25 μm. This is mostly not sufficient to provide slices of reasonable thickness taken from bacterial cells of normal size (1.5×0.5 μm). Yeast cells (approx. 5 μm in diameter) may be sectioned into about ten slices, sufficient to resolve the distribution of cellular components in compartments. Other eukaryotic cells, usually around 30 μm in diameter, allow the generation of numerous sections that may be individually combined to produce convincing images.

Images of 40 serial sections from cells with Mitofluor Green stained mitochondria and Syto 11 stained nuclei, allow an in-focus image to be generated throughout the whole cell (Fig. 1.47 a). Selection of optical sections allows imaging without "overcrowding" the final image with complex structures. Figure 1.47 b depicts cortical microtubules in a projection of nine optical sections taken at 0.5 μm intervals. ER cisternae and the Golgi apparatus may also be visualized in a similar way.

Although visualization of dynamic processes does not require confocal imaging, the extraction of defined focal planes helps for concentration on the relevant image information. Figure 1.48 depicts a time-resolved analysis: the appearance and disappearance of secretory granules during exocytosis at a cell surface. Figure 1.48 b gives an example of the internalization of insulin in endothelial cells. After 2 min incubation, the insulin-FITC signal appears as filaments. After an incubation time of 10 min the distribution pattern is more granular. The filamentous appearance of internalized insulin (and also other ligands) is probably caused by the action of microtubules. Ad-

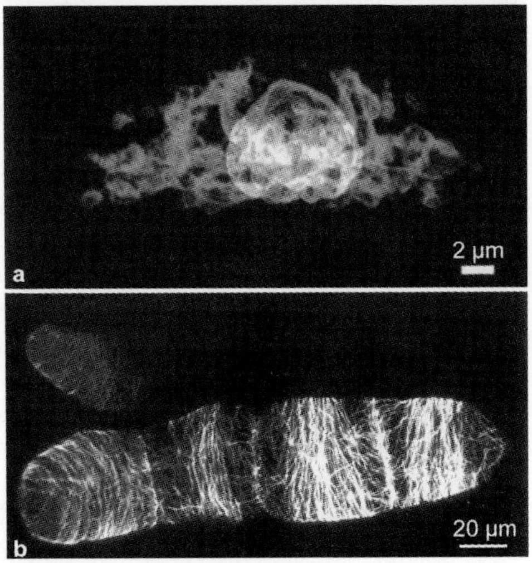

Fig. 1.47 (a) Mitofluor Green stained mitochondria followed by Syto 11 staining of the nucleus (red) in an embryonic chick heart myocyte. Stained mitochondria form a ribbon-like mass surrounding the nucleus. Images were generated from 40 serial sections (0.08 μm) at a step size of 0.49 μm (Bkaily et al., 1999) (b) Immunolocalization of cortical microtubules in the tobacco (*Nicotiana tobacum*) suspension culture BY-2. The cells were taken from a 5-day-old culture that was rapidly dividing, and the microtubules were labeled with YOL 1/34 (Sera-Lab, Crawley Down, UK). Projection of nine optical sections taken at 0.5 μm intervals (Wymer et al., 1999)

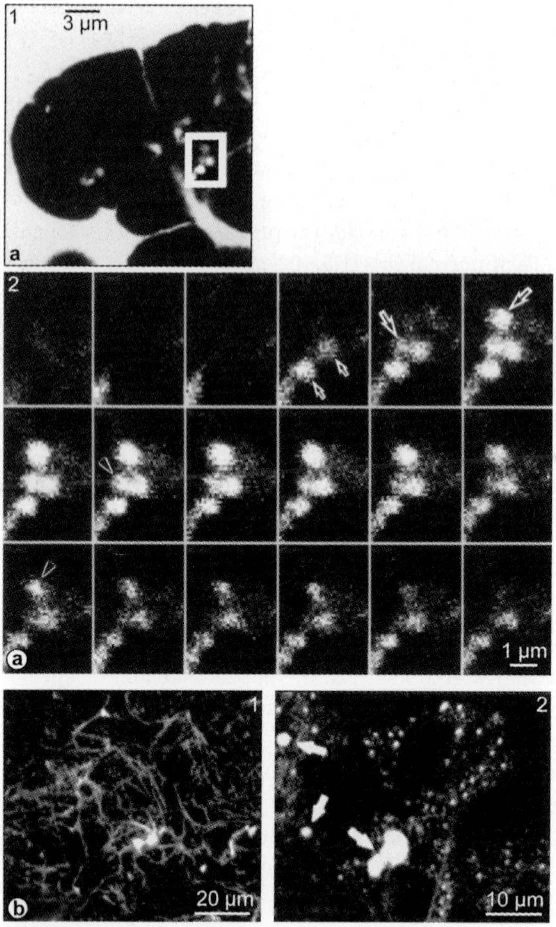

Fig. 1.48 (a) Analytical confocal images showing the exocytosis-removal cycle of single secretory granules. 1, Overall view, 2, Enlarged images, indicated by rectangle in 1, displayed sequentially from the top left to the right bottom. Appearing fluorescent spots are indicated by small and large arrows, whereas disappearing spots are shown by arrowheads (Segawa, 1999)
(b) Insulin-FITC internalization by living endothelial cells. 1, after 2 min of incubation, the insulin-FITC signal appears as filaments. 2, In a distinct experiment, the image was acquired after 10 min of incubation with insulin-FITC; the disruption pattern is now mostly granular. Insulin-FITC also accumulates in large bodies (arrows) (Orlandini et al., 1999)

ditional experiments with specific staining of microtubules revealed co-localizaton of receptor-ligand signals. This feature illustrates the well known function of microtubules in molecular/vesicular intracellular trafficking.

1.3.2
Imaging of Complex Specimens

While serial physical sectioning of a tissue specimen is of course the classical means to reconstruct the three-dimensional distribution, it is time-consuming and requires fixation, embedding, and microtomy prior to inspection. Here, confocal images of thick tissue sections may be an alternative.

Figure 1.49 shows a stereo pair of the vascular net of teeth. Specimens are slices of 0.5 mm, impossible to depict in conventional light microscopy. The stereo pair image

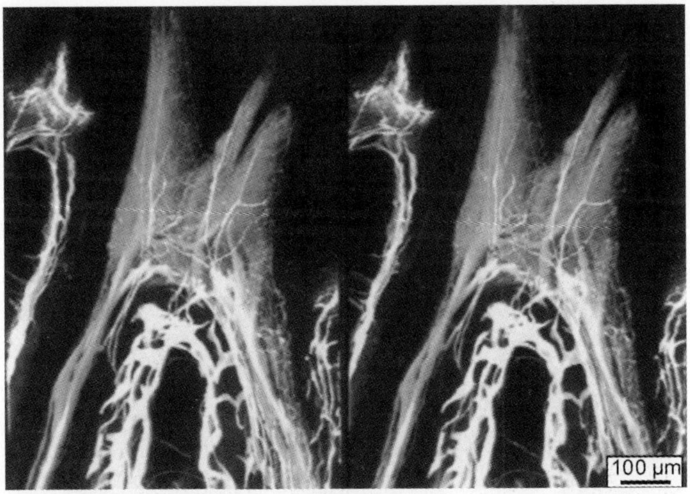

Fig. 1.49 Stereopair image of a vascular net in a mouse tooth. A mouse was perfused with RITC-labeled gelatin and injected with 0.11 ml (net) of FITC-labeled gelatin just before the end of perfusion. The mandible was removed and decalcified with EDTA. The tooth was sliced at 500 μm, treated with sodium deoxycholate, and mounted in Rigolac. Stereopair images were reconstructed from 35 images at 5.81 μm intervals. The perfusion of the circulatory system from the left ventricle with a fluorochrome-labeled gelatin followed by a small amount (0.1–0.6 ml) of gelatin labeled with another colored fluorochrome shows that arterial vessels are distinguished from the venous vessels by the fluorescence from the gelatin filling them (Hashimoto et al., 1999)

of murine decalcified tooth is composed of 35 single slices. The three-dimensional vascular network has been visualized by perfusion with the fluorescent markers RITC and FITC (rhodamine and fluorescein isothiocyanate). The three-dimensional image also allows quantification of the surface areas and volumes of complex structures, as demonstrated for the visualization of mycorrhiza fungal structures (Figs. 1.50 and 1.51). As a prerequisite, the structures of interest should produce a fluorescence signal that allows discrimination from the surrounding plant material. Figure 1.50 a–i depict optical slices through a hyphal coil of *Glomus* sp. in *Allium porrum*. From this data set it is possible to create an extended focus image (Fig. 1.50 j) or a volume-rendered three-dimensional reconstruction, as depicted in Fig 1.51. The data set also allows estimation of the surface area and volume of the depicted specimen.

A whole-mount specimen of animal tissue is depicted in Fig. 1.52. Confocal imaging was used to illustrate the development of a capillary network in stimulated muscle tissue. The small blood vessels are stained with a fluorescent-labeled lectin. Combination of several slices in the extended focus view shows a tight network depicted without out-of focus blur.

In a whole-mount preparation, as depicted in Fig. 1.53, the exclusion of irrelevant information from below and above a desired section of a root tip from an *Arabidopsis* plant is of use. Distracting cell layers are invisible. The specific immunofluorescent

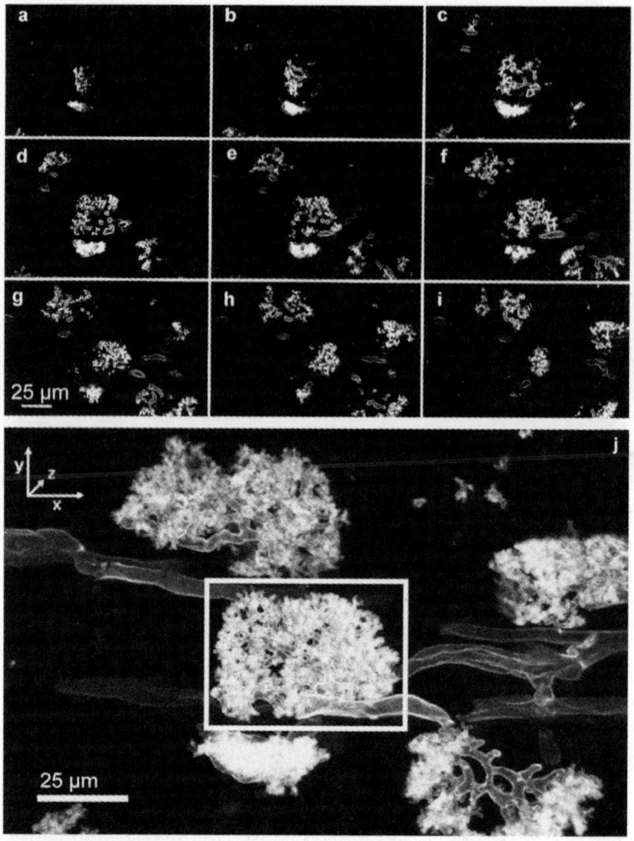

Fig. 1.50 (a–i) Representative optical slices through a group of *Arum*-type arbuscules of *Glomus* sp. in *Allium porrum* shown at 2.1 μm intervals on the z-axis (Dickson and Kolesik, 1999).
(j) An extended focus image of a group of *Arum*-type arbuscules of *Glomus* sp. in *A. porrum* (11 images merged at 2.1 μm intervals) (Dickson and Kolesik, 1999)

stain only shows rows of nuclei with darker nucleoli inside and bright fluorescent spots indicating the presence of a protein specific for small nuclear ribonucleoproteins (U2B", involved in pre-mRNA splicing).

1.3.2.1 **Three-dimensional imaging of microbial communities**

Since chemical fixation of highly hydrated and fragile biofilms is sometimes a delicate task, while monitoring of developing biofilms is possible in continuous flow chambers specially adapted for microscopy, most data concerning the structure and development of biofilms have been obtained with the aid of confocal microscopy. The earliest applications of CSLM to biofilm research were carried out with monoculture biofilms and established the important models of biofilm growth or architecture as already

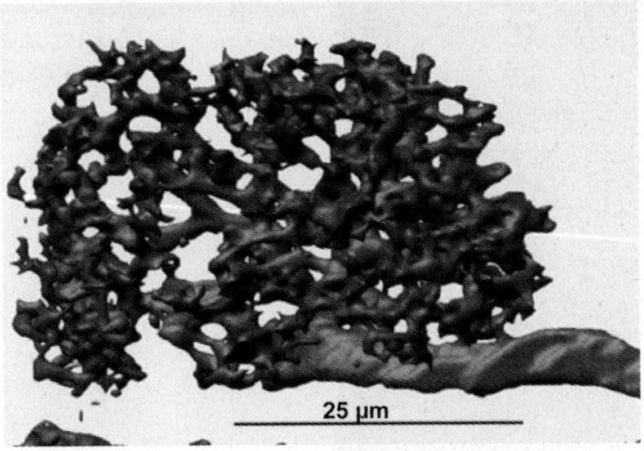

Fig. 1.51 Three-dimensional reconstruction of an arbuscule of *Glomus* sp. City Beach in *A. porrum* from within the group shown in Fig. 1.50j (Dickson and Kolesik, 1999)

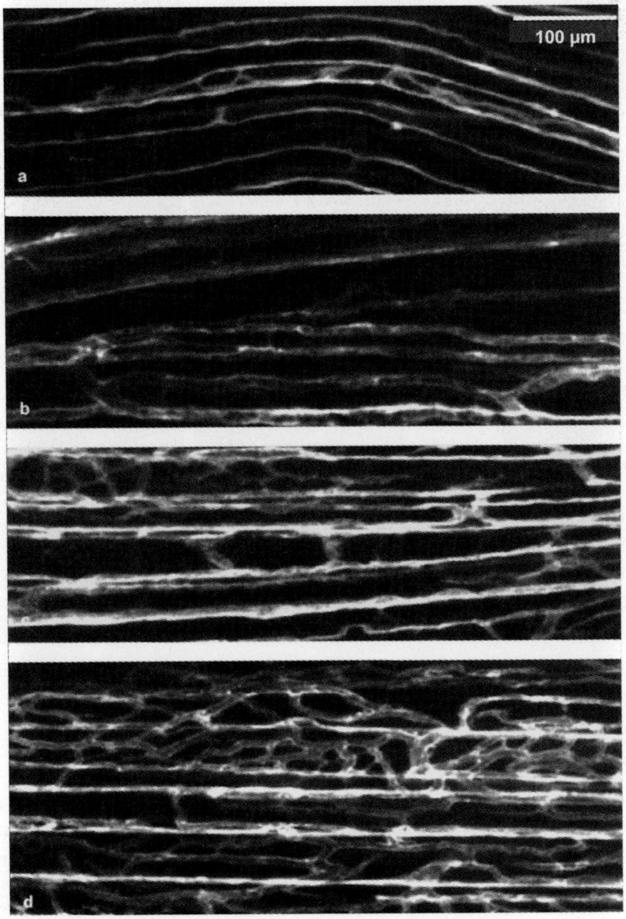

Fig. 1.52 Development of a capillary network in rat skeletal muscle. Control (a), 2 (b), 4 (c), and 7 (d) days stimulated after staining for lectin *Griffonia simplicifolia* I (Hansen-Smith et al., 1996)

Fig. 1.53 Whole mount immunofluorescence labeling of an *Arabidopsis* root with the 4G3 antibody. The 4G3 antibody labels the snRNP-specific protein U2B. It is interesting to note that the label is confined to the nucleus and is excluded from the nucleolus. The round bright spots, often associated with the nucleolus, are coiled bodies (Boudonck et al., 1999; Wymer et al., 1999)

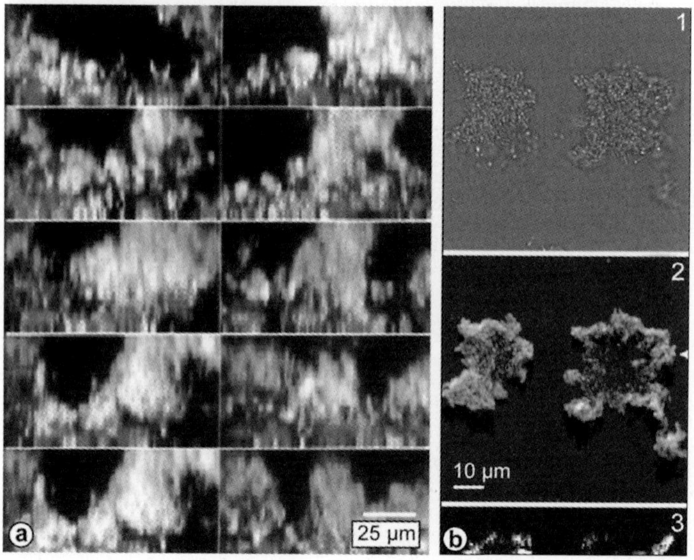

Fig. 1.54 (a) Confocal laser micrographs of fluorescent Gram-stained biofilm showing the distribution of Gram-positive (dark gray) and Gram-negative (light gray) bacteria in a series of *XZ* images through the biofilm (Wolfaardt et al., 1998)

(b) Demonstration of microbial growth activity in a flow chamber monoculture biofilm with the strain SM1639 [*Pseudomonas putida* R1::*Prrn*BP1-*gfp*(AAV)]. 1, bright-field image; 2, CLSM image of the green fluorescent cells presented as a simulated fluorescent projection where a long shadow indicates a large, high microcolony; 3, vertical section through the biofilm collected at the position indicated by the white triangle. The substratum is located at the left edge of panel 3. The vertical section is included to provide information on vertical microcolony size. All images were recorded with the same CLSM settings. The scale bar applies to all of the figures (Sternberg et al., 1999) described above. Studies concerning nutrient supply, nutrient flow, and their effects on biofilm structure, as well as regulation of biofilm development by signal molecules, have been carried out by application of a CLSM to whole biofilms (Palmer and Steinberg, 1999). Fluorescent probes for the detection of bacterial groups, specific cell wall polymers, or molecules of the biofilm matrix are routinely used for imaging of the spatial arrangement of biofilm components. z-Axis sections allow visualization of the vertical distribution of organisms as depicted in Fig. 1.54 a, relating to the distribution of Gram-positive and Gram-negative bacteria. The introduction of GFP has brought about advances in the understanding of gene activity in biofilms. Through the application of unstable GFP, it was possible to monitor dynamic changes of rRNA gene promoter activity (i.e., indirectly the dynamic changes in the rRNA level), allowing determination of growth activity in a complex community

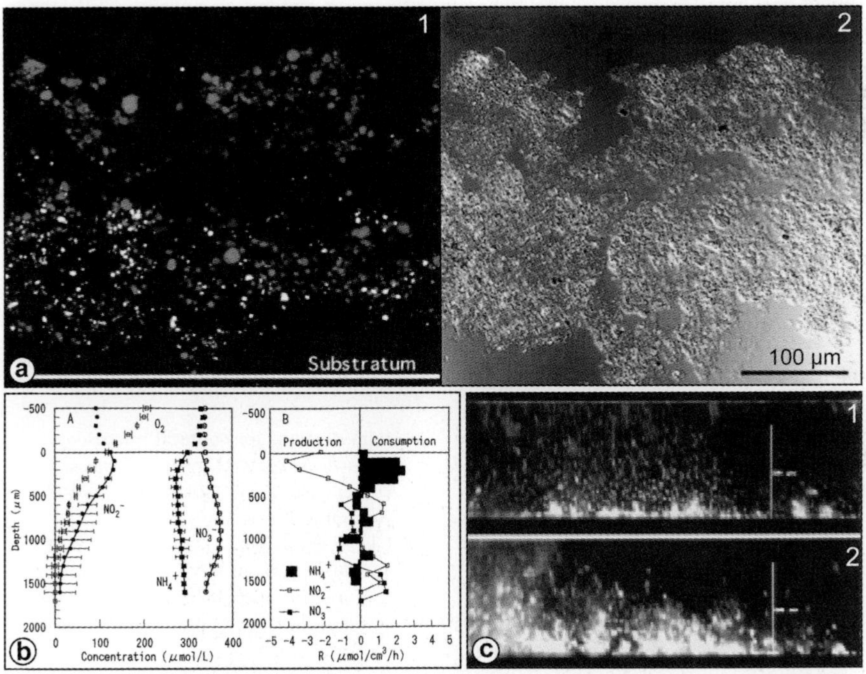

Fig. 1.55 (a) Simultaneous in situ hybridization of vertical sections of an autotrophic nitrifying biofilm with TRITC-labeled probe NEU and FITC-labeled probe Ntsp454. (1 and 2) CSLM projection image (1) and corresponding DIC image (2) of the entire physical biofilm vertical section. The biofilm thickness was about 350 μm (Okabe et al., 1999).
(b) Steady-state concentration profiles for O_2, NH_4^+, NO_2, and NO_3 in a domestic wastewater biofilm incubated in air-saturated medium containing no acetate, 300 μM NH_4^+, 300 μM NO_3, and 100 μM NO_2 (A) and spatial distribution of the estimated volumetric rates of consumption and production of NH_4^+, NO_2, and NO_3 (B). The points are measured means standard deviations, and the lines are the calculated profiles. The biofilm surface was at a depth of zero. R, rate (Okabe et al., 1999).
(c) Most regions in the biofilm matrix in which diclofop accumulation occurred had an overall hydrophobic character. *XZ* sections of the same field in a biofilm show the vertical distribution of binding sites for Nile Red (1; hydrophobic regions) and diclofop (2). Scale bars = 20 μm (Wolfaardt et al., 1998)

(Fig. 1.54 b). With use of these reporter tools, it could be demonstrated that differences in growth activity in individual cells occur in aged *Pseudomonas putida* biofilms. In the initial phases of biofilm development, all cells were highly active, with the ribosomal promoter activity of rapidly growing cells. After primary colonization and formation of small microcolonies, activity gradually decreased, first in the central parts of biofilm microcolonies.

A major approach in biofilm research is to understand the synergistic actions of biofilm organisms in wastewater treatment and decomposition of xenobiotics. The key process for nitrogen cycling in wastewater treatment, microbial nitrification (oxidation of ammonia to nitrate via nitrite), followed by denitrification (reduction of nitrite to nitrogen gas) is easier to understand at the level of the bacterial community by application of CLSM. Characterization of nitrifying bacterial populations in biofilms by culture-dependent techniques is difficult, since neither localization nor determination of the exact number of the naturally occurring species is possible. FISH performed with a set of oligonucleotide probes can reveal symbiotic associations of ammonia- and nitrite-oxidizing bacteria. Combinations of confocal microscopy of labeled cells and microelectrode measurements have demonstrated that the active ammonia oxidizing zone in biofilms is separated from the nitrite oxidizing zone, as reflected by the spatial arrangement of the organisms (see Fig. 1.55 a, b, representing a *physical* section) as well as by microelectrode measurements. The mapping of hydrophobic regions (Nile red staining) and of xenobiotic compounds (e.g., diclofop) is depicted in Fig. 1.55 c (representing an *optical* section). The herbicide could be sorbed by the extracellular polymer matrix, and used as a carbon and energy source.

References

1 Bkaily, G., Jacques, D., Pothier, P. (**1999**). Use of cofocal microscopy to investigate cell structure and function. In: *Methods in Enzymology 307, Confocal Microscopy* (ed. P. M. Conn), pp. 119–135. Academic Press; San Diego.

2 Boudonck, K., Dolan, L., Shaw, P.J. (**1999**). The movement of coiled bodies visualized in living plant cells by the green fluorescent protein. *Mol. Biol. Cell.* 10, 2297–2307.

3 Dickson, S., Kolesik, P. (**1999**). Visualisation of mycorrhizal fungal structures and quantification of their surface area and volume using laser scanning confocal microscopy. *Mycorrhiza* 9, 205–213.

4 Hansen-Smith, S.M., Hudlicka, O., Eggington, S. (**1996**). In vivo angiogenesis in adult rat skeletal muscle: early changes in capillary network architecture and ultrastructure. *Cell Tissue Res.* 286, 123–136.

5 Hashimoto, H., Ishikawa, H., Kusakabe, M. (**1999**). Preparation of whole mounts and thick sections for confocal microscopy. In: *Methods in Enzymology 307, Confocal Microscopy* (ed. P. M. Conn), pp. 84–107. Academic Press; San Diego.

6 Okabe, S., Sathoh, H., Watanabe, Y. (**1999**). In situ analysis of nitrifying biofilms as determined by in situ hybridization and the use of microelectrodes. *Appl. Environ. Microbiol.* 65, 3812–3191.

7 Orlandini, G., Ronda, N., Gatti, R., Gazzola, G.C., Borghetti, A. (**1999**). Receptor-ligand internalization. In: *Methods in Enzymology 307, Confocal Microscopy* (ed. P. M. Conn), pp. 340–350. Academic Press; San Diego.

8 Palmer, R.J. jr., Sternberg, C. (**1999**). Modern microscopy in biofilm research: Confocal microscopy and other approaches. *Curr. Opin. Biotechnol.* 10, 263–268.

9 Segawa, A. (**1999**). Measurement of secretion in confocal microscopy. In: *Methods in Enzymology 307, Confocal Microscopy* (ed. P.

M. Conn), pp. 328–340. Academic Press; San Diego.
10 Sternberg, C., Christensen, B.B., Johansen, T., Nielsen, A.T., Andersen, J.B., Givskov, M., Molin, S. (**1999**). Distribution of bacterial growth activity in flow-chamber biofilms. *Appl. Environ. Microbiol.* 65, 4108–4117.
11 Wolfaardt, G.M., Lawrence, J.R., Robarts, R.D., Caldwell, D.E. (**1998**). In situ characterization of biofilm exopolymers involved in the accumulation of chlorinated organics. *Microb. Ecol.* 35, 213–223.
12 Wymer, C.L., Beven, A.F., Boudonck, K., Lloyd, C.W. (**1999**). Confocal microscopy of plant cells. In: *Methods in Molecular Biology 122, Confocal Microscopy* (ed. S. W. Paddock), pp. 103–130. Humana Press; Totowa, NJ.

1.4 Instrumentation and Preparation Techniques

1.4.1 Basic Features of Lenses, Lens Aberration, and Correction

Only in a (hypothetical) perfect converging lens will parallel light rays be refracted in such a way that they converge at a single point – the principal focus, or focus point. If an object is placed between infinity and the principal focal point of the lens, every point of the original object will be focused onto a plane on the other side of the lens beyond the principal image plane. The image of the object will be inverted relative to the original. Object and image may also be represented by imaginary planes. A plane in front of a lens would be focused onto points along a corresponding plane behind the lens, and vice versa, so one plane is the image of the other. These common planes of focus are referred to as conjugate planes. A real object or an image projected by a first lens (an "intermediate" image) placed between another lens and the focal point of this lens produces an upright and enlarged image (virtual image). This two-lens construction represents a simple compound microscope containing one objective lens and one ocular lens. From this starting point onwards (an original version of a compound microscope was manufactured in 1590), the development towards modern compound microscopes may be viewed as the improvement of the correction of the real lenses (see Spencer 1982, and Meyer 1995 for introductory overviews).

In contrast with their quite simple theoretical characteristics, real glass lenses suffer from a number of aberrations that would massively reduce the actual resolution of the microscope if they were not corrected. The user should be aware of these lens errors, because they are still present in corrected lenses (or lens groups), even though major improvements have been made. The development of achromatic lens in the first half of the 19th century, and especially the work of Ernst Abbe in the late 19th century, prepared the ground for the construction of modern research microscopes. Abbe worked out a comprehensive theory of image formation in light microscopes and constructed apochromatic objectives (see below). Since then, all important lenses have been light microscopes have been combined to lens groups. This mainly had to be done in order to take account of the aberrations of single lenses (Hilger, 1986). Aberrations are grouped into two types: geometrical and chromatic.

Geometrical aberrations are related to the spherical nature of the lens and cause image blur and several types of image distortions (spherical aberration, astigmatism, field curvature), chromatic aberrations refer to the dispersion characteristics of different wavelengths.

All these aberration errors are correctable with respective efforts. Thus, nearly all lenses in light microscope are assembled in lens groups with various correction characteristics. The extent of correction – especially of the objective (see below) – mainly determines the quality class and the expense of a microscope.

1.4.1.1 **Spherical aberration**

Spherical aberration is due to the differential refraction of the light path between the center of a lens and its outer regions. Consequently, the rays emerging from a point source will converge in different planes, and so the point source will not be sharply focused. Spherical aberration is essentially a consequence of the special curvature of a lens. This lens error can be reduced by the use of lenses with ellipsoid or hyperbolic curved faces, the addition of a second lens element with the reciprocal aberrations, and the use of apertures to eliminate the peripheral rays.

1.4.1.2 **Comatic aberrations**

Comatic aberrations are similar to spherical aberrations, but they mainly occur with off-axis light rays and are most severe when the microscope is not properly aligned. The image of a point is focused at sequentially differing heights, producing a series of asymmetrical spot shapes of increasing size that result in a comet-like shape (hence the term coma) in the image of a point. All of the factors that increase spherical aberration are also especially critical to coma. Coma can be compensated by arranging the lens groups symmetrically before and behind an aperture. A lens system such as a microscope condenser or objective that is free of comatic aberration is referred to as aplanatic.

1.4.1.3 **Astigmatism**

Astigmatism aberrations are found at the outer portions of the images produced by uncorrected lenses. Astigmatism is the result of a non-radially symmetrical lens. Light rays in the tangential and sagittal planes of lenses are refracted differently and finally produce a series of elongated images ranging from linear to elliptical, depending on the location of the focal plane. In a zone known as the circle of least confusion, positioned between the tangential line image and the sagittal line image, the major and minor axes of the ellipse are equal and the image approaches a circular geometry. Like coma, astigmatism occurs with off-axis light rays and is most severe when the microscope is out of alignment. Incorrect lens-centering in the objective or insufficient alignment between the objective, the intermediate optics, and the ocular increases astigmatism and coma. Astigmatism errors are usually avoided by precise spacing of individual lens elements, as well as by use of appropriate lens shapes, aperture sizes, and indices of refraction. The correction of astigmatism is often accomplished in conjunction with the correction of field curvature aberrations (see below).

1.4.1.4 Curvature of field

A simple lens focuses image points from an extended flattened object onto a spherical surface, because the greater distance of off-axis points from the lens center results in closer off-axis image points. Curvature of field in optical microscopy is perhaps the most obvious aberration, since the result is an image that is out of focus in a circumferential sector and in focus in the center or vice versa, depending on focusing. Use of an appropriate combination of lens groups eliminates this error.

1.4.1.5 Geometrical distortion

Geometrical distortion arises from differences in focal lengths and magnifications through various parts of the lens. In the absence of other aberrations, geometric distortion results in a misshapen image (often termed pincushion and barrel, respectively), even though each image point is in sharp focus. Distortion aberration is more obvious in specimens with regular periodic features (diatoms or technical objects). Other specimens lacking such periodicity do not appear to be dramatically distorted when viewed under the microscope. In general, thin lenses show little or no distortion, whereas in thicker lenses distortion is more obvious.

1.4.1.6 Chromatic aberrations

Chromatic aberrations arise from variations in the refractive indices of visible light (dispersion). The lens is not able to bring all of the colors into a common focus, which results in a slightly different image size and focal point for each predominant wavelength group. Consequently, a point source of white light will be imaged as a gradient of otherwise identical images of different color along the optical axis. Objects appear blurred, with a halo of a single color. Additional lateral chromatic aberration occurs when the white light source is placed off-axis. A continuous spectrum of images with different magnifications occurs (hence the alternative term "chromatic difference of magnification"). This leads to colored fringes surrounding the image details. Compensation of chromatic aberration is achieved by combining lens elements with different dispersion characteristics (i.e., made from materials with different optical densities). Chromatically separate light rays from the first lens element can then be compensated by the second/third lens element.

1.4.1.7 Numerical aperture and resolution

The most important parameter for microscope objectives is the numerical aperture (NA). The numerical aperture value describes the light acceptance angle of the objective lens, critical for the light-gathering power, the resolving power, and the depth of field of the objective. (It has to be kept in mind that the resolving power of an objective refers to the ability of that objective to yield an image which clearly separates a minimum distance between points or lines, whereas resolution is the actual separation achieved by the microscope system, which is also influenced by other factors, such as the adjustment of illumination and the specimen characteristics.) The light-gathering power is determined primarily by the angle of the cone of light collected by the front element (angular aperture). The angle of the cone is largest when the objective is closest to the specimen, but is also influenced by the refractive index of the medium between the cover glass and the objective front lens (Fig. 1.56 b). It has to be taken into account that the numerical aperture also influences the depth of field (the vertical distance in the specimen measured from above and below the exact plane of focus that still yields an acceptable image). The higher the numerical aperture of an objective, the lower its depth of field.

Theoretically, the highest angular aperture obtainable with a microscope objective would be 180° ($2 \times a$) but in practice the angle does not exceed approximately 144°. Numerical aperture is defined as follows:

$$NA = n \sin a$$

It has to be taken into account that the angle of total light reflection of an air-glass interface is 41.3°. Thus, light coming from the specimen at a greater angle would be totally reflected by the coverslip and not be collected by the objective. The introduction of a medium with a higher refractive index than air (which is approximately 1) would also raise the numerical aperture, as indicated in Fig. 1.56 b. Common immersion media are water (refractive index = 1.33), glycerol (refractive index = 1.47), and immersion oil (refractive index = 1.515, equal to the refraction index of the objective front lens). Since objective lenses are specially designed for the refractive immersion medium (or are used as dry lenses in air), only the appropriate medium should be combined with the respective objective.

In an ideal optical system, the minimum resolvable distance r on an image plane between two self-luminous points depends on the wavelength λ of the emitted light and the numerical aperture of the objective

$$r = 1.22\ \lambda/2NA$$

When the objects are not self-luminescent, but illuminated by a condenser lens (see below), the resolving power is determined both by the illumination angle from the condenser and by the collecting angle of the objective, $r = 1.22\ \lambda/(NA_{obj} + NA_{cond})$. Therefore, if the highest magnification objective is an oil-immersion objective with a numerical aperture of 1.40, then the sub-stage condenser should also have an equivalent numerical aperture to maintain the highest system resolution. Most condensers can be used with immersion oil to achieve their full numerical aperture. This is not recommended unless very demanding photomicroscopy work has to be done.

Illuminated specimens may be regarded as complex diffraction gratings. The smaller the detail structure of the object, the wider the angle of diffraction. Diffracted light is rendered one-half wavelength or 180° out of phase with the undeviated light, allowing destructive interference with the undeviated light. A corresponding diffraction image is projected in the rear focal plane of the objective (it may be observed there when the ocular is exchanged for a phase telescope), brought about by interference of the light waves diffracted by object structures. The dark areas in the image result from destructive interference of waves, the bright spots from the constructively added wavelength. Microscope specimens are irregular, with details and openings of various sizes that do not form an easily observable diffraction pattern (only a central spot is visible), but this complex pattern does contain all the information that is reconstituted into a normal specimen image at the intermediate image plane. Regular objects such as line or cross gratings form a diffraction pattern with the central bright spot (zeroth order spot) and additional spots away from the center. These patterns are easy to observe and contain interpretable and measurable information on the specimen structure. Although direct interpretation of a light microscopic diffraction pattern is of limited use, a corresponding technique is regularly applied to obtain geometrical data from crystalline objects in electron microscopy. Nevertheless, these regular patterns make image formation in the light path of an optical microscope visible. The farther away from the center a diffraction spot is located, the finer the original image detail that is represented by the spot, or in other words, finer details produce higher diffraction angles (which is of course also true for irregular objects). Thus, a large aperture angle of the objective captures finer image details than a small angle. Mathematically, the diffraction pattern is a Fourier transform of the real-space image, which is again transformed at the intermediate image plane.

The diffraction spot itself is an image of the condenser aperture diaphragm, and the spot size varies according to the adjustment of the diaphragm. A small-diameter diaphragm opening produces small, well separated diffraction spots, but also small spots farther away from the center (higher diffraction angle) that cannot be captured by the objective (high contrast, low resolution). A large opening results in large diffraction spots, with more peripheral spots being (partially) captured (low contrast, high resolution). Because light of shorter wavelength is diffracted at a smaller angle than light of longer wavelength, the higher-order diffraction spots all represent spectra of the incident light beam. A lens of a given aperture will capture more shorter wavelengths than longer wavelengths, which explains the influence of wavelength on the resolvable distance between objects (as indicated in the above formula for the minimal resolvable distance of points). From the considerations concerning the microscopical light path discussed above, manipulations in any of the illuminating light path's conjugate plane would have an impact on image resolution. In standard operation procedures, this becomes most obvious through manipulation of the condenser aperture diaphragm (see below).

1.4.2 Microscope Parts

All modern bright-field light microscopes consist of functional parts – illuminating system, condenser, objective, and ocular (eyepiece) – that are briefly described in this Section (Fig. 1.56 a, b). For more comprehensive overviews on this topic, see Abra-

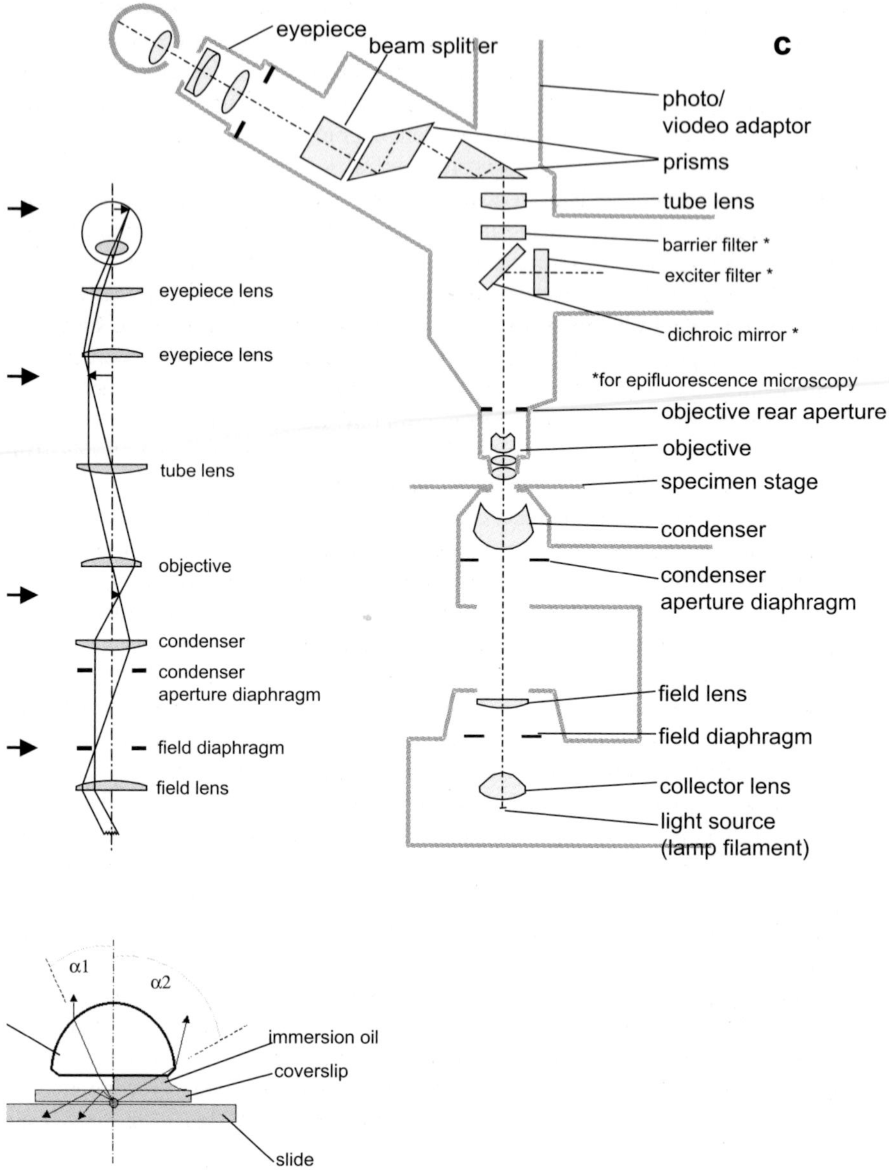

Fig. 1.56 General setup of a standard light microscope:
(a) The image-forming light path. Rays cross at the field diaphragm, the focused specimen, the intermediate image plane, and the retina of the eye or the film plane of the camera (arrows) (Zeiss Inc., with modifications).
(b) The aperture angle α increases when immersion oil increases the refractive index in the space between coverlip and front lens.
(c) Light microscope parts (Zeiss Inc., with modifications)

mowitz, 1994, and Bradbury and Bracegirdle, 1998. Objectives and objective types – as the most important parts of the microscopes – are described in more detail.

The **illuminating system** for bright-field microscopy consists of a light source, usually a (low-voltage) light bulb or halogen lamp, a single collector lens or a collector lens system, and the condenser lens system. A collector lens (system) is placed between the light source and the condenser.

Optimum specimen illumination is achieved by the Köhler illumination technique, which is a standard for all laboratory instruments. In Köhler illumination, only the diameter of the sample area visible through the ocular (field of view) is illuminated. This provides a uniformly bright image, free from glare, and the cone of the illuminating light is optimally fitted to the numerical aperture of the objective. The ray path of the illuminating light in Köhler illumination produces a focused image of the lamp filament on the plane of the aperture diaphragm of a correctly positioned substage condenser, the rear focal plane of the objective, and the eye-point above the eyepiece. These areas are called the conjugate planes of this ray path. An object that is in focus in one plane is also in focus in the other conjugate planes. Similarly, the conjugate planes in the image-forming light path in Köhler illumination include the field diaphragm, the focused specimen, the intermediate image plane, and the retina of the eye or the film plane of the camera (see Fig. 1.56 a). When adjusted according to the Köhler technique, the condenser lenses focus the image of the collector lens (and field diaphragm) onto the specimen plane.

Adjustment of the field diaphragm (which appears focused, together with the specimen, when the conjugate planes are properly adjusted), regulates the field of view, whereas the condenser diaphragm is used to adjust the cone of light emerging from the condenser, which determines the numerical aperture of the condenser/objective system. In general, this diaphragm should be adjusted to 3/4 – 4/5 of the entire light disc size (visible in the ocular tube of the microscope after removal of the ocular). Although the maximum working numerical aperture is not achieved in this position, image contrast is improved. Smaller aperture settings would further improve contrast, but fine image details would become obscured by diffraction artifacts and refraction phenomena.

The **substage condenser** of a microscope is designed to focus the light onto the specimen. The cone of light that illuminates the specimen may be adjusted with the condenser aperture diaphragm opening size. When the light source is not focused at the specimen level, an essentially grainless and even illumination is provided, despite any imperfections on the glass surface of the condenser. This is also the reason why the simple Abbe condenser type in standard microscopes is not corrected for optical aberrations. The primary advantages of the Abbe condenser are the wide cone of illumination that the condenser is capable of producing, as well as its ability to work with long working distance objectives.

Aplanatic or achromatic condensers represent two condenser types corrected either for spherical or for chromatic aberrations. For highest demands, aplanatic-achromatic condensers, corrected for both chromatic and spherical aberrations, are used.

As already briefly mentioned above, a critical factor in choosing substage condensers is the numerical aperture performance necessary to provide an illumination cone adequate for the objectives. The condenser numerical aperture should be equal to or slightly less than that of the highest objective numerical aperture. Therefore, if the highest magnification objective is an oil-immersion objective with a numerical aperture of 1.40, then the sub-stage condenser should also have an equivalent numerical aperture in order to maintain the highest system resolution. In this case, immersion oil would have to be applied between the condenser top lens and the reverse side of the microscope slide to achieve the maximum numerical aperture (1.40) and resolution. Numerous microscopes are equipped with an additional condenser front lens that can be swiveled into the optical axis to select a smaller field of view at high apertures of the objective.

Closest to the specimen, the objective is the first information-gathering lens system of the microscope. It is therefore the most important system, and corrections of aberrations here are of utmost importance for image resolution. A large number of **objective types** are available for a variety of applications and with differences in general optical qualities.

Each objective has all relevant technical information for proper use written on the side of the barrel (Fig. 1.57). For handling, correct interpretation of these data is important. Besides the manufacturer's sign, abbreviations of optical corrections (the most abundant abbreviations used are Acro = achromatic, Apo = apochromatic, Fluar = Fluorite objective, Plano = field curvature corrected objective), as well as specialized optical properties (D = dark-field, DIC = differential interference contrast, Phaco or Ph = phase contrast) are indicated. The choice of these abbreviations and the extent of the information varies slightly between manufacturers. Generally, the magnification is engraved in the largest type font, followed by the numerical aperture value and the immersion medium. Tube length (or the infinity symbol) and recommended coverslip thickness (usually 0.17 mm) are always indicated, while some manufacturers also include the working distance (WD, in millimeters, the distance between the objective front lens and the top of the coverslip for a focused specimen). Sometimes the WD

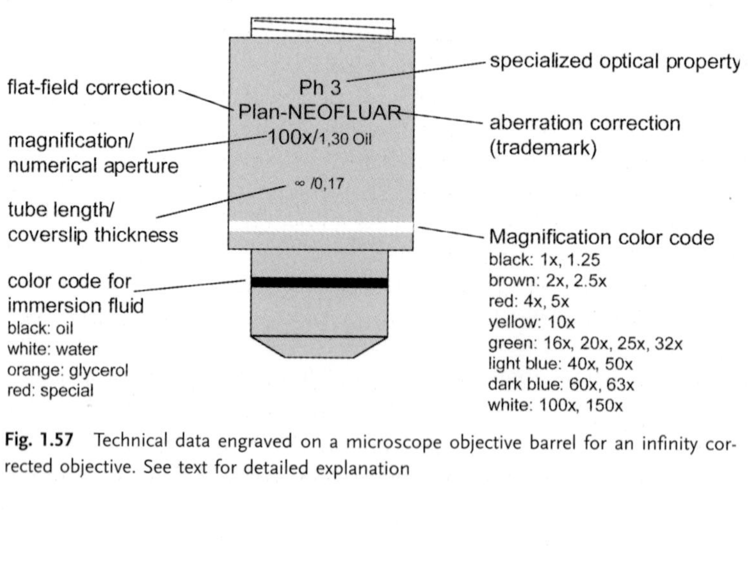

Fig. 1.57 Technical data engraved on a microscope objective barrel for an infinity corrected objective. See text for detailed explanation

values are not included, and the abbreviations used are LWD or ELW to indicate "long" or "extra-long" working distance. To allow rapid identification of essential objective features during operation of the microscope, one or two colored rings are engraved on the objective barrel, indicating the magnification and the immersion fluid to be used.

The most common and inexpensive **achromat objective** is corrected for chromatic aberration in two wavelengths (red and blue) that are brought into the same focus. Because of residual chromatic aberration, yellow-green light is not correctly focused when the red-blue region of the spectrum is focused. This results in a green halo along contours. If focus in the green region of the spectrum is chosen, images will have a reddish-magenta halo. The best results are therefore achieved when the non-focused color is filtered out and black-and-white film is used to record the image. Spherical aberration becomes obvious in high-magnifying achromat lenses. In this type of aberration a curved image plane rather than a flat one does not allow an evenly focused image to be obtained. The image appears sharp and crisp either in the center or on the edges of the ocular viewfield, but not both. Though this may not be a problem for routine inspections, in which continuous focus correction is possible, a portion of a photomicrograph remains out of focus. Objectives based on the simple achromat and corrected for various geometrical aberrations, especially spherical aberration, have been developed. These objectives are also useful for photomicrography at higher magnifications and other applications in routine studies.

In **apochromats**, chromatic aberration is corrected for three to five colors and spherical aberration is corrected for two to three colors. Apochromats are thus best suited for color photomicrography in white light. Because of their high level of correction, apochromat objectives usually have higher numerical apertures for a given magnification than other objective types. **Fluorite** or **Semiapochromat** objectives provide correction of a quality in-between the apochromat and achromat. Without further correction, all these types of objectives would suffer from field curvature. Flat-field or plano objectives compensate for this type of aberration. The flat-field objectives can be optically constructed to be also achromat, semiapochromat or apochromat. In the last case, the lens would be called a plano apochromat, generally regarded as the best lens available. Actually, fluorite or apochromat objectives are regularly flat-field corrected.

The **tube length** refers to the distance between the objective and the eyepiece in millimeters for fixed length, or the infinity symbol for infinity-corrected tube length. At present, nearly all objectives for laboratory microscopes are manufactured with infinity-corrected tube lengths, for which no intermediate image plane directly behind the objective exists any longer because light emerging from the objective is focused to infinity, and an additional lens (tube lens) forms the image at its focal plane. The parallel rays in the "afocal" space between objective and tube lens allows filters, beam splitters, polarizers, and other plane parallel elements to be placed there without additional lenses.

The **ocular** (eyepiece) is basically a projection lens system. Currently available oculars magnify the image projected by the objective lens, mostly by 5–25 times. The simplest types consist of two lenses (Huygens or Ramsden type). The Ramsden eyepiece is further improved by replacement of the simple eye lens by a lens doublet to decrease chromatic aberration. Further variations of this type, more commonly used today, feature a variety of improvements concerning lens aberrations. The increase in magnification allows visual perception of fine image details, but does not improve resolution; details not resolved in the intermediate image, cannot be resolved by an ocular. Visual perception of image details however, requires a certain magnification. A point-to-point distance below 200 μm cannot be resolved by the eye. The numerical aperture of the objective/condenser system defines the range of useful magnification for the objective/eyepiece combination. To exhaust the resolution provided by the objective, the total magnification of the objective/ocular system should be at least 500× NA, but not exceed 1000× NA. Magnifications higher than this value will yield no further resolution of finer image details (empty magnification). As with objective lenses, modern oculars are manufactured of multiple optical components for correction of lens aberrations. Compensating eyepieces are used for further correction of specific object aberrations, especially lateral chromatic aberrations in microscopes with finite tube lengths. Most microscopes with infinite tube lengths do not need compensating eyepieces because all correction elements are implemented in the objective or the tube lens. The ocular type is indicated by engraved abbreviations such as WF for Wide-Field, UWF/SWF for Ultra/Super Wide-Field (referring to the viewfield); HE for High Eye-point (which may be used with spectacles while viewing the samples). Compensating eyepieces are marked with K, C, or comp. Besides the magnification (10×, 20×, etc.) an engraved number (field number) refers to the diameter (in millimeters) of the fixed diaphragm in the eyepiece. The diameter of the field of view (D) is determined by the field number (FN) and the magnification (M) of the objective (D = FN/M).

1.4.3
Standard Features for Enhancement and Modification of Image Contrast in Light Microscopes

1.4.3.1 Phase contrast (Pluta, 1969)

In virtually all laboratory microscopes, illumination may be switched between brightfield and phase contrast, the latter being invaluable for successful imaging of unstained biological objects. In the phase contrast mode, contrast is brought about by the different refractive indices between specimen structures, as well as the specimen and the surrounding medium (water), rather than by use of a colored stain. In terms of wave optics, contrast in stained or "amplitude" objects is brought about by destructive and constructive interference of waves, which becomes directly visible in the diffraction image at the rear focal plane of the objective, whereby the interference between waves shifted by 180° to each other are destructive. The light wave diffracted by an (unstained) specimen becomes retarded in phase by up to 90° of its wavelength. By speeding up the undeviated light by another 90° so that the difference in wavelength between the undeviated and diffracted light for a phase specimen would now be 180°, destructive interference becomes possible – as in stained objects. The phase object is transformed in an amplitude object. This is achieved technically by the separation of the direct zeroth order light spot from the diffracted light at the rear focal plane of the objective. A ring-shaped (annular) aperture is placed in the position of the condenser aperture diaphragm, conjugate to the objective rear focal plane (Fig. 1.58). The undeviated light projects a focused image of the annular aperture into the rear focal plane of the objective, representing the zeroth order maximum. The light diffracted by the specimen is spread over the entire rear focal plane of the objective. A phase plate with a ring-shaped phase-shifter exactly matching the ring of light is installed at the rear focal plane of the objective. The phase-shifter is optically thinner than the rest of the plate. As a result, undeviated light passing through the phase-shifter travels a shorter distance in traversing the glass of the objective than does the diffracted light, which leads to the desired phase shift. The diffracted and direct light can now interfere destructively, so that the details of the specimen appear dark against a lighter background (just as they do for an amplitude specimen). This type of phase contrast (positive contrast) is produced in most available phase contrast microscopes. Alternatively, when the undeviated light is slowed down by 90°, an inverted contrast results. Since the diffracted light is much weaker than the undeviated light, a thin absorptive transparent metallic layer is deposited on the ring-shaped phase-shifter to balance out the light intensities. Phase contrast microscopy is indispensable for imaging of unstained (i.e., living) thin biological specimens, also in combination with – for example – fluorescent images. However, it is necessary to accept some image artifacts such as halos around the outlines of specimen details, which sometimes obscure the boundaries, and a slightly reduced resolution. Phase contrast should not be applied with thick specimens, because shifts in phase occur from areas slightly below or slightly above the plane that is in focus. The halo effect is considerably

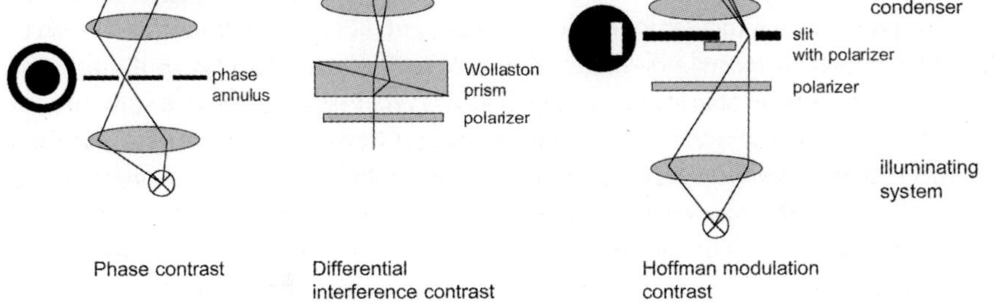

Fig. 1.58 Principle setups of the three most commonly used optical staining techniques for microscopes used in biomedical research: phase contrast, differential interference contrast, and Hoffmann modulation contrast (after Ruzin, 1999, with modifications)

reduced by modified phase plates ("apodized" phase plates) available in objectives manufactured by Nikon Corp.

1.4.3.2 **Dark-field microscopy**

Whereas contrast in phase contrast microscopy is brought about by selective "correction" of the undeviated light wave, this light is completely removed in dark-field microscopes, the zero order spot (visible in the rear focal plane) being abolished completely. This results in an image made up exclusively of higher order diffraction intensities scattered by the specimen, or in other words, the edges of phase objects. Dark-field condensers form an inverted hollow cone of light with a focus centered in the specimen plane. If no specimen is present and the numerical aperture of the condenser is greater than that of the objective, all rays miss the objective and the field of view appears dark.

For low objective apertures, a simple microscope condenser equipped with an annular aperture and an appropriately adjusted iris diaphragm generates a corresponding "hollow" light cone. Higher apertures are achieved with specially designed condensers in which – for instance – a paraboloid-shaped piece of glass in the condenser directs the light in such a way that high-aperture light cones are formed.

Though applications for dark-field illuminations are limited and contrast-enhancement by phase contrast, differential interference contrast, and Hoffman modulation contrast is generally more effective, dark-field illumination is a versatile method with which to observe colloidal suspensions, even when the particle diameter is significantly lower than the limit of resolution for the objective. Particles in the range of some tens of nanometers in diameter diffract enough light to become – indirectly – visible. In this way, the movements of living bacterial flagella, which average about 20 nanometers in diameter, can be observed.

1.4.3.3 **Polarizing microscopy**

The optical anisotropy (birefringence and or dichroism) of crystalline specimens is easily to analyze with a simple variant of a bright-field microscope, equipped with two additional polarizing filters placed in the condenser and close to the aperture diaphragm of the objective (McCrone, 1978). The second polarizing filter (the analyzer) is oriented perpendicular to the first filter (i.e., no light will be transmitted with no specimen in place; crossed polarizing filters). When a (polarized) ray of illuminating light enters the anisotropic specimen (such as a crystalline particle), it is split into two rays. The ray will transverse the crystal by different paths, in which one wave will become retarded with respect to the other (i.e., both rays have two refractive indices). The difference between these refractive indices depends on the orientation of the crystal lattice relative to the ray. At one (or also two in some materials, such as cellulose) positions relative to the light ray, the optic axis of the crystal, the refractive indices are identical (no birefringence occurs). The birefringent crystal transmits the light ray only in two polarization planes, which are perpendicular to each other. When the polarization plane of the incident ray does not meet one of these planes, "partitioning" of the electromagnetic field vector of the incident ray on both "allowed" orientations occurs according to the parallelogram law. This turning angle is greatest when the optic axis of the crystal is oriented at a 45° angle to both the polarizer and the analyzer. No partitioning and no effective turning of the polarization plane occurs when the crystal optic axis is oriented parallel to the direction of the polarizer. At that position, the analyzer would block out every light; at any other position, a portion of light would be transmitted. Rotation of a crystal would show the typical transitions between maximum and minimum brightness against a dark background every 90°.

Because one wave is retarded with respect to the other, interference (either constructive or destructive) occurs between the waves as they pass through the analyzer. The net result is that some birefringent samples acquire a spectrum of color when observed in white light through crossed polarizers. The polarization colors seen in the microscope can be correlated with the actual retardation, thickness, and birefringence of the sample (i.e., the optical path difference of the sample) and are reiterated in multiples of 550 nm. This is visible as a color change every 20–30 nm up to 550 nm, and the colors repeat for an additional 550 nm, and so on. The first set of colors is referred to as first order, followed by second order etc. up to seventh order. The birefringent colors can be correlated to specimen thickness by a reference scale (Michel-Lévy scale).

Specimens exhibiting low birefringence, such as most biological materials, are more difficult to observe by a conventional polarizing microscope than, for example, mineral specimens. Since even slight birefringence induces phase separation of slow and fast waves, this phase difference may be enhanced by a retardation plate or a compensator placed between the specimen and the analyzer. The plate is birefringent and enhances the phase separation by further retarding the phase of one wave relative to the other (i.e., by increasing the optical paths between both rays) by fixed wavelengths (quarter-wave or half-wave plates). Wedge-shaped plates (such as quartz wedges) are adjustable compensators and provide phase retardation from 0–3 wavelengths (0–1650 nm in optical path difference). Compensators are indispensable for quantification of optical path differences in specimens.

The principle setup described above refers to microscopy with the aid of linear polarized light. The periodic change in brightness of a crystal (and extinction at certain orientations) when turning the object in polarized light may be disturbing for practical use. In a specimen composed of numerous randomly oriented crystals, for instance, not all crystals are visible at the same time. Thus, to obtain an optimal "differential stain", circularly polarized light is used for specimen illumination. Circular polarized light microscopy does not produce the characteristic extinction points that plane polarized light does. Instead, all the crystals are illuminated nearly equally, regardless of their orientation.

Circularly polarized light may be best described as the sum of two perpendicular planes of linear polarized light components out of phase by 1/4 wavelength (90°). The resultant of this sum is a rotating vector traveling in the direction of the ray path.

When the resultant of this vector is projected onto the x/y plane, a counterclockwise circular rotation is observed; hence the name of this type of polarization. Circularly polarized light can be created simply by taking plane polarized light and passing it through a 1/4 wave plate oriented at 45° to the light axis. This splits the light into the two perpendicular components, one of which is retarded by 1/4 wavelength. Note that true circular light can be created with this approach only at one wavelength. Light from all other wavelengths will end up being elliptically polarized.

1.4.3.4 **Differential interference contrast**

Similarly to phase contrast, differential interference contrast (DIC) allows effective translation of phase information of objects to amplitude information (Padawer, 1967). A major advantage over phase contrast is that the object appears bright against a dark background, but without the halo effect of phase contrast. In DIC, the optical path difference (the product of refractive index and the geometrical path length) generates contrast (either bright/dark contrast or color). Birefringent specimens may not be suitable for inspection by DIC because of their effect upon polarized light. Similarly, certain types of specimen carriers (birefringent cell culture plates, Petri dishes, etc) may not be suitable. Hoffman modulation contrast (see below) may be a better choice for such specimens.

Before passing the sub-stage condenser, the light traverses a polarizer. The polarized light is then passed through a Wollaston prism. This type of prism is composed of two quartz wedges orientated in a way that their crystal optic axes are oriented perpendicular to each other. The prism separates individual rays of light into pairs of (approximately) in-phase rays but with their electromagnetic field vectors perpendicular to each other. The separation of the two rays (the shearing) is usually about the resolving ability of the objective. After leaving the condenser, the rays of one pair pass through closely adjacent regions of the specimen, which induces an optical path difference due to localized refractive index and thickness variations between the two beams. Above the rear focal plane of the objective, the rays enter a second prism that combines them into a common path. The rays are still polarized and next pass through a second polarizer (analyzer). The analyzer has a polarization plane that is crossed with respect to the first polarizer and is oriented at a 45° angle to the beams exiting the second Wollaston prism. The analyzer recombines the electromagnetic field vectors of ray pairs, allowing interference to occur, which creates the contrast (Fig. 1.58).

The orientation of the specimen can have a pronounced effect on the relief-like appearance, and rotation of the specimen by 180° often changes a hill into a valley or visa versa. The three-dimensional appearance is not a representation of the true geometric thickness but of the optical one, which may sometimes be confusing for interpretation of object structure. The effect may be varied by shifting the second Wollaston prism in a plane perpendicular to the optical axis of the microscope. An additional retardation plate placed in the ray path creates color instead of bright/dark contrast, but contains no relevant additional information.

1.4.3.5 Hoffman modulation contrast

Hoffman modulation contrast (HMC) achieves images of similar appearance to those taken with DIC, but the technique used is here is different. Since no polarizing beam splitters are applied in HMC, birefringent specimens do not cause disturbing image artifacts. Unlike in phase contrast, the modulators used in HMC do not introduce changes in the phase relationships of light passing through the system, but influence the zero order maximum in the diffraction image. An eccentric slit aperture is placed below the condenser, and the slit width is adjusted with a polarizer. A modulator plate is placed at the rear focal plane. This plate consists of three regions of different neutral densities, allowing transmission of 100 %, 15 %, and 1 % light, respectively. The condenser slit is adjusted in such a way that an undeviated light ray traverses the 15 % (gray) region of the modulator. When light rays are deviated by regions of different refractive indices in the specimen, the light rays miss the gray region of the modulator and pass through the other regions of the filter (Fig 1.58). The final image is thus depicted as a pseudo-relief of the original, representing the phase gradients. Generally, HMC is best for specimens with rapid phase gradients, such as large, globular cells (such as plant protoplasts).

Numerous variations of instrumental setups that achieve all these optical staining methods have been developed by the optical industry and are not be discussed in detail here. In a more recent development, variable relief contrast (VAREL, Zeiss, Inc.) is brought about by use of an inclined unilateral illumination. A ring-sector shaped aperture is placed in the sub-stage condenser and may be shifted perpendicular to the optical axis. In this way, several modifications of image contrast may be achieved, such as a “unilateral” dark-field effect, relief contrast superimposed on phase contrast and “inclined” bright-field illumination. The quite simple technique improves the capabilities of phase contrast, since it adds pseudo-relief to phase objects.

1.4.4
Upright and Inverted Microscopes, Epi-Illumination Microscopes

The general descriptions given below refer to the transmitted light microscopes mainly used in biomedical research. These microscopes are available as upright or inverted variants. In upright microscopes, the condenser is situated below the specimen plane, and the objective/ocular system above. Inverted microscopes, with illuminating systems above the specimen plane and objectives below it, provide a long distance between the specimen and the microscope. The bulky objective revolvers will not disturb the handling of large or otherwise complex specimens. This is useful for, for instance, inspection of cell cultures grown at the bottom of culture plates or for introduction of manipulators and additional chambers etc. in the specimen plane. The condenser, rather than the objective above the specimen, has a working distance of up to several centimeters. This depends, of course, on the condenser aperture. Axiovert (trademark of Zeiss, Inc.) inverted microscopes have a condenser working distance of around 7 cm for a condenser aperture of 0.3. For a condenser aperture of 0.55, 2.2 cm are required.

Except for fluorescence microscopy, the transmitted light microscopes are useful for most applications with transmissible thin biological objects. Manufacturers have also designed a whole product family of reflected light instruments for illumination of opaque specimens and inspection of surface features. Both illumination source (and the associated optics) and objective are situated above the specimen, and light is reflected by the specimen. Specialized microscope types (for dark-field, polarization, and DIC) are also available as reflected light instruments. The illuminating light is directed by a beam splitter into the optical axis of the microscope towards the objective (acting at the same time as a condenser). Most objective types are in principle suitable both for transmitted light and for reflected light. In practice, they are specially coated to avoid disruptive light reflections and are corrected for use without a coverslip. This objective type is indicated by the engraved abbreviation "EPI". Dark-field objectives are composed of an outer tube element for illumination of the sample and an inner tube acting as objective. For DIC, the same Wollaston prism first separates the illuminating ray and then combines the reflected ray pairs. Reflected light microscopes may also be build as inverted microscopes. In these, no microscope parts are situated above the specimen stage, which allows working with very voluminous objects, mainly in the field of materials research.

1.4.5
Fluorescence Microscopy

Fluorescence microscopy has become the most rapidly expanding field in light microscopy, with numerous technical variants (Rost, F. W. D., 1995; Tanke and Hermann, 1998). Confocal microscopy, used practically without exception in conjunction with fluorescent techniques, is described in chapter 1.4.5.2.

The most important practical effect of fluorescence microscopy (instead of bright-field microscopy) is a low detection limit for specifically labeled components. Specific labeling may also be achieved under bright-field light microscopes, and detected as a dark stain (i.e., as an amplitude object). To obtain a detectable signal (especially against a corresponding stained background), a considerable amount of stain has to be deposited, which raises several problems. Before the introduction of appropriate fluorescent markers, the application range was limited by inactivation or death of cells, poor resolution, and difficulties in interpretation of stain deposition, as well as by a low detection limit. Fluorescence is – in principle – detectable down to the scale of single emitting molecules (but not, of course, with the corresponding resolution). The use of relatively low concentrations of a fluorescent stain leaves vital function of cells unaffected, which allows visualization of dynamic processes. Autofluorescent fusion proteins further widen the application range. Fluorescence techniques are therefore most powerful tools for elucidation of structure-function relationships in all fields of cellular biology.

In fluorescence microscopes, the incident light excites light emission by the irradiated specimen. Transmitted light microscopes are thus of only limited suitability for use as fluorescence microscopes, because the directions of the excitation light and the emitted light are identical, orientated towards the objective. Detection of faint fluorescence requires expensive blocking of the excitation light. Thus, only microscopes with transmitted dark-field illumination were of practical use in the past. Reflected light fluorescence (commonly referred to as epifluorescence) is now the method of choice for all applications in biomedical research. As in other reflected light illumination techniques, the exciting light is reflected by a beam splitter into the objective, acting as a condenser. The fluorescent light is then collected in the objective, filtered, and guided to the ocular. Clear separation of exciting and fluorescent light is essential in fluorescent microscopy, and selection of light sources, filters, beam splitters, and fluorescent dyes has to be performed with due regard to this problem.

1.4.5.1 **Fluorescence microscopes**

Based upon the setup of a standard transmitted light microscope, fluorescence microscopes have become indispensable tools for sensitive detection of fluorescent markers. Most research microscopes are now available with (optional) equipment for fluorescence (Fig. 1.56).

Normally, the **light source** is not a halogen lamp, which provides a continuous spectrum, but rather a Hg, Xe, or Hg/Xe arc lamp. These light sources each emit a distinct line spectrum. The main peaks of the emitted line spectrum in the widely used Hg arc lamps are at approximately 365, 404, 434, 546, 577, and 613 nm. For several specialized setups, lasers, which offer monochromatic light of high intensity, are most suitable.

With a line spectrum rather than a continuous spectrum, it is easier to select a distinct wavelength with an **excitation filter**, which is necessary for excitation of a marker. These filters are of either band pass or short pass filter type. Band pass filters allow only a short range of wavelengths to pass and block out all shorter and longer wavelengths. Short pass filters allow all wavelengths shorter than a certain value to pass. These filters are marked BP or SP, respectively. For BP filters, the transmitted wavelength peak and the respective bandwidth at the 50 % level of maximum transmission are indicated. For SP filters, the indicated number refers to the wavelength at 50 % of the maximum transmission.

The transmitted light is then reflected by the **beam splitter** into the optical axis of the microscope, orientated towards the objective. The beam splitter must allow passage of the emitted light of the specimen, which is of longer wavelength than the excitation light. The beam splitter acts as a long pass filter that allows the passage of wavelengths longer than a certain value, but reflects all shorter wavelengths. Thus, the exciting light beam is reflected, together with any excitation light back-scattered by the specimen.

A third filter, the **barrier filter**, is another long pass or band pass filter used to perform additional filtering of the emitted light. The filters further suppress transmittance of the shorter-wavelength exciting light and are also necessary when unwanted fluorescence has to be blocked out, either autofluorescence of specimen components or another specific fluorescent signal that has to be recorded separately (with another filter combination). The three filters are integrated in a module, the filter cube, that also allows quick exchange of filter combinations. The coordinated use of fluorescent dyes, excitation filter, beam splitter, and barrier filter is, of course, essential to obtain an optimum signal output. The numerous possible combinations, together with application data, as well as suppliers of fluorescent dyes, are provided by microscope suppliers.

Limitations involving lens aberrations also apply to fluorescent images, of course. Although, in principle, all objective types used for bright-field microscopy can also be applied for fluorescence, the special features of fluorescence microscopy (i.e., epi-illumination with light of a selected wavelength, possibly ultra-violet light, and the necessity to capture low light intensities) restricts the use of some types. Generally, objectives that provide high numerical aperture should be selected, in order to capture the maximum amount of light from faintly emitting fluorescent specimens. In these situations, auto-fluorescence and/or internal reflection inside the objectives can also interfere with imaging of small structures and low-fluorescence targets. The objective should transmit a wide range of wavelengths, from the ultraviolet (down to 340 nanometers) through the infrared regions of the electromagnetic radiation spectrum. These requirements are generally best met by (plan-corrected) fluar objectives, which provide large apertures and also allow transmittance of ultraviolet light. Older apochromat objectives may not be the best choice. Because of their numerous lens elements they may absorb ultraviolet light, and transmittance is also diminished. For standard applications that require only a limited range of exciting light wavelengths and when a clear fluorescent signal is expected, (corrected) achromats may also be sufficient.

The use of electronic cameras in concert with image processing has circumvented the limitations of direct observation or photographic emulsions (dim light and small differences in light intensity against a bright background). With highly light-sensitive cameras (such as SIT- or CCD-cameras; see below) it is possible to observe cells for long periods at very low light levels, avoiding the damaging effects of prolonged bright light – and heat – exposure. Such image intensification systems are especially important for viewing fluorescent molecules in living cells.

Subsequent computer-based image processing makes it possible to compensate for various optical faults and to attain the theoretical limit of resolution. Moreover, by linking video systems to image processors, the signal-to-noise ratio, and thus the contrast, can be greatly enhanced by processes such as frame accumulation, thereby overcoming the eye's limitations in detecting small contrast differences. Although this processing also enhances the effects of random background irregularities, this 'noise' can be removed electronically by background subtraction. Small transparent objects previously impossible to distinguish from the background then become visible.

1.4.5.2 **Confocal microscopes**

The idea of confocal microscopy, originally developed over forty years ago, has become especially feasible now that affordable data acquisition and storage techniques have become available. Confocal microscopy is now the most widely used technique for optical image slicing (Fig. 1.59). For biomedical applications, confocal microscopes are almost exclusively used for imaging of fluorescent specimens, and so the description presented here mainly refers to fluorescent specimens and fluorescence micro-

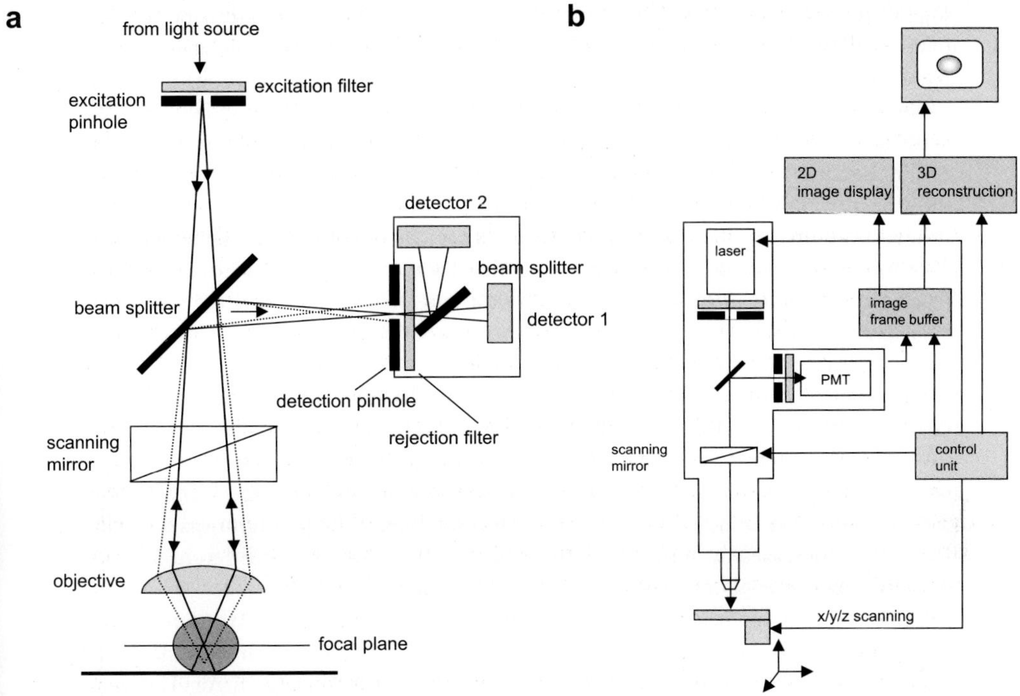

Fig. 1.59 Confocal microscopy:
(a) Simplified version of a principle setup and the optical ray path of excitation and emission light.
(b) Essential control circuits for confocal microscopes (PMT: photomultiplier tube)

scopy. See Pawley, 1995 and Sheppard and Shotton, 1997 for comprehensive surveys. The image obtained by a standard (fluorescence) microscope includes the in-focus portion and the out-of-focus portions above and below the focal plane. The blur of the out-of-focus portions in standard microscopes reduces contrast and the effective resolution. This is especially critical for sections exceeding 2 μm in thickness.

"Out of focus" light is reduced in confocal microscopes, essentially by spotwise illumination and by imaging of the specimen. The spot of light is focused by the objective lens at the focal plane of the specimen. The spot is produced either by imaging of an illuminated small aperture or, more usually, by focusing a parallel laser beam. With only one single spot illuminated, illumination intensity falls off rapidly below and above the focal plane as the beam converges and diverges, thus reducing fluorophore excitation out of the focal plane. The size of the spot cannot, of course, be reduced to an ideal point. Its size and shape is mathematically described by the point spread function (PSF), in which more than 80 % of the light energy is concentrated in a space roughly described by a rotational ellipsoid. The actual size of the spot is determined by the wavelength of the illuminating light, the size of the illuminating pinhole, and the aberration of the microscope optics. Light reflected back from this spot (or light from an excited fluorophore) is collected by the objective and reflected to a pinhole positioned in front of a detector. Light rays from below the focal plane (see

Fig. 1.59 a) are focused before reaching the detector pinhole. The light cone then expands so that most of the light is blocked. In the same way, light from above the focal plane focuses behind the pinhole. Only "confocal" light from the conjugated optical plane is not rejected. The light detected after passage through the pinhole is also described by a point spread function, similar to the "illumination" PSF. The resulting PSF is hence a product of the illumination and detection PSFs, which results in an improvement of the lateral resolution by, at best, a factor of 1.4 relative to conventional microscopes (which is not quite essential), but which also results in a defined optical slice thickness – which is impossible to define for a conventional microscope. Axial resolution depends, much more strongly than lateral resolution, on the numerical aperture of the objective, and is, through the use of suitable objectives, 2–3 times lower than lateral resolution.

The detection point spread function (i.e., the final resolution) may be regulated by changing the size of the pinhole. Opening the pinhole to a larger diameter produces a thicker optical section and reduced resolution, but is often necessary in order to include more specimen detail or to increase the amount of light striking the detector. As the pinhole diameter is reduced, the optical section thickness and brightness decrease. Resolution increases until a certain minimum pinhole diameter – different for each objective aperture – is reached.

Since the confocal microscope only illuminates and collects light from one specimen spot at a time, an image has to be built up by scanning the specimen in the focal plane. Most confocal systems now employ a galvanometric scanning mirror. The beam scans the specimen at a rate of one image frame per second. To achieve faster scanning for real-time imaging, some systems use an acoustic-optical device.

Tunable by ultrasound, acoustic-optical devices diffract, deflect, and focus light beams. Ultrasound waves propagating through solid crystals (such as TeO_2) induce alternating regions of rarefaction and compression – areas of lower and higher refraction index – in the material. The material thus acts as a tunable diffraction grating, which can also be used to deflect an incident light beam by changing the acoustic frequency.

Faster scanning rates can also be achieved by simultaneous illumination by multiple spots focused in the object plane with a Nipkow disk. These are spinning disks with a large number of pinholes. To avoid most of the illuminating light being lost, an instrument carrying a micro-lens in each pinhole has been developed (Ultraview Live Cell imager High speed confocal imaging system with Yokogawa scan head, from Perkin Elmer Corp.), and is now readily applicable for real-time imaging of cellular events at sufficient light intensities.

Since most biomedical applications require fluorescence imaging, all types of confocal microscopes for life sciences are equipped with lasers as illumination sources (CLSM, confocal laser scanning microscope). Although the illumination in the typical laser scanning confocal system appears to be extremely bright, the average illumination at any given point on the specimen is relatively moderate due to the fast scanning of the specimen, the actual illumination at any given point generally being less than in a conventional epifluorescence microscope. This also reduces the problem of fluorophore bleaching (see below). A number of monochromatic and multi-line lasers are now available to meet the requirements for excitation of a combination of more than one fluorophore at a time. Usually, three to five laser sources are implemented in the CLSM system. They may be used simultaneously or sequentially to collect multi-channel images. Laser lines are band-pass-filtered, or for highest demands, selected by acoustic-optical tunable filters, which allows fast switching of the excitation wavelength. Excitation light is split and filtered by dichroic mirrors and barrier filters into separate channels for two to five detectors, whereby several channels may be recorded simultaneously.

Some advanced instruments apply spectrometric detection strategies. Here the fluorescent light is dispersed by a prism, and rays of different wavelengths are guided to separate detectors. In advanced detector designs, such as the LSM META series (Zeiss, Inc), the entire emission spectrum of the fluorescent specimen is analyzed simultaneously. This feature is especially versatile for experiments with multiple probes of similar emission spectra, which are difficult to separate with conventional modules. For detection, an optical grating projects the entire fluorescence spectrum onto 32 channels of a multiple detector. In this way, a spectral signature for each pixel of the scanned image – or in other words, an image stack comprising the entire emission spectrum – is acquired and can subsequently be used for digital separation into component dyes. An alternative system for spectral detection has been developed by Leica. The signals from each fluorescent pixel are compared with reference spectra of fluorescent markers previously taken under similar experimental conditions. This allows clearer separation of largely overlapping signals. The potential of this technique also improves signal analysis capabilities in fluorescence resonance energy transfer.

Though the generation of one optical slice through the region of interest often provides sufficient information for interpretation of a result, an important feature of confocal microscopes is the generation of image stacks from consecutive focal planes. After a stack of consecutive planes has been collected, individual slices can be examined and merged to provide an image of defined optical thickness. Volume-rendered views of the whole specimen can also be digitally reconstructed. Scanning in one to three dimensions of space can also be combined with signal acquisition over time. Thus, with appropriate fluorophores, dynamic changes in cells are also detectable. As in all fluorescent microscopic techniques, however, the successful outcome of the experiment also strongly depends on the longevity of the fluorescence signal upon illumination.

Improvement of both LSM and wide-field images may be performed by use of a mathematical method to remove out-of-focus haze by image deconvolution. Though less effective than image acquisition by confocal microscopy, the image quality of wide-field images is improved remarkably. In addition, software packages for LSM contain deconvolution algorithms to remove residual out-of-focus blur from microscope images.

The big advantage of deconvolution is that almost every laboratory microscope may be upgraded with the system. A precise stepper motor for focus control, a high-resolution camera, and the computer hardware and software are necessary. The camera captures and digitizes the images from the microscope, which are then processed by the deconvolution algorithms. In principle, deconvolution regenerates one focal plane from a wide-field image that contains information "convoluted" from all focal planes. Deconvolution software requires an image stack consisting of a through-focus series of the specimen. Included in every single image is the in-focus and out-of focus information. In-focus images are generated by correction for the blurring effect of the PSF and reassignment of out-of-focus intensity information to its appropriate focal plane. The point spread function of the specific optical system is experimentally determined with fluorescent beads or is calculated by the software, utilizing the data of the optical system. Deconvolution in wide-field microscopy may have advantages over LSM. Deconvolution systems costs less than laser scanning confocal microscopes because a standard microscope can be equipped with the deconvolution package and no additional optical equipment is required.

Acquisition of an image can be faster than with confocal microscopes. Including the time to move the stage, three images can be captured in about 3 seconds. Because of the scanning procedure, confocal microscopes can be slower, especially when each specimen plane is scanned more than once with reduced laser power and integrated multiple times to reduce fluorophore bleaching effects.

The principal limitation of the digital deconvolution approach has been the amount of computer time required for processing. A single slice could require considerable time to deconvolute on a personal computer. A large data set could take an entire day to process.

1.4.6
Stereo Microscopy

Useful for inspection of surfaces and large objects in three dimensions, stereo microscopes are effective instruments for large samples or for specimen preparation.

Stereo microscopes of the (older) Greenough type use double inclined body tubes to produce the stereo effect. The type more frequently used at present is a common main objective that utilizes a single large objective shared between a pair of eyepiece tubes and lens systems. Both types are usually equipped with continuously variable zoom-type magnification system.

Both types have their advantages and disadvantages. Since Greenough-type microscopes essentially consist of two tubes of a "classical" compound microscopes, the tubes are arranged in a way that the eyes view the objects from slightly different perspectives, at a convergence angle of 10° to 12°. The Greenough design, introduced by Zeiss at the beginning of the twentieth century, consists of two identical (and symmetrical) optical systems each containing a separate objective arranged in accurate alignment within a single housing. A major advantage of this design is the high numerical apertures that can be obtained, since the objectives are very similar in design to those utilized in classical compound microscopes. Because of the convergence angle, the left eye views the object from the left-hand side while the right eye views the same object from a slightly different perspective on the right-hand side.

A prism or mirror system is necessary to rotate and invert the magnified image received from the objectives, so that the original orientation of the specimen can be viewed in the oculars.

In common main objective stereo microscopes (CMO), the two divergent light rays are collected by a common infinity-corrected objective. The intermediate image is taken up by two parallel arranged optical trains of the zoom lens system and the oculars. The infinity-corrected optics allows the introduction of filters and other accessories above the common main objective. Because the optical system produces a parallel bundle of light rays, these accessories do not introduce significant aberration or shift the position of images observed in the microscope. It is obvious that these accessories are difficult to integrate into the Greenough microscope optical trains. This type is thus predominantly used as a routine inspection and dissecting microscope, whereas the CMO is a more flexible instrument for research purposes.

Opaque and sculptured specimens are usually illuminated from above by external light sources such as fiber optic illuminators. Oblique light sometimes creates distracting shadows in highly sculptured specimens. Here, ring light sources (arranged around the objective) are used to provide equal illumination from a high angle. Translucent specimens are illuminated from below the specimen stage. Here, besides conventional bright-field illumination, dark-field and oblique illuminators are also available. Dark-field and asymmetrical oblique illumination of a specimen provide enhanced image contrast. Some oblique illumination techniques give rise to a relief contrast effect with a similar appearance to differential interference contrast images.

Fluorescence illuminators for CMO stereomicroscopes are also available, especially in conjunction with detection of green-fluorescent protein (GFP). In principle, the fluorescence illuminator works as described for compound microscopes, with a common ray path of emitting and fluorescent light through (one channel of) the zoom lens system and the common main objective. Alternatively, microscopes with an external pathway for the excitation light – that does not utilize the microscope imaging system – are also available. Here, an additional lens system directs the excitation light onto the specimen. This leaves the illumination independent of the magnification zoom position.

1.4.7 Image Acquisition and Processing

During the past years, due to the need to record low light intensities, more and more ultra-sensitive cameras have been adapted for use with fluorescence microscopes. Generally, a signal from the camera is digitized and then further analyzed by use of an image processing system; further analysis (measurement of dimensions, image reconstruction) may also be applied. Thus, in most imaging systems now available for microscopes, the final image is not directly transferred to a monitor, but routinely stored, archived, displayed, or printed by use of a personal computer system.

Currently, the standard image acquisition tool is a CCD camera. CCDs are arrays of dense matrices of photodiodes that operate by converting light energy in the form of photons into an electronic charge; the stored charge is sequentially read out and transformed into a digital signal. CCDs (other than photomultiplier tubes or avalanche photodiodes) do not intensify the incoming signal, but they store the photon signal electronically over a certain exposure time. This reduces the frame rate (the rate at which a new still image is made available), but allows a measure of adaptation to low light signals. Since CCDs detect the whole image at a time, they are also referred to as area detectors, like CID (charge injection device) or CMOS (complimentary metal-oxide semiconductor) detectors or vidicon cameras.

Commonly used light detectors lacking spatial discrimination are the photomultiplier tube and the avalanche photodiode. Photomultiplier tubes (PMTs) are widely used detectors in confocal microscopes. Thanks to the signal enhancement process (a photon liberates a photoelectron from a photocathode, the electron liberates secondary electrons in a cascade-like process) an electron gain by a factor of 10^7 is achieved. The advantages of photocathodes over other low-light intensity detecting devices are their quick respond to changes in light input and their low dark current or background noise. Avalanche photodiodes enhance an incoming photon signal by a factor of 1000. Their modest gain and high background noise makes them inferior to PMTs, but they are also less expensive due to the simplicity of their construction, and also widely used as detectors. To produce an image with a CLSM, the image is built up in synchronization with the raster scan of the laser beam.

The spatial resolution of a digital image is determined by the distance between pixels (sampling interval). The sampling interval should be one half the size of the smallest resolvable feature of the optical image, or smaller. At a resolution limit of 0.22 μm for the light microscope, sampling intervals should be 0.11 μm or less. Smaller sampling intervals do not add any relevant spatial information from the object (oversampling), but can often help to improve the appearance of a final image after further image-processing steps. Thus, an interval of 2.5 to 3 samples for the smallest resolvable feature is suggested. It is often difficult for the user to match this criterion, because most digital cameras used for microscopy have a fixed minimum sampling interval, which cannot be adjusted to match the specimen's spatial frequency. Below a critical level, undersampling produces severe artifacts, known as spatial aliasing. These artifacts are obvious as repetitive specimen patterns looking like Moiré fringes.

A digitized (grayscale) image consists of picture elements (pixels) with a digital brightness value. Analog-digital converters operate with a certain bit-depth and should have at least 8- to 10-bit resolution, which corresponds to 256 or 1024 gray levels. The grayscale of an image is also referred to as dynamic range. Higher numbers of gray levels represent a greater dynamic range. Though nearly all final images in printing processes or for presentation are reduced to 256 gray levels, it is better to work with a higher bit depth during processing (as long as the hardware capacity allows it) and to discard the additional gray levels in a later processing step. A small number of gray levels results in contouring or posterization: a stepwise rather than a smooth gray-level gradient in the image. All operations that are performed on grayscale images can be extended to color images by applying the algorithms to each color channel separately, and then combining the channels. Thus, each color channel is processed at a resolution equivalent to the bit depth utilized in a grayscale image (e.g., 8 bits). The resulting 8-bit components are then combined to produce 24-bit pixels (referred to as true color), although some applications may require higher bit depths. Commonly, red/green/blue color output signals are converted into HSI or HSL (hue, saturation, intensity/lightness) signals, which in the end allows more intuitive signal correction by the user.

A raw image recorded by the camera is often unsuitable for further analysis or presentation. Bright-field images may be blurred, or show low contrast, false colors (due, for example, to improper color balance of the CCD sensor), illumination gradients, or distracting speckles. Contrast between bright fluorescent particles and a fluorescent or dark background can pose additional problems. Commonly, at least simple image processing software is included as a part of digital imaging packages for microscopy. More complex packages such as Image-Pro (Media Cybernetics), but also the freeware NIH image (developed by the Research Service Branch of the National Institute of Health), are comprised of several modules that may be combined individually.

Generally, image processing is of the following types: image enhancement for improvement of the overall image appearance or highlighting of certain details, image segmentation for categorization of elements or structures within an image, and image manipulation (i.e., the geometric alteration of an image), which is of minor importance for processing of microscopic images.

The algorithms for image processing comprise point-processing, area-processing, and frequency domain-processing algorithms. For **point processing**, each pixel is replaced by a new pixel, the value of which depends only on the old pixel value, irrespective of the values of other, neighboring pixels (typically brightness and contrast adjustment or replacement of colors). In most algorithms, relevant information for point processing is displayed by image histograms and lookup tables. An image histogram is a graph of pixel intensity distribution, showing the dynamic range of pixel values for that image. The plot displays the number of occurrences of each pixel value in the image against pixel value. Image histograms provide the essential information for the user to change the dynamic range of pixel values, such as contrast stretching. With look-up tables, pixels of one gray level or color are mapped to another gray level or color. They are thus the basis for contrast modification or false color display.

Area processes transform an image on a pixel-by-pixel basis, giving each pixel a new value depending on the values, and/or positions, of its neighbors. The neighborhood of a pixel forms an N by N grid around the pixel, where N is an odd number. The brightness trend or spatial frequency information may thus be described and modified. A number of spatial filtering processes (removal of noise, image smoothing, image sharpening, edge enhancement and detection) are used for spatial filtering. Through the replacement of each pixel in an image by the median value from that pixel's neighborhood, abrupt changes in pixel intensity – random noise – are smoothed out. Low-pass spatial filters attenuate the high-frequency content of an image (characterized by rapid changes in brightness and/or color), while low frequencies are left intact. This can also help to remove noise from an image, or to allow the low-frequency portion of an image to be examined more closely. High-pass spatial filters operate vice versa: low-frequency image content is attenuated, and the high frequencies are passed. By shifting the image by one pixel, and subtracting from the original, the slope or change in spatial frequency is found. In an area of constant pixel intensity, this slope is small or zero, while large slopes indicate rapidly varying pixel intensity. Areas of the former type are replaced by darker pixels, whilst areas of the latter type are replaced by lighter pixels. This shows edges or color change boundaries more clearly.

Area-processing methods dealing with spatial frequency can also be viewed as conversion from the spatial domain to the frequency domain and back again. In the strict sense, this is only true when the size of the filter matches the size of the image. For many applications it may be (computationally) easier to transform the image by **frequency domain processing** by use of the Fourier transformation technique, and to work on the data directly in the frequency domain. All processes concerning spatial frequency enhancement or removal may be very effectively corrected in the frequency domain. Key features of Fourier transformation are discussed in conjunction with processing of electron microscopic images (see chapter 2.4.9.2).

Comprehensive overviews in relevant fields of image acquisition and processing are presented in Ray, 1994, Russ, 1999, and Wilkinson and Schut, 1998.

1.4.8 Visualization of Living Unicellular Organisms

Numerous samples, single cells or small tissue fragments, may be visualized by brightfield or phase contrast microscopy after suspension in buffer without any further treatment. Microorganisms need no elaborate embedding and sectioning techniques for inspection (see, for example, Bast, 2001). They are visualized as whole cells, suspended in aqueous solution, such as the original culture medium or a suitable buffer (such as

Tab. 1.1 Living unicellular specimens for light microscopy

For culture suspensions:

1.) Dilute or concentrate a cell suspension to faint turbidity.
2.) Place approximately 20 – 50 μl onto a microscope slide and cover with a coverslip.

For solid or semi-solid material (e.g., taken from agar plates or a pellet):

1.) Place a drop of liquid (20 – 50 μl) onto a coverslip.
2.) Transfer a barely visible quantity of cells onto the surface of the coverslip with the aid of a (sterile) needle, carefully moistening the specimen with the drop if required, and smear the cells onto the coverslip. Dilute the homogeneous suspension of cells with the drop, mix the diluted suspension and cover with a coverslip.

Hanging drop procedure:

1.) Line the edge of the depression in the deep-well slide with a thin layer of sealing material (Vaseline, grease).
2.) Place a small drop of the liquid containing the specimen to be examined in the center of a coverslip positioned on a flat surface.
3.) Invert the deep-well slide and position it over the coverslip so that the drop of liquid is directly in the center of the well.
4.) Press the deep-well slide firmly onto the cover glass.
5.) With a quick, smooth motion, invert the slide. If this step is performed properly, a hanging drop will be formed on the coverslip.
6.) The slide is now ready for microscopic examination.

Tab. 1.2 Agar slide cultures

1.) Select an Agar medium according to the growth requirements of the organism.
2.) Dip clean, sterile microscope slides into the molten nutrient agar (0.7–2.0% w/v agar in nutrient solution). Dip two or more times for thicker layers. Remove the agar from the underside of the slide.
3.) Inoculate the agar surface using a finely drawn out glass fiber (prepared from a Pasteur pipette): moisten the tip with a sterile buffer solution or sterile liquid medium, lightly touch it to a colony liquid culture of the organism and spread it on the fresh agar surface.
4.) Trim away the agar around the inoculated area, place a coverslip on top, and seal the edges with wax.
5.) If the square of nutrient agar is smaller than the coverslip, there is a supply of air and minimal drying. Incubate according to conditions appropriate for the organism.

Tab. 1.3 Immobilization of motile organisms

Use agarose that is free of disturbing insoluble particles; alternatively, 15–20% gelatin solution may be used. The high optical density of gelatin reduces the disturbing halo effect in phase contrast.

1.) Melt 2% agarose in distilled boiling water.
2.) Transfer 2 ml of the molten agar solution onto microscope slides. After solidification of the agar, dry the slide at 50 °C in a dust-free drying oven.
3.) Store the slides (usable for several months) in a dry and dust-free place.
5.) Apply a drop of suspended cells onto a coverslip; turn the coverslip over and place it on the agar-coated face of the slide. The swelling agar squeezes the organisms against the coverslip and immobilizes the cells, but keeps the surroundings hydrated.

50–100 mM potassium phosphate buffer, perhaps supplemented with 0.9%, w/v sodium chloride – "phosphate-buffered saline"); see Table 1.1 for details.

For some sensitive organisms, mechanical squeezing by a coverslip may be deleterious. The hanging drop method (Table 1.1) should be used as an alternative. Especially useful for observation of vegetative growth of fungal filaments are agar slide cultures (Table 1.2). Observation of highly motile cells sometimes requires immobilization. If it is required that organisms should stay alive (not necessary in, for example, simple cell counting), the viscosity of the embedding medium is raised by addition of gelatin or preparation of agar-coated microscope slides (Table 1.3).

1.4.8.1 Bright-field staining techniques for unicellular organisms

"Classical" staining techniques brought about by use of colored chemicals enhance the amplitude contrast in bright-field microscopy (see, for example, Baker, 1958; Habs and Seeliger, 1967; Lillie, 1977). They are used for general enhancement of organisms against a background (simple stains), differentiation of groups (differential stains), or for selective visualization of internal cell structures (cytological stains). Negatively charged cellular components are best stained by cationic dyes (Crystal Violet, Toluidine Blue). Basic cellular components bind anionic dyes such as Acid Fuchsin. Contrary to these ionic interactions, some dyes bind covalently to cellular structures. The color-rendering component (chromophore) is coupled to a reactive component that usually binds to hydroxy, amino, or thiol groups. The Procion dyes, with dichlorotriazine as reactive group, are commonly used representatives of these agents.

All these dyes are applied in solution form, and the choice of solvent can significantly influence the activity of the dye. The duration of exposure to the specimen and the ambient temperature affect the deepness of staining. Since most dyes are salts, a polar solvent (an aqueous or ethanol solution) is used. A non-polar solvent, usually in conjunction with a mounting medium, is applied to stop the staining process and to keep the staining permanent.

Extraction of dye from an overstained specimen is also achieved by use of a polar solvent (ethanol, acetone), acetic acid (0.5 – 1 %, v/v), or diluted mineral acids. An effective destaining solution is one composed of three parts HCl (25 %) and 97 parts ethanol (70 %, v/v).

Chelating metals (usually Al^{3+}, Fe^{2+}, Fe^{3+}, Cr^{3+}) are often required as additional components to the staining solution to enhance the staining properties of a dye (mordants). pH and certain electrolytes can also alter staining characteristics. These accentuators are commonly applied to intensify staining. Trapping agents (such as iodine in the Gram staining procedure) are chemicals that inhibit or slow down removal of dyes from tissues.

Numerous staining methods for bright-field microscopy of microorganisms are commonly used in medical diagnostics, in order to detect organisms in smear specimens (sputum, stool, tissue, etc.). They are also used, in conjunction with other diagnostic tools, to identify a pathogen and to estimate its spread in the host's body. Since these techniques are well established and reliable, they are still highly relevant for routine diagnostics. For in situ identification of organisms other than in biomedical applications, these techniques are of limited use, because no classical identification scheme for non-pathogenic organisms exists. Stains for bright-field microscopy are thus commonly used for visualization of general features such as the bacterial capsule and storage material and for Gram classification.

General staining of organisms, which in the past had to be applied to obtain sufficient amplitude contrast for observation, has, however, been replaced by optical staining: phase contrast (for smaller organisms such as bacteria and yeasts), or DIC (for larger cells). Cell suspensions and large inclusions (endospores, storage granules) need no further staining unless further differentiation of their chemical nature becomes necessary. The application of commonly used stains for the identification and visualization of relatively large features in unfixed microorganisms (storage granula and extracellular capsules) are described in Table 1.4.

Generally, cells are fixed prior to application of a specific cytological stain. For this purpose, a thin layer is smeared onto the coverslip surface, air dried, and chemically or heat-fixed. Fixed bacteria are firmly attached to the microscope slide surface, but protoplasts become coagulated. Only large and rigid cellular structures (endospores, storage granules) remain preserved, and become visible after application of an appropriate stain (the effect of fixatives is discussed below in more detail). The most widespread staining technique for heat-fixed cells is the Gram differential stain (Table 1.6). Both Gram-positive and Gram-negative bacteria take up crystal violet. The outer membranes of Gram-negative bacteria are disrupted in the acetone rinse step, allowing

Tab. 1.4 Staining of storage granules and capsules

Polysaccharides:

1.) Apply a drop of Lugol's solution (0.3 % iodine w/v, 0.6 % KI w/v in distilled water) to one side of the coverslip of a prepared specimen suspension (see Tab. 1.1) and soak it through the suspension by applying a piece of filter paper at the opposite side of the coverslip.
2.) Observe in bright field (no phase contrast). Starch granules appear black/blue, granulose blue-violet/brown-violet, glycogen reddish-brown.

Lipids:

1.) Place a drop of Sudan Black solution (0.3 % Sudan Black B w/v in aqueous ethanol solution, 70 %, v/v) onto a microscope slide.
2.) Suspend a small amount of cells (pellet or concentrated cell suspension) in the drop, place a coverslip on the drop, and incubate for 15 min.
3.) Observe in bright field.

Capsules:

1.) Apply a drop of diluted Indian ink (25 – 50 % v/v in water) to one side of the coverslip of a prepared specimen suspension (see Tab. 1.1) and soak it through the suspension by applying a piece of filter paper on the opposite side of the coverslip.
2.) Observe in bright field (no phase contrast). remove excess liquid; only a thin film (approx. one diameter of cells in thickness) is allowed, to obtain a visible effect. Capsules exclude Indian ink particles and appear bright against a dark background of ink particles.

the crystal violet to be differentiated out (i.e., selectively destained). This allows the Gram-negative bacteria to take up the basic fuchsin stain. Gram differentiation should be performed with at least one known control strain processed on the same microscope slide.

Results of the Gram staining procedure are often ambiguous because they depend on incubation times, rinsing steps, and the growth phase of the organisms themselves. Thus, numerous modifications of Gram stain protocols may be applied, optimized for differentiation among certain bacterial groups.

Among numerous stains used for detection and differentiation of pathogenic organisms, some – such as the Neisser stain for detection of polyphosphate-granules – are of relevance for non-medical applications. Phosphate is stored under unfavorable nutrient conditions as polyphosphate-granula. Polyphosphates form linear polyanionic chains with Mg, Ca, or K as counterions. The classical Neisser stain is sometimes used for identification of such granula monitoring of phosphate removal in activated sludge processes.

Tab. 1.5 Fixation of small unicellular organisms (bacteria, yeasts)

Heat fixation:

1.) Smear a diluted cell suspension (see Tab. 1.1) on the surface of a microscope slide and allow to air-dry thoroughly.
2.) Fix the smear on the microscope slide by passing the slide over a Bunsen burner flame three times (smear directed upwards). Avoid overheating; the slide should be touchable after having passed through the flame.

Tab. 1.6 Gram staining procedure

Crystal Violet solution: Crystal Violet (1 % w/v); phenol (2.5 % w/v); ethanol (10.0 % w/v) in distilled water
I/KI-solution: see Tab. 1.4

1.) Use heat-fixed smears of cells: the unknown organism and two (Gram-positive and Gram-negative) reference organisms.
2.) Cover with Crystal Violet solution, incubate for 2 min and rinse in a water bath for 5–10 s.
3.) Incubate with I/KI solution for 2 min and rinse in a water bath for 5 s.
4.) Destain for 1 min in 1-propanol and rinse with water for 5 s.
5.) Cover with safranin solution (0.5 %, w/v) and rinse with water for 5 s.
6.) Cover the wet specimen with a coverslip and observe in bright field.

Tab. 1.7 Neisser stain procedure for polyphosphate staining

Methylene Blue solution: Methylene Blue (0.1 %); ethanol (2 % v/v); acetic acid (5 % v/v), dist. water to 100 %.

Crystal Violet solution: saturated Crystal Violet solution (> 13.87 g in 100 ml 96 % ethanol), diluted to a 2 % (v/v) working solution with distilled water.

KI/I solution: see Tab. 1.4

Chrysoidin solution: 0.7 % (w/v) Chrysoidin G in distilled water

1.) Mix 2 parts Methylene Blue solution and 1 part Crystal Violet solution prior to use.
2.) Use heat-fixed cells on a microscope slide. Cover the slide with the Methylene Blue/Crystal Violet mixture. Incubate for 20–30 s. Remove the solution, rinse under a mild jet of tap water.
3.) Cover with KI/I solution, incubate for 5 s. Remove the solution and rinse with tap water.
4.) Cover with chrysoidin solution, incubate for 10 s. Remove the solution, do not rinse. Polyphosphate granula appear as dark blue particles in yellow-brown colored cells.

For preservation of smaller structures, fixation is carried out by application of aldehydes or osmium tetroxide. Aldehyde fixation in particular is most widely used for fluorescence microscopy.

1.4.9
Embedding and Sectioning – a Quick Start

The structures of complex specimens are only maintained if the organisms (or cells as parts of a tissue) are held in place, especially when sectioning of specimens is to follow. Optical sectioning by confocal microscopes and other devices does not replace embedding and sectioning procedures, because all these microscopes need translucent specimens. Only some tomographic techniques provide insight into (intact) specimens, but still at lower resolutions than a histological section.

Sectioning is performed either by conventional processing at room temperature or by cryopreparation. In the conventional process, cells and tissue are chemically fixed to withstand the subsequent preparation procedures. Embedding in paraffin or embedding resin and thin sectioning by microtomy follows. Embedding is an essential prerequisite for numerous cytochemical staining methods. Comprehensive overviews on

Tab. 1.8 Gelatin embedding of biofilms

1.) Fix samples with 2 % w/v formaldehyde solution and 0.2 % glutaraldehyde solution.
2.) Prepare a mixture of 0.2 g gelatin in 1 ml of a 50 % v/v glycerol/ 50 % v/v water mixture. Incubate at 95 °C to obtain a homogeneous solution. Cool down to 40–45 °C.
3.) Incubate the specimen in the gelatin-glycerol solution for 5 min.
4.) Freeze the specimen at –20 °C. After solidification, perform razor blade or microtome/vibratome sectioning.
5.) Microscope on a slide with a cover slip in a drop of water.

the respective methods can be found in Gray (1964), Berlyn and Mischke (1976), and Ruzin (1999). Cryopreparation of specimens needs a cryostat (cryotome) as special equipment, but allows the whole specimen preparation procedure to be shortened to hours rather then days. Cryostat sectioning is widely used for animal and human tissue (especially as a rapid procedure for routine clinical diagnostics), but may also be applied to plant tissue (for applications in biomedical research, see Bancroft and Gamble, 2002). For preparation of objects of low mechanical stability (such as biofilms), embedding in gelatin is a simple and straightforward alternative, as long as sections of 100 μm (and thicker) are tolerable (Table 1.8).

1.4.10
Tissue Sectioning with Vibratomes

Vibratome sectioning allows preparation of sections from material without embedding or freezing. Chemical fixation may also be omitted. Since the sample is prepared in its native state, sections are suitable for subsequent immunolocalization techniques (see below) and localization of enzyme activities. The principle of the vibratome is a vibrating blade (rapid lateral movement and slow forward movement in the cutting direction) that produces uniform slices of a specimen. To facilitate the cutting process, specimen and knife are submerged in a (cooled) bath of liquid. The knife may be a razor blade, but sapphire or glass knives are also available. The vibrating movement of the knife enables cutting of non-stabilized specimens, which would be impossible with a sliding microtome. The big advantage of vibratome sectioning is obvious: artifacts due to fixation and embedding may be avoided. The cells in tissue slices may even be kept alive for further studies after the sectioning procedure. The technique is therefore often used in conjunction with fluorescence microscopy for dynamic processes. Nevertheless, in comparison with sections stabilized by an embedding medium, several concessions have to be made. The mechanical properties of the specimen itself greatly influence the quality of vibratome sections. In embedded material, mechanical stability is mainly determined by the embedding medium, and is also easily controllable by slight variation of the embedding procedure. In vibratome sectioning either of native or of just chemically fixed material, every specimen type needs individual parameter settings. Generally, vibratome sections are 3–5 times thicker than paraffin-embedded sections. The actual thickness greatly depends on the mechanical properties of the specimens and the knife itself, as well as the instrument settings listed below.

Tab. 1.9 Basic vibratome sectioning operation procedures

1.) If necessary, embed the specimen in 4–6% w/v molten agarose – at a temperature slightly above the gelling point, so as not to heat-damage the specimen. Gelatin-embedded specimens (Tab. 1.8) are also appropriate.
2.) Cut the specimen to an appropriate size and then mount it on the vibratome base plate with the aid of appropriate glue (tissue glue or cyanoacrylate-based glue). Pieces of wood, plastics, or agar may be glued onto the base plate to stabilize the specimen further. Allow the glue to dry.
3.) Place the base plate in the vibratome chamber.
4.) Adjust the knife relative to the specimen: advance the knife to just in front of the specimen; raise or lower the specimen so that the top is just below the knife edge.
5.) Perform initial sectioning to obtain a flat specimen surface. Set (for example) the advance speed to a low value, the amplitude at a medium value, and section thickness to 50–100 μm until whole sections are produced.
6.) Optimize control settings and perform sections of appropriate quality.
7.) Use a small paintbrush to pick up the floating sections and transfer them to a microscope slide. Sections may be observed directly, stained, or processed by localization procedures.

Though sapphire blades might cut sections even down to 1 μm thickness, this only applies for ideal specimens. With standard steel blades, sections in the 20–50 μm range may usually be achieved. Although the basic operation procedures of vibratome sectioning are quite simple (Table 1.9), some experience is required for optimization of the sectioning process.

Section quality is influenced by five basic parameters under the control of the operator:

- section thickness,
- nature of the bath (usually a physiologically compatible buffer or distilled water),
- temperature of the bath (the softer the specimen, the colder the bath should be),
- speed of the sectioning blade (the softer the specimen, the slower the speed),
- amplitude of the lateral excursion of the sectioning blade vibratory movements (the softer the specimen, the higher the amplitude),
- blade angle (the softer the specimen, the steeper the angle of attack required).

An example relating to brain tissue (soft) may illustrate the setting parameters:

low bath temperature (2–6 °C), slow sectioning speed (setting 1–2), high amplitude (> 8), steep blade angle (25° or more). Optimization of the parameters may be elaborate for specimens with high internal variations of mechanical density, or when especially thin sections are required.

1.4.11
Fixation, Embedding, and Sectioning of Specimens

Although vibratome and cryostat sectioning are much quicker procedures for sample processing and are advantageous in conjunction with subsequent application of immunolocalization and other techniques with specific probes, conventionally embedded specimens are still useful. Specimens embedded in resin or paraffin may be easily handled and stored at room temperature for long periods of time without loss of quality. Thus, numerous experiments may be performed with the same specimen.

Embedding procedures vary greatly concerning fixation, dehydration, and embedding media, but all of them follow a standardized scheme (see also, for example, Steward and Pearse, 1992, and Sanderson, 1994):
- fixation,
- dehydration of fixed objects,
- embedding in paraffin (or another appropriate wax),
- mounting and trimming of the specimen,
- sectioning,
- transfer of the sections onto microscope slides.

1.4.11.1 **Fixation**

The purpose of fixation is to preserve samples permanently, especially against the subsequent dehydration and embedding procedures, in which the perturbing chemical and morphological alteration of the specimen should be as little as possible (Steward, 1973). For light microscopy, a number of different modes of chemical fixation for numerous purposes have been described. All procedures, however, affect the structures of objects to a greater or lesser extent. Fortunately, most of these alterations affect cellular and sub-cellular organization rather than tissue organization. These artifacts may be tolerated when only gross morphological features are of interest. Localization experiments, however, require a more careful selection of fixatives, or omission of chemical fixation.

The action of a fixative on the chemically highly heterogeneous specimen is complex, and for numerous fixatives the chemical details of the interaction with the specimen are still obscure. Moreover, fixatives with different effects on the specimen are often used as mixtures or in subsequent steps. The effects on the preservation of enzyme activities or antigenicity are thus difficult to predict and must often be determined empirically.

Essentially, numerous fixatives transform the **cytosol** into a **gel**, the specimen thereby becoming stabilized against extraction of components by subsequent procedures. Depending on the preservation characteristics of the fixatives, three types can be distinguished. Cytological fixatives preserve cellular structures. They are often sub-classified into cytoplasmic fixatives (which preserve the components of the cytoplasm) and nuclear fixatives (preserving the nuclear structure, but not explicitly the other components). Microanatomical fixatives achieve preservation at tissue level, but intracellular components are destroyed. Histochemical fixatives especially preserve the antigenicity of proteins and activity of enzymes.

Fixatives cannot be cleanly separated into these three groups; most of them meet more than only one of these features. Another classification scheme, by mechanism, is also used. By the action of **coagulating** fixatives, proteins become precipitated, which generally massively affects the intracellular structure. Crosslinking fixatives are retained in the tissue (and so they are also referred to as **additive**), chemically interconnecting structures. This is generally more suitable for maintaining the cytoplasm in a "life-like" structure. These fixatives are thus also widely used for electron microscopy. Fixatives of this type are (bifunctional) aldehydes and osmium tetroxide. Alternatively, fixatives may be **non-additive** and/or **non-coagulant** (acetic acid).

Typical coagulating fixatives are chromic acid, mercuric chloride, picric acid, and alcohols. Chromium salts (or the respective acids) dissociate in aqueous solution to give Cr-O-CR complexes, which react with most cellular components. They are especially good fixatives for chromosomes. Mercuric chloride as a fixative agent strongly coagulates protein and is an excellent mordant. A disadvantage is the appearance of precipitates throughout the tissue; these must be removed by a post-fixation treatment as described in Table 1.10. Picric acid is a component in numerous fixatives. Picric acid acts in both coagulating and additive fashion (protein crosslinker). The extreme shrinkage of tissue is counteracted by inclusion of acetic acid in the fixative. Alcohols are protein denaturants and are not used routinely for embedding of animal tissues because they cause too much artificial shrinkage, distortion, and brittleness of the specimen. However, they are very good for cytological smears because they act quickly and give good nuclear detail.

Non-coagulating agents are aldehydes, acetic acid, and potassium dichromate. Aldehydes include formaldehyde (formalin) and glutardialdehyde (glutaraldehyde). Tissue is fixed by (cross-) linkages formed in the proteins, particularly between lysine residues. Formaldehyde is generally recommended as an appropriate, all-purpose fixative. This cross-linkage does not harm the structure of proteins greatly, so activity and antigenicity is not lost, and so formaldehyde is good when subsequent immunological techniques are to be used. Nucleic acids are not affected. Formaldehyde is therefore used, often in conjunction with alcohols, as an appropriate fixative for subsequent in situ hybridization techniques. Glutaraldehyde gives best overall cytoplasmic and nuclear detail, but penetrates tissue slowly. It is thus often used in a mixture together with formaldehyde. Glutaraldehyde/formaldehyde fixatives penetrate tissue quickly and are best for preservation of fine structure down to electron microscopic levels. Unlike formaldehyde, however, glutaraldehyde may mask antigenic sites even at low concentrations. It may also suppress in situ hybridization signals. Autofluorescence of aldehydes, especially glutaraldehyde, though, is sometimes troublesome in fluorescence microscopy. Additional steps for reduction of autofluorescence prior to fluorescence detection are therefore necessary.

Most fixation recipes have been determined empirically, and so numerous combinations of fixatives and buffers in wide concentration ranges exist, with quite similar results for a wide variety of tissues. However, thanks to the long experience collected in the fields of histology and pathology of human tissue, well established optimized procedures exist for fixation of virtually all tissue types, often in conjunction with an appropriate stain. Table 1.10 presents only the most widely used fixatives, suitable for animal as well as plant tissue, together with general remarks on tissue type and handling procedures. The fixation process is affected by a number of factors, such as pH, penetration time, volume of the fixative vs. specimen volume, temperature, concentration of the fixative, and time interval of fixation. Certain combinations of physiological buffers with fixative should be avoided, due to formation of precipitates or partial inactivation of fixative by buffer components. The fixatives listed in Ta-

Tab. 1.10 Common fixatives for use in light microscopy

Fixative (100 ml)	Features	Use
FAA/FPA – ethanol (96 %) 50/70/88 ml – glacial acetic acid/propionic acid 5 ml – formaldehyde (37 %) 5 ml – distilled water 20/10/2 ml	Most commonly used general purpose fixative for plant specimens; rapid tissue penetration; higher ethanol concentration for more rigid, lower concentration for delicate specimens; stable working solution	Fixation time 8 – 24 h, usually no overfixation (specimen may be kept in the solution for longer time periods)
Carnoy's fixative + chloroform – ethanol (100 %) 60 ml – glacial acetic acid 10 ml – chloroform 30 ml	General purpose cytological and histochemical fixative for animal and plant specimens, rapid penetration of objects lower acetic acid concentrations for delicate objects; stable working solution	Fixation time for thin, easily penetrable specimens about 1 min; for rigid specimens, 1 – 4 h, wash in 50 %–70 % ethanol*)
Bouin's fixative – formaldehyde 37 % 25 ml – picric acid 1.3 % (saturated) 70 ml – glacial acetic acid 5 % 5 ml	General purpose cytological (esp. nuclear) fixative for animal and delicate plant specimens (anthers, root tips), chromosomes; stable working solution	Fixation time 2 – 24 h usually no overfixation wash in 50 %–70 % ethanol (several exchanges) for up to 2 days to remove surplus picric acid*)
Allen's solution – formaldehyde 37 % 23 ml – glacial acetic acid 4.6 ml – urea 1.4 g – chromium(IV) oxide 1.4 g – picric acid 1.3 % (saturated) ad 100 ml	Cytological (esp. nuclear) fixative for animal specimens, chromosomes	Dissolve chromium oxide and then urea in the liquid components after gentle warming; fixation time 2 h at 37 – 38 °C; wash in 50 %–70 % ethanol – overfixation leads to tissue hardening
Hollande's fixative – copper acetate 2.5 g – picric acid 4 g – formaldehyde 37 % 10 ml – glacial acetic acid 1.0 ml – distilled water to 100 ml	A modification of Bouin's fixative; mild decalcifyer	Fixation time for small specimens 2 – 4 h, application of phosphate buffers after fixation causes precipitation
Zenker's fixative (stock solution) – mercuric chloride 5 g – potassium dichromate 2.5 g – sodium sulfate 1.0 g – distilled water ad 100 ml **additional solutions:** I/KI–solution – iodine 1 g KI 2.0 g – 70 % (w/v) ethanol ad 100 ml sodium thiosulfate 5 % (w/v)	Efficient routine microanatomical fixative; also nuclear fixative and mordant stable stock solution (stored dark)	Add 1 part of acetic acid to 19 parts of the stock solution just prior to use. After fixation, incubate specimen in 70 % I/KI solution to remove mercury pigments. Decolorize the iodine by titration with 5 % (w/v) sodium thiosulfate

Tab. 1.10 continued

Fixative (100 ml)	Features	Use
Pfeiffer's fixative formaldehyde 37 % 33 ml glacial acetic acid 13 ml methanol 33 ml distilled water 21 ml	Fixative for micro-algae, fungi, mosses; stable working solution	Fixation time 30 – 60 min, no overfixation (specimen may be kept in the solution for longer time periods)
Karnovsky's fixative – paraformaldehyde 2 g – glutardialdehyde 25 % 10 ml – distilled water (or appropriate buffer) to 100 ml	Fixative for best preservation of cytological features (also suitable for electron microscopy), antigenicity and enzyme activity can be retained	Dissolve paraformaldehyde in slightly basic aqueous solution (use NaOH solution for titration) under gentle heating (50 – 80 °C) fixation time 1 h–24 h. After fixation, wash in (buffered) aqueous solution
Formalin and other formaldehyde solutions (37 % formaldehyde solution or higher dilutions)	Cytochemical fixative (antigenicity is mostly retained, for in situ hybridization)	Add approximately one volume of formalin to ten volumes specimen-containing buffer fixation time 1 h–24 h. After fixation, wash in (buffered) aqueous solution. For subsequent immunohistochemistry, use lower formaldehyde concentrations, if possible
Glyoxal-based fixatives (8 %, v/v in bicarbonate-NaOH buffer, 0.5 M, pH 8.9)	Similar to formalin	Fixation time 1 – 2 h
Zinc fixative – 0.5 g zinc acetate – 0.5 g zinc chloride – 0.05 g calcium acetate Tris buffer (0.1 M, pH 7.4) to 100 ml	Mild fixative; antigenicity is mostly retained, as an alternative to formaldehyde fixatives for immunohistochemistry	Fixation time 8 – 12 h
Commercially available fixatives	Special purpose fixatives (usually mild fixatives suitable for subsequent immunohistochemical procedures and in situ hybridization), of various, partially unknown, modes of action	To be applied according to supplier's instruction, often used in conjunction with immunohistochemical kits. – GlyoFix (Shandon-Lipshaw), glyoxal-based fixative – Mirsky's fixative (National Diagnostics) carrier-coupled dialdehyde – NoToX (Earthsafe Industries) (Prento and Lyon, 1997)

*) when embedding procedures are to follow, washing is carried out by dehydration in the ascending ethanol series

ble 1.10 are commonly used without buffering. Paraformaldehyde/glutardialdehyde fixatives, however, may be used very well in physiological buffers, as long as the buffer chemicals do not contain reactive amino groups.

Generally, fixation should be performed as soon as possible after the collection of the specimen, or as necessary for larger organs, in situ, by perfusion fixation to avoid artificial rearrangement of structures. The diffusion time of the fixative in the specimen therefore also has to be considered. A small, thin specimen allows quicker penetration than a large cube. Especially in plant specimens, diffusion may also be impaired by hydrophobic surfaces or air-filled spaces. Penetration often has to be facilitated by vacuum infiltration (especially important for tissue) in the presence of air-filled spaces such as aerenchyma in plant tissue. A one-stage vacuum pump produces a suitable vacuum for this purpose. The open vial containing the sample submerged in the fixative should be placed in a suitable desiccator connected to the vacuum. The air-filled plant tissue will often float on the liquid surface and then sink to the bottom of the vial during infiltration. Some recipes include organic solvents that facilitate the passage of fixatives. Among the fixatives listed in Table 1.10, the best penetrating agents are formaldehyde and alcohol. Application of microwave radiation also facilitates fixation. A short heating pulse (20 s to 1 min) generally accelerates the reaction of a fixative. In order to obtain equally fixed specimens, microwave fixation should be applied after the specimen is completely soaked with the agent. Instruments especially developed for histology for microwave fixation and other microwave-facilitated processing steps (see below) are available from several suppliers.

There should be a 10:1 v/v ratio of fixative to specimen and/or change of the fixative at intervals to avoid exhaustion. Mild agitation of the specimen will enhance fixation. Moderate heating may, but does not necessarily have to, facilitate fixation. Concentration of fixative should be adjusted down to the lowest level possible. Despite possibly facilitated fixation, the specimen may be damaged in several ways: besides greater artificial swelling or shrinkage, more damaging precipitation may also be introduced. Excessive concentrations of aldehyde fixatives (especially glutardialdehyde), will disturb antigenicity and catalytic activity. Table 1.10 shows the most widely used concentrations.

Most of the solutions contain highly toxic, sometimes volatile, agents. Preparation requires special safety measures. Waste must be collected and disposed of separately. Most of the stable working solutions are available from a number of suppliers, which reduces danger of exposure during use.

1.4.11.2 **Dehydration and embedding**

Various embedding media, based on paraffin, carbowax, or synthetic resins, are routinely applicable in light microscopy (Table 1.11). The embedding medium should be selected with regard to the post-sectioning procedures. The harder paraffin and resin mixtures are more suitable than the standard paraffin mixture for sections of less than 5 μm in thickness. Not all embedding media are equally suitable for certain staining procedures (especially with fluorescent stains), for detection of enzyme activities, in situ hybridization, and immunolocalization techniques.

Table 1.12 shows a commonly used procedure using paraffin-based embedding media (e.g., Paraplast, Histowax, or Histosec). These media include additives as listed in Table 1.11. Paraffin embedding requires, after dehydration in an ascending ethanol series, transfer to a paraffin-miscible solvent (transition fluid). Whereas xylene or methylbenzoate/benzene are very suitable transition fluids, they are highly toxic and have been replaced by other, non-toxic substances. In particular, terpene mixtures such as the limonene-based Histo-clear, Histo-clear II, or Rotihistol (Roth, Karlsruhe) are widely used. Alternatively, preparations using acetone as transition fluid or an acetone instead of an ethanol series for dehydration have been proposed. Table 1.12 summarizes a standard procedure for dehydration and paraffin embedding. Incubation times refer to thin and small tissue specimens (either animal or plant). Large or hard (plant) tissues may need as much as 24 h per step.

Infiltration and embedding may be markedly accelerated by application of microwave procedures. Under strict temperature control (to avoid boiling of the solvents),

Tab. 1.11 Embedding media for light microscopy

Embedding medium	Constituent(s)	Use
TissuePrep, TissuePrep 2	Purified paraffin/synthetic polymers; melting temp. approximately 56 °C	Section thickness down to 2–4 μm
Paraplast, Paraplast plus	Purified paraffin/synthetic polymers/DMSO; melting temp. approximately 56 °C	Section thickness down to 2–4 μm
Paraplast X-tra	Purified paraffin/synthetic polymers; melting temp. approximately 50–54 °C	Section thickness down to 2 μm, low viscosity
Steedman's wax	9 parts PEG 400 distearate 1 part 1-hexadecanol; melting temp. approximately 38 °C	Section thickness down to 8 μm, softer than paraffin; improper sectioning with rigid specimens; immunolocalization
Glycol methacrylate (Historesin)	2-Hydroxyethyl methacrylate; additional catalyst and accelerator required	Section thickness down to 0.5 μm, use with ethanol (no absolute dehydration required; final step: 95 %); immunolocalization
Butyl/methyl methacrylate	*n*-Butyl and methyl methacrylate mixture; additional catalyst required	Section thickness down to 0.5 μm, desired hardness is obtained by adjusting the relative proportion of *n*-butyl and methyl methacrylate; immunolocalization, in situ hybridization
embedding resins for electron microscopy	Spurr's resin Epon resins Lowicryl resins (see section 2.4.6.4)	Section thickness up to 1 μm with glass knives, for combined light/electron microscopy

Tab. 1.12 Dehydration and paraffin embedding

Dehydration:

1.) Dehydrate in an ascending ethanol series (10 % v/v, 15 %, 20 %, 30 % ... 70 %, 1 h per step; 80 %, 90 %, 96 % 4 h per step, 100 % 2 h) with gentle and continuous shaking. Incubation times for the steps up to 96 % may be prolonged, especially for rigid materials (such as woody plants). Upon dehydration, the specimens become very hard in 100 % ethanol, especially when the incubation times are extended. Isopropyl alcohol may be used instead.
2.) Replace the 100 % ethanol by methylbenzoate. Exchange methylbenzoate twice (incubation time: 2 h per step).
3.) Replace methylbenzoate by benzene. Exchange benzene twice (incubation time: 30 min per step).

Paraffin embedding:

4.) Prepare a solution of benzene and molten paraffin (equal volumes), maintain at 30 – 40 °C. Replace the benzene solution from step 3, incubate for 60 min, then replace the mixture by pure paraffin (melting point 56 – 58 °C), incubate for 2 h. The samples may be stored in the solidified paraffin.
5.) Use embedding molds of appropriate dimensions for the final embedding step. Coat the molds with a thin glycerol layer (this will later facilitate the removal of the paraffin bloc). Transfer the object to the embedding mold; cover the sample with molten paraffin. Orientate the sample with warm preparation needles or forceps. When a thick surface layer of the paraffin is solidified, the sample may be submersed in cold water to speed up solidification.
6.) Remove the solidified block from the mold.

Tab. 1.13 Dehydration and glycolmethacrylate resin (GMA) embedding

The procedure refers to Technovit kits (trademark Heraeus Kulzer, Germany, and available from several suppliers). Other kits are used in a similar way, although the composition of the resin kit may differ slightly. Refer to the suppliers' instructions.

1.) Prepare the infiltration solution. Add 1 g of Hardener I (benzoyl peroxide) to 100 ml of 2-hydroxyethyl methacrylate (Technovit-GMA). Mix using a magnetic stirrer until the benzoyl peroxide is completely dissolved. Store refrigerated for up to 2 months.
2.) Dehydrate a fixed specimen in an ascending ethanol series (70 % v/v, 96 % ethanol for 2 h, repeat the step with 96 % ethanol; 100 % ethanol for 1 h). Place specimens in vials or appropriate embedding molds.
3.) Prepare a mixture of infiltration solution (step 1) and 100 % ethanol (equal volumes), incubate for two hours. Exchange for 100 % infiltration solution and incubate for 8 – 16 h. Replace the incubation solution with fresh solution and incubate for another 1 – 3 h. The incubation time with infiltration solution should be prolonged with large, rigid specimens. Air-filled spaces in specimens may infiltrated under vacuum (one-stage rotary pump).
4.) Mix 1.15 parts of infiltrating solution and 2.1 part of Hardener II. Use the solution within 10 minutes, before polymerization occurs. Cover the specimens in embedding molds or fill gelatin capsules with the solution. Close or seal the molds/capsules to exclude oxygen. There is a varying sensitivity of the GMA resin mixtures to oxygen, which prevents polymerization; sometimes a short degassing step to remove disturbing air bubbles and/or polymerization in a vacuum desiccator may also be necessary (refer to suppliers' instructions). Allow polymerization to take place overnight at 4 °C (the polymerization process heats the specimen, which may induce damage). If the block is still too soft for sectioning after polymerization, harden it by heating at 60 °C for 1 – 12 h.

Tab. 1.14 Dehydration and butyl/methyl methacrylate embedding for subsequent in situ hybridization or immunolocalization

1.) Fix the specimen for 2–6 h in formaldehyde/glutardialdehyde mixtures (e.g., 2.5 % formaldehyde – freshly prepared from paraformaldehyde – 0.5 % glutardialdehyde) in an appropriate buffer at pH 7.0.
2.) Remove the fixation solution by several washing steps in buffer.
3.) Dehydrate the specimen in an ascending ethanol series (e.g., 30 % v/v, 50 %, 70 %, 90 %, 95 % ethanol*))
4.) Infiltrate with butyl/methyl methacrylate. Use a mixture of 1 volume of resin and two volumes of ethanol for at least 2 h, then a mixture of equal volumes for at least 2 h, finally a mixture of 12 volume ethanol and two volumes of resin. Then infiltrate with pure resin (including 0.5 %, w/v, benzoin methyl ether w/v and 1 %, w/v, benzoyl peroxide) for at least 12 h (up to several days).**)
5.) Place the specimen in appropriate embedding containers (see Tab. 1.13, step 4) containing the infiltration medium (this table, step 4).**)
6.) Polymerize in the cold room or refrigerator for 18–72 hours under UV light.
For subsequent in situ hybridization of mRNA, the following treatments are necessary:
– All glassware should be baked at 180 °C or treated with 5 % diethyl pyrocarbonate (DEPC) in 100 % ethanol. All solutions should be DEPC-treated overnight prior to being autoclaved. This has to be done to inactivate RNase. Care has to be taken not to re-contaminate the RNase-free solutions.

*) Addition of 1 mM dithiothreitol to all solutions
**) Addition of 10 mM dithiothreitol to all solutions

dehydration time may be reduced to minutes. Facilitated paraffin infiltration by microwaves is especially useful for rigid specimens, reducing infiltration times roughly by a factor of 10. Automated histoprocessing for small, soft tissue specimens (e.g., needle biopsies) based upon the application of microwaves can reduce the time for specimen preparation to 30 min.

Polymerized resin ("plastic") is a harder, denser embedding matrix for tissue, improving support and preservation of fine structural features. Sections from plastic-embedded specimens may be cut considerably more thinly than those from paraffin blocks. Some of the resins are explicitly suitable for subsequent localization procedures. Two embedding procedures for commonly used resins are summarized in Table 1.13and Table 1.14.

Sectioning of the harder embedding resins (also for electron microscopy) often requires the use of a harder glass knife instead of the steel knife sufficient for paraffin sectioning. The embedding procedure is analogous to the respective procedure described in chapter 2.4.6.2. An advantage of embedding in harder resins is also that ultrathin sections, suitable for transmission electron microscopy, may be obtained. This allows direct comparison between semi-thin light microscopic sections and ultrathin sections cut from the same specimen.

1.4.11.3 **Thin sectioning**

Operation procedures for commercially available ultramicrotomes differ and are described in detail in the respective users' manuals. The procedures included here refer to those common to all standard instruments, but do not supplant the manuals. Thickness of microtome sections typically range between 5 and 50 μm, glass or sapphire knives allow thinner sectioning. Sections of up to 1 μm in thickness may also be cut with an ultramicrotome (for preparation of ultrathin sections for use in electron microscopy). Standard microtomes used for histology are rotary (with a specimen moving up and down and a static knife) or sliding microtomes (with a knife mounted on a movable sledge and a static specimen).

Microtome knives

Paraffin sectioning may be performed with a steel knife, or for numerous routine applications, with a disposable blade. Steel knives are manufactured in four different profiles designed for cutting material of different hardnesses. Plane concave knives (type A) are designed for very soft materials, slightly plane concave (type B) for soft materials. Type C is plane on both sides and used for paraffin or frozen sections, while type D has a steep cutting edge on one side for hard materials. The knife angle is typically 15° but larger at the cutting edge itself (30°) This larger angle (facet) stabilizes the knife.

Knives for sectioning polymerized synthetic resin are made of tungsten carbide, sapphire, diamond, or, most simply, glass. The glass knife may be broken from glass strips, as usual for ultrathin sectioning (see Section 2.4). Since only small cutting edges may be obtained with these triangle-shaped knives (approximately 2 mm in diameter), glass knives for histology are regularly broken in a different way. The commonly used Ralph-type knives have a broad cutting edge suitable for obtaining large sections. When supported in an appropriate way, and with pressure applied at a certain distance from the support, a scored glass strip breaks in such a manner that a well shaped knife edge is formed (see Fig. 1.60). Special knife-makers are used for a highly reproducible scoring and breaking procedure, but the knives may also be broken from a thin glass strip (such as a microscope slide), which needs some experience.

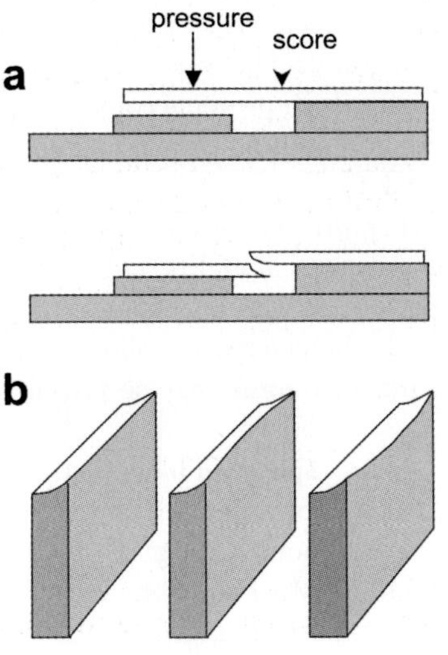

Fig. 1.60 (a) Principle procedure for making glass knives for histology (after Ruzin, 1999, with modifications).
(b) Ideally, a knife with a straight cutting edge is achieved (left). Convex or concave deviations from the straight line (center, right) due to misalignment of the knife-maker are acceptable within certain tolerances

Tab. 1.15 Operation of a standard rotary microtome

1.) Mount the paraffin block on an appropriate carrier, such as a wooden block, most easily by gluing the molten paraffin surface onto the piece of wood.
2.) Trim the cutting face of the specimen to the form of a flat-topped pyramid. The flat top should have a trapezoidal shape.
3.) Insert the knife and adjust to an appropriate clearance angle, and fix the specimen carrier to the specimen arm.
4.) Bring the specimen arm towards the knife. Then align the cutting face of the embedded specimen to the knife:
   - bring the trapezoid face level with the cutting edge by adjusting the microtome arm height
   - adjust the specimen in a way that the edge of the cutting face is oriented parallel to the knife edge and that the knife edge is at an equal distance to the cutting face during the cutting movement.
5.) If necessary, bring the specimen arm closer to the knife. Pre–select an intermediate section thickness. Begin sectioning by using the fine advance mechanism. Use the fine advance manually first, until the first section is produced.
6.) When complete sections are produced, switch over to automatic, motor–driven advance. Ribbons of sections should result.
7.) Pick up the ribbon with a thin paintbrush and mount the sections to microscope slides.

For specimen sectioning, rotary microtomes operate with the highest precision and flexibility of instrument settings. The instruments are manufactured for the softer embedding media and thicker sections commonly used in light microscopy, as well as for the harder resins and ultrathin sections found in electron microscopy. The latter instrument type may also be used, with some restrictions, for cutting semi-thin sections for light microscopy. The specimen arm is connected to coarse and fine advance mechanisms: controlled hand wheels or automatic mechanisms. The coarse advance mechanism advances and retracts the specimen horizontally. When turned by 360°, the fine advance wheel provides downward movement of the specimen arm just behind the knife (cutting movement), mechanical retraction of the arm during the upwards movement, and a subsequent horizontal advance of the specimen arm according to the pre-selected section thickness. Retraction prevents contact between the specimen and the knife during the return path of the specimen, which prevents contamination of the knife or damage of the specimen surface. An outline procedure for standard rotary microtome operation is summarized in Table 1.15.

Paraffin sections are routinely cut down to 5 μm in thickness. The very soft Steedman's wax should be cut at about 4 °C to provide best results. This is most simply performed with a cryostat: a microtome build in a thermostatically controlled refrigerated cabinet (cryostats allow temperature control down to –40 °C). Alternatively, a standard microtome may be placed in cold room for processing.

1.4.11.4 **Specimen mounting**

Usually, specimen sections are mounted on microscope slides. To facilitate adhesion of the sections to the glass surface, the slides are coated with an appropriate "subbing" solution. Most solutions for subsequent routine staining, such as the chrome alum subbing solution, contain 0.5 – 1.0 % (w/v) gelatin.

Tab. 1.16 Adhesives for mounting of sections on microscope slides

Working solution	Slide preparation	Use
0.25 – 10 g/l aqueous gelatin solution (traditionally often used with a bactericidal protective: 0.03 % chromium potassium sulfate, 0.5 % sodium benzoate–chrome alum subbing solution)	Slides are washed in 100 % ethanol, dried, and immersed in gelatin solution. Slides are dried at room temperature. Freshly prepared gelatin solution should be used without protective. The dried coating will be not affected	For subsequent standard staining procedures
0.5, 1.0, or 2.0 % triethoxy-3-aminopropyl silane in acetone	Slides are washed in distilled water and oven-dried, then immersed for 10 – 15 s in the triethoxy-3-aminopropyl silane solution. The slides are then dried, rinsed in three changes of distilled water and dried at 50 °C	For subsequent in situ hybridization and immunolocalization experiments
100 μg/ml–1 mg/ml poly-L-lysine (molecular weight: 47,000 Da) aqueous solution	Slides are washed for 2 min each in distilled water and 100 % ethanol and oven-dried, then immersed for 20 s in poly-L-lysine solution. After drying at room temperature, dipping is repeated. The slides are then finally air-dried	For subsequent in situ hybridization and immunolocalization experiments
Vectabond (Vector Laboratories; solution in acetone prepared according to manufacturers instructions)	Slides are washed for 5 min in acetone, immersed in a Vectabond/acetone solution for 5 min, rinsed for 1 min in distilled water, and dried	For subsequent in situ hybridization and immunolocalization experiments

For subsequent in situ hybridization, however, silane, polylysine or Vectabond (Vector Laboratories, Inc.) are more appropriate, because the sections stick better to the glass surface for the localization procedure and possible non-specific binding of markers by gelatin, which may sometimes occur, is avoided (Table 1.16).

The stability of the sections on the coated slides varies with the embedded specimen, the embedding matrix, and the subsequent procedures. Adhesives such as gelatin may not be stable enough to withstand the washing procedures for in situ hybridization or immunolocalization. Background signals caused by gelatin have also been observed. The use of additives in gelatin solution is often only suitable when the solution itself has to be protected against contamination (which occurs within hours in a warm gelatin solution), but the additives often interfere with reagents used in subsequent procedures.

1.4.12
Cryotechniques in Light Microscopy

Numerous freezing techniques are applied in conjunction with light and electron microscopy. Freezing provides hard, readily sectionable specimens without the application of time-consuming embedding procedures, but with inferior preservation of morphological details.

Freezing of unfixed or fixed tissue is applied when rapid processing of tissue is required (such as for rapid diagnosis as part of an intra-operative consultation), or for subsequent histochemical procedures. However, sections of soft unfixed tissue are difficult to handle when thawed (necessary, of course, to process sections by staining or by application of cytochemical markers). Also, even minimal fixation (i.e., with fixative concentrations below the recommended values) improves preservation of fine structure as compared to unfixed specimens.

During the freezing process, however, specimen structures are affected by growing ice crystals. This may in principle be overcome by use of ultra-rapid and high-pressure freezing techniques, but only for a very limited specimen size (see Rasmussen, 1982, and Barnard, 1987). The cooling rate in the deeper parts of the sample are still slow enough for formation of large hexagonal ice crystals with subsequent tissue damage. It is possible to reduce the level of ice crystal formation by use of cryoprotectants, which reduce the rate of ice crystal nucleation by freezing-point depression. Cryoprotectants may either be non-penetrating macromolecules such as polyvinyl pyrrolidone (PVP), dextran, hydroxyethyl starch, and sucrose, or they may be penetrating agents such as dimethyl sulfoxide (DMSO), glycerol, ethylene glycol, and dimethyl formamide (DMF). A solution of 0.5 mol/l sucrose in 3.5 mol/l DMSO is a versatile standard mixture. The commercially available OCT solution (for specimen mounting, see chapter 1.4.12.1) may be also used as cryoprotectant (Table 1.17).

Whereas specimen damage due to ice crystal formation in non-cryoprotected specimens is most critical for electron microscopy, for which excellent preservation of intracytoplasmic structures down to nanometer scales may be required, limited specimen alterations may not be of relevance at the light microscopic level. Thus, a variety of cryogens and various freezing methods may be used, especially when cryoprotection is applied. Liquid nitrogen (–196 °C) is readily available and easy to handle, but overcooling can cause cracking and brittleness of the specimen. A layer of gaseous nitrogen surrounding the specimen reduces heat transfer considerably. This effect is of no relevance for cryoprotected specimens, and so slow cooling rates (to avoid cracking of the specimen) may also be tolerated in liquid nitrogen. To overcome the disadvantages of liquid nitrogen as a direct cryogen, some other, more suitable substance – such as propane, isopentane, or hexane – cooled by liquid nitrogen may be used. Counter-heating may help to keep a desired temperature. This is best provided in devices normally used for freezing specimens in electron microscopy (cryoimmersion fixation apparatus, see chapter 2.4.5).

Since the temperatures provided by liquid nitrogen are not usually necessary for effective freezing, carbon dioxide (–70 °C) is also used as an alternative. Freezing attachments connected to a cylinder of carbon dioxide rapidly freeze a sample by heat exchange from the expanding and vaporizing gas. Similarly, solid chunks of carbon dioxide (dry ice) held against the specimen can be effective, and are acceptable for routine freezing. Again, combination with a hydrocarbon may be effective, pellets of solid carbon dioxide being mixed with isopentane, hexane, acetone, or ethanol. The specimen may be plunged directly either into this mixture or into a container of pure hydrocarbon cooled in the hydrocarbon-carbon dioxide slurry. Commercially available thermoelectric cryobaths, which cool isopentane, provide temperatures down to about −60 °C. Thermoelectric cooling (−40 °C) employs the Peltier effect and is used as part of the stage of some freezing microtomes.

Several techniques to freeze an exposed specimen in situ are also available. This is most simply done by use of a short spray burst of – for example – tetrafluoroethane, which produces temperatures as low as −45 °C, but only in a small area close to the surface. A hand-held cryofixation unit ("cryogun") is an advanced version of a freezing device for in situ applications.

1.4.12.1 **Freeze sectioning**

The various forms of microtomes (rotary microtome, sledge microtome) have all been modified in ways to produce sections from frozen specimens. Piped cryogen (usually carbon dioxide), refrigerated coils, and thermoelectric devices have been applied to cool down the relevant microtome parts (i.e., the knife and the specimen holder). In a simple freezing microtome, the knife is moved horizontally in a curved path over a chilled specimen holder. The holder also acts as freezing stage on which the specimen is mounted with the aid of mounting solution. Alternatively, freezing stages based on refrigerated coils or thermoelectric modules are available. Mostly, however, specially adapted rotary microtomes placed in a cryo cabinet (cryostats) are used for light microscopy. Unlike the simple versions of freezing microtomes, the cryostat provides a refrigerated environment, facilitating routine work and minimizing problems with condensing water vapor.

When frozen tissue is to be sectioned, the sample must be firmly attached to the specimen holder. This is achieved by use of a mounting solution that surrounds the specimen and first acts as a sticky glue that allows orientation of the specimen before freezing occurs. After freezing, the specimen is firmly attached to the holder, brought about by the frozen mounting solution. Cryoprotectants of higher viscosity are appropriate mounting solutions. The most widely used mounting solution is an aqueous solution of glycols and resins, providing an inert matrix for sectioning, commercially available as "OCT" or "Tissue Freezing Medium" from various suppliers (Ishii et al., 1990). Alternatives to OCT are bovine albumin solution (30 %, w/v), Apathy's mounting medium (50 g gum arabic and 50 g sucrose in 50 ml of distilled water) or methylcellulose solution. Mounting on an appropriate holder may be performed before freezing, but the frozen specimen may also be mounted in a cryo cabinet (see Table 1.17), necessary when freezing has been applied in situ. Care has to be taken that the applied mounting solution works well at the desired cutting temperature.

A disposable blade or a type C knife with a wedge-shaped profile is used for the sectioning procedure. The knife should be cleaned with xylene followed by alcohol after each use. To reduce friction during sectioning, a Teflon spray or application of a metal polish is recommended. A spare microtome knife should always be stored in the cryostat, so that it is maintained at cabinet temperature.

Frozen sections are more difficult to handle than paraffin- or resin-embedded sections. Since sections tend to curl at the knife edge, most cryostat microtomes are fitted with an anti-roll plate. The plate is situated just above the knife surface so that a cut section passes between it and the knife. Application of Teflon spray reduces friction and section adhesion to the anti-roll plate.

After the sections have been placed on an microscope slide and are at room temperature they are less stable than paraffin or resin-embedded sections to all subsequent procedures. A chemical fixation step stabilizes the sections and enhances binding to gelatin-coated microscope slides. Mounted and fixed sections may be stored temporarily in an airtight plastic container at 0–4 °C, which minimizes loss of enzyme or antigenic activity.

Unfixed sections for fluorescence microscopy may also be stored temporarily in this manner with little deterioration. Longer term storage, however, requires lower tem-

Tab. 1.17 Specimen freezing and freeze sectioning

1.) Fix a specimen in 4–8 % w/v formaldehyde solution (freshly prepared from paraformaldehyde; Tab. 1.10) in an appropriate buffer (if there are no special requirements needed, 50 mM potassium phosphate, pH 7.5, is recommended) for 1–24 h min at 0 °C.

For cryoprotection prior to freezing, incubate the specimen in a solution of 1.6 M sucrose and 25 % (w/v) polyvinylpyrrolidone for 8 h (for small unicellular organisms) up to several days (for tissue specimens, change the solution several times).

2.) Place the specimen in a drop of a mounting solution (e.g., OCT, Tissue Tek Inc. or methylcellulose solution) on a mounting stub.
3.) Freeze in a cryogen (liquid nitrogen, propane, isopentane, or hexane). This step may also be performed before Step 2. Then mount the frozen specimen on a cold mounting platform (at best in a cryostat), on which the mounting stub is placed during preparation. While the mounting solution starts to freeze, orient the specimen.
4.) If necessary, trim the specimen to a flat-topped pyramid with a cold razor blade. The flat top should have a square or trapezoidal shape.
5.) Transfer the mounted specimen to a cryostat and allow the specimen to equilibrate to the working temperature in the cryostat cabinet (commonly between –14 and –20 °C; harder tissues produce better sections at lower temperatures and vice versa).
6.) Proceed, in principle, as described in Tab. 1.15, from Step 4 onwards. Set section thickness between 15–50 μm. To avoid rolling of sections, an anti-roll plate is placed just parallel to and just below the knife edge.
7.) Pick up the sections with microscope slides. Use plain slides for unfixed tissue and slides coated with gelatin or poly-L-lysine for fixed tissue.
   - Rotate the anti-roll plate out of the way (this step may be critical for specimen types, which tend to flip to tight rolls. Variation in the cutting temperature and cutting speed may help; some skill is needed to overcome this problem)
   - Swivel the slide down onto the sections and hold it there for a few seconds to allow the sections to melt onto the slide.

7a.) Alternatively, sections may be picked up with cold forceps or a cold paintbrush and then transferred to objective slides.

peratures of –80 °C or less. However, it is recommended that the specimen is stored in this way before cutting and the sections processed soon after sectioning.

1.4.13
Staining of Sections

Selected staining procedures for microorganisms have already been described in Tables 1.4, 1.6, and 1.7. However, hundreds of stains and staining methods have been applied, especially in conjunction with histopathology of (human) tissue. For staining of more than single slides at a time, staining cuvettes of various sizes are available. Automated tissue-staining systems are also used for routine diagnostics in pathology laboratories.

Paraffin-embedded tissue may be stained with any dye, but the paraffin is almost always removed prior to processing by incubation of the mounted sections in an organic solvent. A generalized staining procedure is outlined in Table 1.18.

Only selected stains and staining methods are presented in Tables 1.19 and 1.20. Since the numerous staining techniques are all well described in the literature (McManus and Mowry, 1964; Gray, 1973; Pearse 1980, 1985) the reader is referred to those sources. For routine stains, ready-to-use kits are available from various suppliers. Numerous procedures are also adapted to a microwave-enhanced staining method.

1.4.14
Preparation of Specimens for Fluorescence Microscopy

New classes of microscopic and analytical techniques based upon the application of fluorescent instead of bright-field stains have now become established. Technically, the excitation of fluorescence by an incident beam rather than absorption of transmitting light has allowed instruments to be created with analytical features that cannot be

Tab. 1.18 General staining procedure for paraffin sections

1.) Use a Coplin jar as container for all procedures. Deparaffinization: incubate three times for 5 minutes in xylene, then in an descending ethanol series: twice for 2 – 3 min in 100 % ethanol; then for 2 – 3 min in 90 % (v/v), 80 %, 70 %, and 50 % ethanol. Avoid excessive carry-over of the solutions used. Allow excess liquid to drain between the steps, but do not allow the sections to dry out. Finally transfer to distilled water.
2.) Apply staining procedure in aqueous solution (Tab. 1.20).
3.) Destain differentiate if necessary (Tab. 1.20).
4.) Dehydrate in an ascending ethanol series: 50 %, 70 %, 80 %, 90 %, and 100 % ethanol, for 2 – 3 min per step.
5.) Stain in a non-aqueous solvent if desired.
6.) If step 5. has been applied, destain and clear in an appropriate organic solvent (e.g., xylene).
7.) Clear in xylene (two changes).
8.) Mount with an appropriate mounting resin (some resins also work without application of step 7).

Tab. 1.19 Common stains for use in histology. The compositions of working solutions refer to the procedures given in Tab. 1.20

Stain	Preparation of stock solution	Remarks
Alcian Blue	Alcian Blue 8GX 1.0 in 100 ml acetic acid solution (3 %, v/v in distilled water), pH 2.5 (stock solution is working solution)	Blue stain of mucosubstances and acetic mucins in animal tissue (general cytoplasmic stain in plants when applied as described in Tab. 1.20)
Hematoxylin	Hematoxylin (1.0 g) is dissolved in 50 ml isopropanol (96 %) and 50 ml glycerol, 1.5 g potassium aluminum sulfate, and 5 ml glacial acetic acid are added, counterstain Eosin γ (1 g in 100 ml 95 % (v/v) ethanol) (stock solution is working solution)	Most widely used nuclear stain; numerous variations of staining characteristics when used with various metal salts
Safranin	Safranin (2.0 g) is dissolved in 50 % (v/v) ethanol in distilled water (stock solution is working solution)	Predominantly for plant tissue, generally used as or in conjunction with a counter-stain
Neutral Red	Neutral Red (1.0 g) in 100 ml acetic acid solution (3 %, v/v in distilled water), pH 2.5. The hydrophobic powder is slowly applied onto the water surface during continuous agitation, larger particles are filtered	Nuclear stain, counter-stain; 0.01 % working solutions used as vital stains (accumulates actively in, for example, vacuoles)
Giemsa	Azure A (0.15 g), Methylene Blue (0.30 g), Eosin Y (0.36 g), Phloxine B (0.04 g) added to a 60 ml glycerol/ 40 ml methanol solution, (working solution: 10 % of stock solution in distilled water or 50 mM phosphate buffer)	General purpose stain for animal tissue, routine chromosome analysis, bacteriological stain
Methyl Green	Methyl Green (1 g) in 100 ml heated distilled water (extraction of crystal violet impurities with chloroform prior to use) (stock solution is working solution)	Nuclear stain; differentiation of DNA and RNA with Pyronin G
Pyronin G	Pyronin G Methyl Green (1 g) in 100 ml distilled water (stock solution is working solution)	Counter-stain with Methyl Green
Nuclear Fast Red	Nuclear Fast Red (0.1 g) in 100 ml 5 % potassium aluminum sulfate (stock solution is working solution)	Nuclear stain, counter-stain
Toluidine Blue	Toluidine Blue (0.1 g) in 100 ml distilled water or neutral buffer (working solution: 10 % of stock solution in distilled water)	General purpose stain for animal tissue

Tab. 1.20 Staining procedures for tissue sections (see Tab. 1.19 for the working solutions)

Alcian Blue-Neutral Red

1.) Deparaffinize and rehydrate sections (Tab. 1.18).
2.) Stain in Alcian Blue solution for 5 min.
3.) Rinse in running tap water for 5 min.
4.) Counterstain in 1 % neutral red for 1 min, rinse in tap water.
5.) Dehydrate, clear, and mount (Tab. 1.18). Acid mucopolysaccharides appear blue, nuclei red.

Hematoxylin

1.) Deparaffinize and rehydrate sections (Tab. 1.18).
2.) Stain in Haematoxylin solution for 2 – 15 min.
3.) Rinse in running tap water for 10 min.
4.) Differentiate in 0.1 % HCl (control microscopically); stop process by neutralization with 0.1 % ammonia solution.
5.) Rinse in running tap water for 10 min.
6.) Dehydrate, clear, and mount (Tab. 1.18). Nuclei appear dark blue.

Hematoxylin-Eosin

In conjuncton with Hematoxylin staining: proceed after step 5 of the Hematoxylin staining procedure.
1.) Stain with Eosin solution for 4 min.
2.) Rinse in running tap water for 10 min.
3.) Dehydrate, clear, and mount (Tab. 1.18). Nuclei appear dark blue, cytoplasm light red.

Hematoxylin-Safranin

In conjunction with Hematoxylin staining: proceed after step 5 of the Hematoxylin staining procedure
1.) Stain in Safranin solution for 12 h.
2.) Rinse briefly in running tap water. Sections appear red.
3.) Differentiate in 96 % ethanol until excess color is washed away (control microscopically if necessary).
4.) Rinse in running tap water for 10 min.
5.) Dehydrate, clear, and mount (Tab. 1.18). Wooden cell walls red, other tissue features in grades of blue, nuclei dark blue.

Giemsa

1.) Deparaffinize and rehydrate sections (Tab. 1.18).
2.) Stain in Giemsa solution for up to 1.5 h.
3.) Rinse briefly in 0.5 % (v/v) acetic acid (solution in distilled water).
4.) Dip in 96 % (v/v) ethanol.
5.) Dehydrate, clear, and mount (Tab. 1.18). Cytoplasm appears light blue, nuclei deep blue/violet, bacteria deep purple.

Methyl Green-Pyronin

Prepare mixture prior to use: Methyl Green working solution (9 ml), Pyronin G or Y working solution (4 ml), glycerol (14 ml), 0.1 M acetate buffer, pH 4.8 (23 ml).
1.) Deparaffinize and rehydrate sections (Tab. 1.18).
2.) Stain in Methyl Green-Pyronin solution for 10 s to 5 min.
3.) Rinse in running tap water for 5 min.
4.) Dehydrate, clear, and mount (Tab. 1.18). Nuclei appear green or green blue, RNA (nucleoli), cytoplasm, bacteria red.

Tab. 1.20 continued

Nuclear Fast Red
1.) Deparaffinize and rehydrate sections (Tab. 1.18).
2.) Stain in Nuclear Fast Red solution for 5 min.
3.) Rinse in running tap water for 5 min.
4.) Dehydrate, clear, and mount (Tab. 1.18). Nuclei appear red.
Toluidine Blue
1.) Deparaffinize and rehydrate sections (Tab. 1.18).
2.) Stain in Toluidine Blue solution for 10 s to 5 min.
3.) Rinse in running tap water for 5 min.
4.) Dehydrate, clear, and mount (Tab. 1.18). Nuclei appear deep blue violet, cytoplasm in a lighter tone.

achieved by transmitted light microscopy, such as the selection of a certain volume in a specimen in which excitation should occur (Table 1.21). The low detection limit of fluorescent dyes has enormously broadened the applications of specific stains in light microscopy, but most fluorescent dyes are unsuitable for showing all the cytological features of an object, and so fluorescent images are often combined with bright-field images. Staining is mostly achieved by optical methods (DIC or phase contrast images), but some (fluorescent) stains (such as Neutral Red) also provide bright-field contrast. Specific probes coupled to dyes are often used in fluorescent techniques in order to detect and localize sub-cellular components – organelles inside cells or biological macromolecules (protein, nucleic acids) – but numerous more or less specific dyes may be used to detect sub-cellular compartments such as liposomes or mitochondria. Numerous dyes may be used without (obviously) damaging effect on the living cell, allowing the observation of dynamic processes in living systems. Heterologous expressed proteins may be tagged with autofluorescent protein as an intrinsic marker system.

Fluorescent markers are not used only for microscopic imaging techniques. Detection and quantification of fluorescence changes are used in numerous applications such as binding studies between molecules, conformational fluctuations of molecules, etc. (see Table 1.21 and Slavik, 1998; Mason, 1999).

1.4.14.1 **Fluorescent dyes**

Numerous chemicals, especially those containing highly conjugated aromatic systems, are fluorescent. Some of these are suitable fluorescent dyes (fluorophores) for microscopy. A steadily growing number of fluorescent dyes for single or multiple labeling experiments are offered (an extended and periodically updated review is provided by Molecular Probes Inc.).

For a given fluorescent dye, the manufacturers indicate the wavelengths of the excitation and the respective emission spectrum. The emission spectrum is shifted to a longer wavelength than the excitation spectrum (Stokes' shift), the emission intensity peak systematically being lower than the excitation peak. The greater the Stokes' shift, the easier it is to separate excitation from emission light. To achieve maximum fluorescence intensity, the fluorophore is usually excited at the wavelength at the peak of the excitation curve, and the emission is selected at the peak wavelength of the emission curve, selection of the wavelengths being brought about by an appropriate combination of filters as mentioned above.

The emission efficiency of a fluorophore is essentially determined by the quantum yield Q, the ratio between emitted and absorbed photons, which ranges from zero to 1. The amount of fluorescence is determined by the quantum yield, the concentration of the dye, the intensity of the incident light beam, the thickness of the fluorescent layer penetrated by the beam, and the efficiency with which the dye absorbs the excitation light (extinction coefficient).

Tab. 1.21 Applications in fluorescence microscopy and related analytical techniques (Hovius et al., 2000)

Technique	Application
IF (Immunofluorescence) FISH (Fluorescence in situ hybridization) GFP (Green fluorescent protein)	Identification, localization, and visualization of cells and specific tissue components with fluorescent probes
FP, FA (Fluorescence polarization or anisotropy), TRFA (Time-resolved fluorescence anisotropy) Phosphorescence anisotropy	(Time-resolved) rotational mobility changes of fluorescent/phosphorescent probes due to molecular interactions (information on local viscosity, flexibility, and dimensions of the fluorescently labeled molecule)
FRAP (Fluorescence recovery after photobleaching) or FPR (Fluorescence photobleaching recovery)	Destruction of fluorescence by a short, intense photobleaching pulse and measurement of recovery time of fluorescence after diffusion/transport of neighboring unbleached fluorophores into the bleached area
	Lateral diffusion of labeled molecules over small (2–5 μm) distances
TIRFM (Total internal reflection microscopy)	Excitation of fluorophores attached or close to a quartz surface with an evanescent wave produced by a totally internally reflected laser beam
FRETM (Fluorescence resonance energy transfer microscopy)	Short distance (1–10 nm) energy transfer between an excited donor fluorophore molecule and a respective acceptor without the emission of a photon
	Interaction between molecules (e.g., receptor-ligand studies, membrane fusion events)
FLIM (Fluorescence lifetime imaging microscopy)	Measurement of the characteristic fluorescent lifetime of a fluorophore
FCS (Fluorescence correlation spectroscopy)	Fluorescence intensity fluctuations due to chemical reactions, diffusion, receptor-ligand binding, etc. in a small observed volume

The sensitivity of fluorescence detection is severely reduced by background signals originating from constituents of the specimen and the embedding media, as well as from unbound or non-specifically bound fluorescent dyes. Flavins and flavoproteins absorb at 488 nm and emit in the same spectral region as the classical dye fluorescein. Aldehyde fixatives, especially glutardialdehyde, induce fluorescence that spans the visible spectrum, although it is especially bright in the yellow range. Sometimes a generalized fluorescence of all cellular features may be useful, especially when the fluorescent image cannot be combined with a bright-field image. Formaldehyde vapor is a suitable inducer, but so are Hematoxylin, Aniline Blue, acridine, and other general purpose stains.

Absorption of light by the fluorophore may also cause damage (covalent bond breakage) of the molecule. The process, called photobleaching, is common to all fluorophores and produces nonfluorescent products. Thus, the observed effect, fading of the fluorescent signal, always occurs and depends on the intensity of the excitation light and the duration of illumination. The application of less photolabile fluorophores, antifade reagents, and a high detection sensitivity of the fluorescence signal are means by which fading problems may be overcome. The photobleaching effect may also be used to determine the diffusion characteristics of molecules used in conjunction with the FRAP technique (Table 1.21).

Several environment effects may reduce the quantum yield of fluorescence. These effects may be undesired, but may also be useful for special applications. After interaction of the excited fluorophore with another molecule, non-radiative relaxation (quenching) may take place. Self-quenching between fluorophores may occur, especially at high fluorophore concentrations. Quenching may weaken the signal of the fluorescent dye but also of undesired background signals, and so appropriate quenchers may be used to improve the signal-to-noise ratio. The binding of fluorescent dyes to a particular target (a probe or a target molecule) may also affect fluorescence, as may solvent polarity and pH.

Cytochemical fluorescent dyes

Both bright-field stains and fluorescent dyes have specificities for certain cellular components, and several manufacturers provide detailed data on the approximately 500 fluorescent dyes suitable for fluorescence microscopy. Table 1.22 therefore shows only a small number of some of the more widely used dyes: consult Lillie (1977), Ruzin (1999), and Haugland (2002) for further detailed procedures. An introduction into the application of laser dyes for CLSM has been presented by Maeda (1984). The excitation/emission wavelengths presented in Table 1.22 do not refer to in situ conditions, which are, of course, highly variable. The shift of emission maxima to longer wavelengths (and decreasing quantum yield) with increasing solvent polarity may be a disturbing effect, but also offers a way to obtain information on the polarity or polarity changes in the immediate environment of the dye (see also Czarnik, 1993). In a similar way, the pH-dependent change of emission spectra may be useful. The selective ion sensitivity of chelating dyes may be used for the determination of calcium ion, magnesium ion, or zinc ion concentrations.

Tab. 1.22 Some unconjugated fluorescent dyes and their preferred targets (Ruzin, 1999)

Dye	Excitation/emission wavelengths (nm)	Working solution	Target
DAPI	358/461	1/100 Dilution of a 1 mg/ml stock solution in water	DNA/nucleus adenine/thymine selective, *Mycoplasma* detection, chromosome/nucleus counter-stain (semi-permeable)
Ethidium homodimer	492/627	1/1000 Dilution of a 1 mg/ml stock solution in water	DNA/nucleus (impermeable)
Hoechst 33342	350/461	1/500 Dilution of a 1 mg/ml stock solution in water	DNA/nucleus adenine/thymine selective (permeable)
Hoechst 33358	352/461	1/500 Dilution of a 1 mg/ml stock solution in water	DNA/nucleus (permeable)
SYTOX Green	504/523	1/1000 Dilution of a 5 mM stock solution in DMSO	DNA/nucleus (chromosome banding)
Propidium iodide	493/630	1/1000 Dilution of a 1 mg/ml stock solution in water	DNA/nucleus dead-cell dye, chromosome counter-stain (impermeable),
DiOC6	484/519	1/500 Dilution of a 0.5 mg/ml stock solution in ethanol	Endoplasmic reticulum and mitochondria (permeable)
Rhodamine 6G	547/571	1/100 Dilution of a 1 mg/ml stock solution in DMSO	Endoplasmic reticulum (permeable)
ER-tracker Blue-white DPX	374/460	1/ 10,000 – 1/1000 Dilution of a 1mM stock solution in DMSO	Endoplasmic reticulum (permeable)
Lyso-Tracker Blue DND-22	373/422	1/100,000 Dilution of a 1 mM stock solution in DMSO	Acidic organelles, lysosomes (permeable)
Calcofluor White	350/440	1/100 Dilution of a 10 mg/ml stock solution in water	Cellulose and chitin (i.e. cell walls)
Mito-Tracker Red (CMXRos)	460/534	1/1000 Dilution of a 1mM stock solution	Mitochondria (permeable)
Acridine Orange	500/526 (DNA) 460/650 (RNA)	1/1000 Dilution of a 1 mg/ml stock solution	DNA/RNA discrimination, lysosomes (permeable)

Tab. 1.22 continued

Dye	Excitation/emission wavelengths (nm)	Working solution	Target
SNARF-1	Acid 518/548–587 Base 579/640	10–50 μM working solution	pH indicator
SNAFl-1	Acid 479/508–543 Base 537/623	10–50 μM working solution	pH indicator
p-Phenylenediamine	(Antifade reagent)	20 mg/ml in glycerol containing buffer*)	Effective antifade reagent, toxic, photosensitive, heat labile
n-Propyl gallate	(Antifade reagent)	50 mg/ml in glycerol containing buffer*)	Less effective than *p*-phenylenediamine, but may be used for in vivo studies
1,4-Diazabicyclo (2,2,2)octane DABCO	(Antifade reagent)	Approximately 25 mg/ml in glycerol containing buffer*)	Less effective than *p*-phenylenediamine
Ascorbic acid	(Antifade reagent)	1 mg/ml in glycerol-containing buffer*)	For in vivo studies
Prolong antifade kit (Molecular Probes)	(Antifade reagent)	Product of two components to be mixed just prior to use (polymerizing agent!)	Especially useful for FISH

*) for a 10 ml volume, dissolve the respective quantity in 0.8 ml distilled water, and add 0.2 ml Tris-HCl-solution, pH 8.0 and 9 ml glycerol

The first prerequisite for use as a vital dye is membrane permeability, whereas impermeable dyes require permeabilized membranes for penetration and binding to their targets. This also allows discrimination between living or dead cells, with intact or damaged membranes. These dyes find use in discrimination between cells for flow cytometry as well as fluorescence microscopy. In this way, for example, the efficiency of fungicidal or bactericidal agents may be checked in situ.

Because of the sensitivities of the reagents, most staining procedures need more optimization than simple bright-field stains, especially for combinations of several dyes. Thus, the remarks given in Table 1.22 and the procedures in Table 1.23 give only general information on staining and fixation conditions. The procedures are recommended for cytological nuclear (counter-)staining (Molecular Probes, Inc.) but are also useful for staining other cytological features with fluorescent stains. Whereas permeable dyes penetrate the membranes of living cells (and may be used as vital stains), cell membranes must be permeabilized to allow introduction of membrane-impermeable dyes, and are only suitable for staining dead cells.

In spite of the high autofluorescence of glutardialdehyde, this fixative is often preferred because of its rapid tissue penetration and good preservation of cytological features. Control of autofluorescence is not a trivial problem, though, because it also depends on the specimen type, the pretreatment, or the quality of the fixatives. A combination of the fixative and a 1 mg/ml solution of sodium borohydride is used to suppress background fluorescence. However, problems with induction of background fluorescence by application of borohydride has also been described (Baschong et al., 2001). Mixtures of glutardialdehyde at concentrations below 0.5 % (v/v) with formaldehyde are also used. The application of the less fluorescent formaldehyde alone may also give sufficient fixation, though it does not provide the extensive cross-linking of aldehydes. Classical alcohol and glacial acetic acid fixatives have also been described for application in conjunction with fluorescent dyes, but this procedure is not suitable for studies at the sub-cellular level, due to the coagulation of cellular components.

After the fixation steps as described in Table 1.23, the specimens may be directly applied to staining. Some nucleic acid stains such as propidium iodide also stain RNA. RNase treatment before staining enhances the selectivity of DNA staining.

The application of vital stains is of course useful for the observation of numerous dynamic events. One has to be aware, however, that these stains may interfere with vital functions. Nuclear stains in particular have turned out to be deleterious to vital

Tab. 1.23 Basic fixation procedures for cytological stains in fluorescence microscopy

Hanks balanced salt solution (HBSS; 8 g NaCl, 0.4 g KCl, 185 mg $CaCl_2 \times 2\ H_2O$, 98 mg $MgSO_4$, 48 mg Na_2HPO_4, 60 mg KH_2PO_4, 1 g D-glucose, in 1000 ml distilled water, pH 7.4).
Phosphate buffered saline (PBS; 1.44 g Na_2HPO_4, 0.24 g KH_2PO_4, 8 g NaCl, 0.2 g KCl, in 1000 ml distilled water, pH 7.4).

Glutardialdehyde fixation (also suitable for cell monolayers attached on poly-L-Lysine coated slides)

1. Rinse cells in HBSS at 37 °C.
2. Incubate for 10 min in a 0.5 % (w/v) solution of glutardialdehyde in HBSS.
3. Add 1 ml of sodium borohydride (1mg/ml $NaBH_4$ in PBS), incubate for 5 min, this step may be repeated 2–3 times. (Alternatives: incubate in 50 mM ammonium chloride or 100 mM glycine solution in PBS for 5–15 min)
4. Remove the sodium borohydride solution and incubate in PBS for 10 min at room temperature.

Formaldehyde fixation (also suitable for cell monolayers attached on poly-L-Lysine coated slides)

1. Rinse cells in HBSS at 37 °C.
2. Incubate for 10 min in 1–4 % (w/v) solution of formaldehyde in HBSS.
3. Rinse 3–4 × in PBS (one minute per rinsing step). If rinsing does not sufficiently reduce autofluorescence, refer to step 3 of the glutardialdehyde fixation procedure.

If cell permeabilization is required, proceed as follows:

4. Incubate in 0.2 % w/v solution of Triton X-100 in PBS for 5 min.
5. Rinse 3–4 × in PBS (one minute per rinsing step).

Alternatively, acetone may be used for permeabilization:

4. Incubate in acetone at –20 °C for 10 min.
5. Remove the acetone solution and incubate in PBS for 10 min at room temperature.

Methanol/glacial acetic acid fixation for cells in suspension

1. Add one volume of methanol/glacial acetic acid (3/1, v/v) to one volume of growth medium. Incubate for 5 min at room temperature.
2. Remove the mixture, rinse in methanol/glacial acetic acid, and incubate for an additional 10 min in fresh methanol/glacial acetic acid.
3. Rinse 2 × in distilled water.

Tab. 1.24 Basic protocol for application of fluorescent dyes

Unfixed specimens

1.) Prepare a dilution of the fluorescent dye in a solution appropriate for the living specimen (buffered solution, growth medium). The appropriate concentration has to be found by trial and error; use data in Tab. 1.22 and manufacturer's instructions as a first guide. To avoid artificial overloading, begin with the lowest concentration.
2.) For cells grown on coverslips in a Petri dish: remove the culture medium and replace by rewarmed (37 °C) culture medium containing the fluorescent dye. For other specimens: replace the medium after centrifugation of cells or add the diluted fluorescent dye to a thin tissue section.
3.) Incubate with the dye solution for 5–60 min (optimization necessary, start with short incubation times) and replace by fresh medium without dye.
4.) Observe cells under a fluorescence microscope fitted with the appropriate filter set providing the excitation and emission wavelengths given in Tab. 1.22 (recommended filter sets are listed by and available from the manufacturers of fluorescence microscopes).
5.) If fluorescence signals are too low, increase the concentration of the dye (step 1) or the incubation time (step 3).

Fixed specimens

1.) Perform fixation (see Tab. 1.23).
2.) Dilute the fluorescent dye in an appropriate buffer (e.g. PBS).
3.) Rinse cells in PBS and incubate with the dye solution for 10–20 min. Rinse at least once in PBS. Note that the loading concentration is lower and staining time is shorter than in unfixed specimens.
4.) Observe cells under the fluorescence microscope.

functions, and most of the antifade reagents are also toxic. Several vital stains are also used with fixatives: the specimens may be fixed after vital staining and then further processed (i.e., with an additional stain).

The staining or "loading" procedure itself is performed with the appropriate dye (Table 1.24) after fixation or with unfixed specimens. Some dyes with tendencies towards rapid photobleaching should always be used in conjunction with an appropriate antifade agent. Ambient light should routinely be excluded during and after application of the staining procedure.

1.4.14.2 **Fluorescent probes**

A fluorescent probe consists of the fluorescent molecule and a covalently bound (conjugated) molecule that selectively binds ("labels") a specific specimen structure (Kasten, 1993; Johnson, 1998). Thiol or amine reactive dyes are most often used to prepare conjugated probes. Numerous types of conjugated probes (oligonucleotides, antibodies, lectins, avidin/biotin are most frequently used) are also commercially available and are directed towards diverse cellular targets, for detection of whole organelles and specific proteins. Fluorescein and its derivatives, tetramethylrhodamine, and Texas Red are some of the standard fluorophores for coupling (see Table 1.25, Hermanson, 1996; Haugland, 2000). Several classes of fluorophores (such as the Alexa Fluor dye series, Molecular Probes, Inc.) offer improved properties (such as high photostability) and broad variation of excitation/emission spectra.

Often, several fluorescent dyes (such as a specific fluorescent probe and a free dye as "counterstain") are combined in a single experiment. In addition, numerous endogenous fluorophores (flavins, flavoproteins, reduced pyridine nucleotides, porphyrins, chlorophylls) cause autofluorescence in biological specimens. Though sometimes useful as an additional "counterstain", the autofluorescence signal often interferes with the signal from the probe. The specimens therefore have to be checked with the filter sets intended for use with the fluorophores prior to experiments with specific probes. Autofluorescence is usually most intense with excitation by violet-blue to blue light (400 to 450 nm) and decreases with increasing wavelength. Cells containing significant amounts of porphyrins and chlorophyll will exhibit far-red autofluorescence.

Before a multiple labelling experiment is started, it is recommended that the specificites in single labelling experiments are checked. When specimens labeled with one probe are then examined by use of the filter sets for each of the fluorophores, each specimen should be visible only with the appropriate filter set. Several manufacturers recommend appropriate dye/filter combinations for multiple labelling experiments.

When both signals are bright and their fluorescence spectra are clearly separable, excitation and emission at the same time with one filter set is possible. Different signal intensities, however, should be better recorded with an appropriate filter combination providing a clear separation of excitation and emission signals. In specimens with a FITC and DAPI combination, for instance, an FITC excitation filter (465 – 495 nm) and DAPI UV excitation filter (340 – 380 nm) are used to obtain two separated fluorescence signals that are processed separately. The final image is a combination of the separate channels. Kits with appropriate dye combinations and the recommended filter sets are available for numerous applications in routine and research.

As a demonstration for the application of three fluorescent dyes, the ViaGram Red+ staining procedure (Molecular Probes, Inc.) is presented in Table 1.26. After the procedure, living cells show blue fluorescence (exclusively stained by cell-permeable DAPI) and dead cells green (stained by DAPI and SYTOX Green).

Tab. 1.25 Some frequently used dyes for conjugation with marker molecules

Dye	Excitation/emission	Reactive group
FITC Fluorescein isothiocyanate	490/520	thiocyanate group
Coumarin maleimide	385/465	thiol group
Cy3	552/565	amino group
Cy5	650/667	amino group
Texas Red	596/620	sulfonyl group
TRITC tetramethylrhodamine isothiocyanate	541/572	thiocyanate group

Tab. 1.26 Double staining of bacterial specimens with the ViaGram Red+ staining procedure

The following procedure describes the preparation of enough staining solution for up to 20 slide preparations. If there is a problem in observing a difference between Gram-negative and Gram-positive cells, then it may be necessary to reduce the amount of WGA conjugate added to the cells.

1.) Prepare a 2 mg/ml stock solution of Texas Red-X conjugate of wheat germ agglutinin (WGA).
2.) Prepare a working solution of the vital stains by adding 3 μl of DAPI stain stock solution and 3 μl of SYTOX Green nucleic acid stain stock solution to 54 μl of water.
3.) Centrifuge a small volume of bacterial cells (50 – 100 μl of a well grown culture; about 5×10^7 cells) and wash the cells in an equal volume of 100 mM phosphate-buffered saline (PBS; pH 7.5 containing 0.1 % w/v BSA). Recentrifuge and resuspend in an equal volume of PBS-BSA.
4.) Add 2.5 μl of the WGA conjugate stock solution to 50 μl of the washed suspension and mix by pipetting up and down several times.
5.) Incubate for 5 – 15 minutes at room temperature.
6.) Centrifuge at 2000 rpm for 1 – 2 minutes to remove the WGA staining solution and resuspend in 50 μl PBS-BSA.
7.) Add 2.5 μl of the DAPI stain/SYTOX Green working solution, mix, and incubate at room temperature for 10 minutes.
8.) Transfer about 10 μl of the sample to a slide, apply a glass coverslip, and immediately observe under the fluorescence microscope.

The actual appearances of bacteria in stained preparations vary according to several factors: differences in the relative brightness of the fluorescent reagents, the abundance of target sites for dye-binding, and the spatial localization of the fluorophores. Recommended filter sets include Chroma 31000 for DAPI fluorescence (exciter filter with a center wavelength at 360 nm and an emission filter with a center wavelength at 460 nm), Chroma 31001 for Sytox Green (exciter filter 480 nm, emission filter 535 nm) and Chroma 31004 for Texas Red (exciter filter 560 nm, emission filter 630 nm) – filters from Chroma Corp.

The specific probe is a lectin (wheat germ agglutinin), that binds to the wall of Gram-positive cells (red fluorescence), Gram-negative cells being colorless. Four color combinations are thus possible: blue, green, blue/red surface, and green/red surface. The suppliers recommend the use of three filter sets to separate the three color channels.

1.4.14.3 Immunofluorescence techniques

Together with in situ hybridization (see below), the most widely used application with conjugated probes is fluorescence detection with immunomarkers. Though immunotechniques are also applicable as bright-field methods, fluorescence techniques are again far more sensitive, and multiple labelling with combinations of antibody-labeled fluorescent markers is more easily to achieve. Immunocytochemistry at the light microscopic level is therefore exclusively presented here for fluorescence microscopy (Table 1.27).

The probe consists of the dye coupled to a protein ("secondary" antibody, Protein A or Protein G) that binds to the "primary" marker, the antibody (mono- or polyclonal). Ideally, the primary marker binds specifically to the target, and the fluorescent probe binds exclusively to the primary marker. Controls have to be performed to exclude misinterpretations: to prevent the occurrence of false positive signals, it is necessary to verify that:

1) the combination of fluorophores can be distinguished with the available filter sets,
2) secondary antibodies specifically recognize the primary antibody – in multiple labeling experiments only one antibody, and
3) the primary antibodies specifically recognize their antigens.

Specificity of secondary antibodies

Commercially available secondary antibodies should have been checked for cross-reactivity and specificity to the primary marker. The secondary antibody should not react with any component of the specimen, and so a negative control by omission of a specific primary antibody has to be performed.

For double labeling experiments, the primary antibodies from different animal species (rabbit/mouse) are used. Secondary antibodies with specificites for either of these primary antibodies are used as markers. The commercially available (Molecular Probes, Inc., for example) secondary antibodies are extensively adsorbed against solid phase-immobilized immunoglobulins and serum proteins of other species to minimize cross-reactivity with them.

Different binding affinities to other secondary markers are also used. Protein A and protein G, for instance, bind IgG antibodies from different animal species with different affinities. However, cross reactivities are also not unlikely in these systems. Combinations of three antibodies are possible and are applied in microscopy (standard fluorescence as well as CLSM) and quite often in flow cytometry (see, for instance, Lanier and Recktenwald, 1991; Schubert, 1991; Brouns et al., 2002).

Specificity of primary antibodies

Primary antibodies should be extensively tested to exclude unwanted reactivities with cell components. Antibodies against cellular proteins should routinely be checked by Western blotting for specificity of binding to a certain antigen protein. Nevertheless, Western blotting does not rule out unwanted binding of the primary antibody to other

Tab. 1.27 Outline procedure for immunofluorescence

1.)	For cells grown on microscopic slides: perform formaldehyde fixation, permeabilization and rinsing according to Table 1.23 (paragraph “Formaldehyde fixation“). For paraffin sections, apply deparaffinization and rehydration according to Tab. 1.18, step 1 and finally transfer slides to 100 mM phosphate-buffered saline (PBS; pH 7.0). For butyl/methyl methacrylate sections (Table 1.14), remove resin by incubation in 100 % acetone for 15 min (not necessary for glycol methacrylate). Rehydrate according to Table 1.18, step 1 and finally transfer to 100 mM PBS. For frozen sections refer to chapter 1.4.12.1.
2.)	Incubate the specimen on the slide in a drop (approx. 100 μl) of a primary antibody solution (serum, purified antibody diluted in PBS) for 60 min at 37 °C. A first experiment may be performed with a 1/100, 1/1000, 1/10000 (...) dilution series.
3.)	Wash the slide 3 times for 5 min per step by incubation in PBS.
4.)	Incubate with a secondary antibody tagged with a fluorescent marker for 60 min at 37 °C. Wash with PBS 3 times for 15 min.
5.)	Wash the slide 3 times for 5 min per step by incubation in PBS.
6.)	For double labelling: Repeat steps 2–5 with different primary/secondary antibodies.
7.)	Apply other stains (see Table 1.22), if necessary.
8.)	Mount in 90 % glycerol in PBS, add antifade agent, if necessary (Table 1.22).

cellular components, such as polysaccharide cell wall components. An important control experiment is to block out the binding of the antibody by application of the pure antigen to the specimen and checking by fluorescence microscopy. Controls with similar specimens that do not contain the relevant antigen are also extremely valid, these controls often being performed with the respective antigen-lacking mutants.

Extended incubation times, too high concentrations of either primary or secondary antibodies, and also contaminating unconjugated fluorophore molecules increase non-specific labeling of the samples. Background staining is reduced when optimized antibody concentrations and incubation times are used. Addition of serum proteins, either in solution or in solid phase, are useful to reduce unspecific adsorption.

Repeated freezing and thawing affects the reactivity of antibodies by denaturation or aggregation of antibody molecules, resulting in more background staining. Larger volumes of antiserums should therefore be aliquoted before freezing.

Green fluorescent protein as a genetic reporter system for in situ analysis

In expression studies, easily detectable (either biochemically or by their color or fluorescence) reporters for gene expression are widely used. Several fluorescent and non-fluorescent probes, based on the activities of, for instance, luciferase, β-galactosidase, and β-glucuronidase, have been widely used as reporters for gene expression in plants (Jefferson et al., 1987; Mantis and Tague, 2000).

Green fluorescent protein (GFP) offers features that overcome many of the problems of other reporter systems (Chalfie et al., 1994; Chalfie and Kain, 1998; see Conn, 1999 for a methodological overview). Thanks to the intrinsic fluorescence of GFP, addition of external substrates is not necessary. GFP produces bright green fluorescence in prokaryotic or eukaryotic systems, independent of the location of the protein. Several genetically engineered derivatives of GFP with different emission spectra are available, allowing multiple-marker experiments (Patterson et al., 1997; Falk and Lauf, 2001).

The wild-type green fluorescent protein consists of 238 amino acids, and has a cylindrical structure with the fluorophore element located in the center. Once expressed by the cell, autocatalytic oxidation and cyclization of the GFP amino acids at positions 65 to 67 results in the formation of the fluorophore. The fluorescence also requires further interactions of the fluorophore with other parts of the protein. The oxidation reaction only requires the presence of molecular oxygen, and the subsequent fluorescence does not require any additional gene products, substrates, or other factors. The bright green fluorescence emission from GFP (maximum: 508 – 515 nm) is induced by illumination of the molecule with blue light (maximum: 470 nm) or long-wavelength UV (395 nm). Hence, GFP can be monitored in viable cells and organisms in real time. Its widespread use as a reporter has considerably broadened the application potential of qualitative and quantitative fluorescence microscopy and flow cytometric techniques (Cubitt et al., 1995; Gerdes et al., 1996; Misteli and Spector, 1997; Gerdes and Kaether, 1996).

GFP is utilized for localization studies at cellular and sub-cellular levels. When genetic fusions between the relevant promoters (e.g., for expression of catabolic genes) and the GFP are performed, it is possible to monitor the respective expression in microorganisms for localization of cells in microbial biofilms (Unge et al., 1999). In eukaryotes, the expression of GFP is utilized to localize transfected cells and tissues in vivo, but also compartmentation of proteins, fused to GFP, and dynamic processes within a single cell (c.f. Klonis et al., 2002). Because of the high background fluorescence of the plant pigments, detection of GFP fluorescence in thick green tissue may be difficult. Non-pigmented tissues such as roots exhibit lower autofluorescence. Since flavin coenzymes have absorption and emission maxima (450 and 515 nm, respectively) very similar to those for wild-type GFP and the red-shifted GFP variants, autofluorescence of non-pigmented and animal tissue may also obscure the GFP signal. The application of GFP variants with shifted excitation/emission spectra should be checked as alternatives when auto- or background fluorescence interferes with the specific signal.

Expression of wild-type GFP has caused several problems. In bacterial expression systems, the wild-type variant was found in insoluble, non-fluorescent inclusion bodies. Mis-splicing in plants (such as in *Arabidopsis thaliana*) and inappropriate codon usage had to be overcome to obtain high and reproducible expression rates in organisms. Numerous GFP variants are now available, suitable for a variety of purposes, for expression in bacteria, plants, *Drosophila*, or mammal cell lines (see e.g., Atkins and Izant, 1995; Brand, 1995; Haseloff and Amos, 1995; Hodgkinson, 1995; Pines, 1995; Gage et al., 1996). In practice, however, the GFP variants have to be selected carefully; not all combinations are useful and some variants are not yet optimized for expression in all types of organisms. GFP chromophore variants with different excitation and emission spectra have allowed monitoring of gene expression from multiple promoters in the same organism or analysis of mixed cell populations. In practice, separation of the fluorescence signals by standard filter sets is often difficult, because excitation/emission wavelengths of most variants are still quite similar and difficult to separate. Here, the advanced techniques for analysis of overlapping emission spectra in some CLSMs (such as the Zeiss LSM Meta series) allow clear signal separation.

Wild-type GFP is resistant to heat (but fluoresces poorly above 25 °C), mild denaturants (1 % SDS or 8 % urea, fixatives in low concentrations), and also several organic solvents. In vivo, GFP appears to be stable for 1 day or longer. The protein is therefore not suitable for expression studies in which rapid turnover of the marker is necessary (such as the measurement of fast changes in promoter activity). Addressing this obstacle, GFP variants with more rapid turnover rates have been developed. Variants with stabilized fluorescence at elevated temperature are also available. In yeast, wild-type GFP fluorescence is strongest when cells grow at 15 °C, but decreases to about 25 % of the original value when the growth temperature is raised to 37 °C. Fluorescence-enhanced GFP variants show little difference in fluorescence when expressed at either 25 °C or 37 °C Some variants (available from Clontech Laboratories, Inc.) are listed in Table 1.28.

Fixation of cells expressing GFP fusions has varied results, which seem to depend on the specific protein being used. Both methanol and aldehyde fixation have been applied successfully. In the presence of formaldehyde, the GFP signal is quenched considerably, but it does recover to some extent when fixative is washed away. Glutar-

Tab. 1.28 GFP and GFP variants

GFP variant	Max. fluorescence intensity relative to wild-type GFP	Excitation/ emission wavelengths	Codon optimized	Host cells	Protein half-life
EGFP	35×	488–509	human	mammalian	> 24 h
Destabilized EGFP variants	35×	488–509	human	mammalian	1–4 h
EYFP	35×	513/488–527	human	mammalian	> 24 h
EBFP	1×	380–440	human	mammalian	> 24 h
GFPmut3.1 and its descendants	20×	470/395–509	none	bacterial	40 min to >24 h
GFP (wild type)	1×	470/395–509	none	bacterial	> 24 h

aldehyde fixation with GFP fusion proteins is not usually recommended, although low doses of glutaraldehyde (e.g., 0.1 %) may be tolerated. Loss of fluorescence in ethanol-treated cells has been observed.

Like other chromophores, the fluorescence of GFP will also suffer from photobleaching, which should be reduced by minimizing exposition to short-wavelength excitation light (especially UV).

GFP may be cloned form available plasmids in various expression vectors for monitoring promoter activity, for fusion of heterologous proteins to N- or C-termini and with various sub-cellular location markers. When GFP is used in a bidirectional vector, a cloned target gene and GFP may be simultaneously expressed. Thus, GFP may be used as a reporter to screen the positive clones expressing the gene of interest. The appropriate vectors are transformed or transfected into their target cells by a variety of procedures.

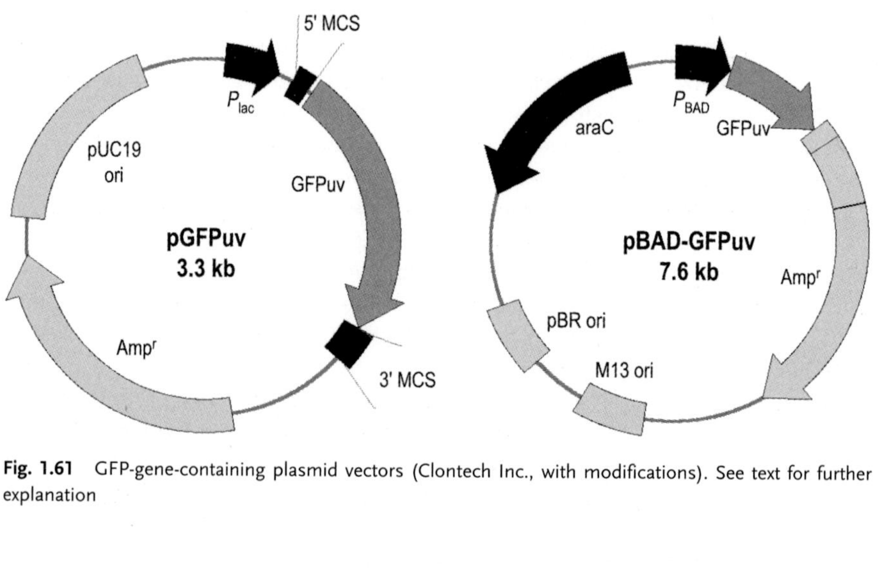

Fig. 1.61 GFP-gene-containing plasmid vectors (Clontech Inc., with modifications). See text for further explanation

Especially in eukaryotes, FACS analysis is often necessary to select and concentrate a small number of transfected cells. Transfection efficiency varies greatly with the cell line used and so it is difficult to put forward a representative procedure. GFP fusion tags at N- or C-termini of proteins have worked in numerous pro- and eukaryotes, but most important is how the fusion affects the function and localization of the tagged protein. The protein chimerae may not act exactly like their wild-type counterparts, and so localization purely by detection of the GFP fusion protein may be misleading.

For regulated expression in Gram-negative bacteria (*E. coli* and other species), the lactose or arabinose promoter-operator system in high-copy number standard plasmid vectors is frequently used. Regulation of the expression system is useful to adjust the level of GFP expression to individual experimental requirements. Moreover, possible toxic effects of constitutively expressed GFP on cells may be avoided.

As an example, the pBAD-GFPuv vector, available from Clontech, is depicted in Fig. 1.61. The plasmid contains, besides a modified GFP gene, an ampicillin resistance gene as a selection marker, the P_{BAD} promoter, and its regulator gene, *ara*C. GFPuv represents an GFP variant optimized for bacterial expression systems (preferably *E. coli*) and excitation by UV light. The expression is regulated by the arabinose promoter P_{BAD} (i.e., induced by arabinose and repressed by glucose). Expression of GFP in colonies of plate-cultivated transformed cells is visible upon UV irradiation. The fluorescent clones are useful markers for re-detection of the cells in, for example, artificial biofilms or to visualize phagocytosis of the fluorescent organisms.

Other vectors are suitable for excision of the GFP gene (such as the pGFPuv plasmid, also available from Clontech). The *gfp* gene may be excised due to the flanking of multiple cloning sites (see Fig. 1.61) or amplified by PCR. The gene will be then fused to the C- or N-terminus of the target gene, and transformed by means of an appropriate plasmid vector.

Fluorescence in situ hybridization (FISH)

Specific detection of nucleic acid sequences with tagged probes, such as Northern or Southern blotting belongs to standard laboratory techniques. These techniques require the isolation of the nucleic acid, separating it on a gel, blotting it onto nitrocellulose, and probing it with a complementary sequence. This principle is also applied for (fluorescence) in situ hybridization, in which the probe is used in situ to detect specific nucleotide sequences within cells and tissues. Four types of probes may be used for in situ hybridization: single- or double-stranded DNA probes, RNA probes, and synthetic oligonucleotide probes.

Single-stranded DNA probes are several hundreds of basepairs in length and are produced by reverse transcription of RNA or by PCR with a single antisense primer. Once the sequence of interest is amplified in the first round, a subsequent round is carried out with the first PCR product as template by use of the anti-sense primers, which yields single-stranded DNA.

Double-stranded DNA probes are generally less sensitive than single-stranded DNA. They may be prepared by reproduction of the sequence of interest (usually on a plasmid) in a bacterial host and excision of the sequence after purification of the (plasmid) DNA from the organism. The sequence is now commonly amplified by PCR, which is less time-consuming and affords very pure samples. The advantage of the bacterial preparation, however, is the availability of large quantities of the probe sequence in question. Because the probe is double-stranded, denaturation or melting has to be carried out prior to hybridization.

RNA probes have the advantage that RNA-RNA hybrids are very thermostable and are resistant to digestion by RNases. This allows the possibility of post-hybridization digestion with RNase to remove non-hybridized RNA and therefore reduces the possibility of background staining. Especially in plants, RNA may be used instead of DNA for probing, due to the high background signaling of DNA probes. The handling of these probes, however, is much more complicated because of the ubiquity of highly stable RNases, which have to be excluded from all material by sterilization or chemical treatment to avoid rapid decomposition of RNA.

These probes are prepared by an RNA polymerase-catalyzed transcription of mRNA in the 3' to 5' direction. Alternatively, in vitro transcription of linearized plasmid DNA with RNA polymerase can be used to produce the RNA probes.

Synthetic oligonucleotide probes are also single-stranded. The probes are produced synthetically, and so the target sequence has to be known. The probe size (40–50 base pairs) is ideal for in situ hybridization, because cells or tissue of interest are easily penetrated. Since they are synthetically designed, they may be "custom-tailored" for the target sequence. In a short oligonucleotide probe, a small number of wrong bases would considerably reduce binding affinity to the target sequence. Since G/C base pairs bond more strongly than A/U base pairs, differences in GC content would require different hybridization conditions. For a set of comparable experiments, it is advantageous to make a series of probes that have the same GC content. Hybridization procedures can thus be standardized for many different probes irrespective of the target genes being measured. Synthetic oligonucleotide probes often represent a reasonable choice for the experimenter, because commercial suppliers offer ready-to-use custom designed probes.

There are various methods by which to attach the fluorescent marker to the nucleotide probe. Fluorescent dye molecules may be directly bound through an amino linker at the 5'-end of the probe or, in an enzymatic reaction, to the 3'-end. An increase in signal intensity is observable, when the fluorescent dye is coupled through a spacer molecule (e.g., 18 carbon atoms in length) or has labeling probes at both ends. It is also possible to enhance the fluorescent signal by indirect detection with a fluorescently labeled antibody. The specific antibody is targeted to the oligonucleotide-bound marker (such as digoxigenin; DIG). This system also allows the fluorescent marker to be exchanged without the need to generate a new oligonucleotide. Enzymatic signal amplification may also be applied, when an anti-digoxigenin antibody is coupled to alkaline phosphatase that converts the non-fluorescent substrate HNPP (2-hydroxy-3-naphthoic acid-2'-phenylanilide phosphate) to the dephosphorylated form, which generates fluorescence in conjunction with the Fast Red TR dye. Enzymatic amplification may reduce the spatial resolution of the signal, however, due to the formation of fluorescent precipitates.

Nucleotides labeled with tags such as fluorescein, cyanine dyes (Cy3, Cy5), Oregon Green dyes, and Texas Red, and also biotin or digoxygenin, are available from various suppliers and may be used for construction of the respective DNA, RNA, or oligonucleotide probe by techniques such as random primed labeling, PCR labeling, nick translation labeling, or oligonucleotide tailing. These procedures are described in detail in molecular biology sourcebooks (Andreev and Pinkel, 1999; Roche Applied Science, 2000; Lloyd, 2001). All major suppliers also offer synthesis of labeled custom oligonucleotides.

It is widely known from DNA hybridization in blotting techniques that specific binding of probes is determined by a variety of parameters. In situ, after the varying conditions of fixation and embedding, these parameters have to be optimized for virtually every type of specimen. A huge number of individual hybridization protocols have therefore been described. A description of the general steps for tissue-embedded specimens is presented in Table 1.29. Widely used applications of FISH relate to the detection of (bacterial) microorganisms in samples and expression studies in plant tissue. These two aspects are therefore discussed here more intensively in the following.

Detection of bacterial microorganisms by FISH

The most commonly used target molecule for detection and differentiation of bacterial microorganisms by FISH is 16S rRNA (18S rRNA for eukaryotes). As a unit of the ribosome, 16S rRNA is present in every living cell in a high copy number. Its structure with conserved and variable regions allows classification of the organisms at various taxonomic levels. Since the 16S rRNA sequences are the most widely used taxonomic markers, growing datasets and supporting software tools for the construction of marker molecules are available (Ludwig and Klenk, 2001; Hugenholtz et al., 2002). Appropriate detection systems based on rRNA-specific oligonucleotide probes for several diagnostic purposes in medicine and industry have been made commercially available (e.g., by Vermicon AG, Munich). These markers are ready to use for hybridization. Known sequences, available from existing databases, allow direct synthesis of labeled probes. Analysis of so far unknown environmental samples is processed by amplification of 16 sRNA sequences with appropriate universal primers, sequencing, and selection of an appropriate oligonucleotide probe, which is then synthesized and used for hybridization (Amann and Ludwig, 2000; Moter and Göbel, 2000; Amann et al., 2001, and references therein).

Prior to hybridization, the specimens have to be fixed and permeabilized. Fixation is generally performed with ethanol and/or paraformaldehyde. For Gram-negative bacteria, 3–4 % formaldehyde (prepared from paraformaldehyde or formalin) solutions are most commonly used. To enable penetration of the marker in Gram-positive bacteria, ethanol/formaldehyde mixtures are used. Some organisms require additional treatment after mounting. After fixation, specimens are mounted onto pre-coated slides. The subbing solution should not, of course, induce autofluorescence or non-specific binding of the marker system (see Table 1.16). More complex specimens (biofilms, cell cultures) may require the use of coverslip-seal combinations that attach to the microscope slides to form microwells. Solutions are then added or removed through inlet and outlet ports (HybriWell coverslips, Molecular Probes, Inc.). For some Gram-positive bacteria, enzymatic digestion of the cell envelope with lysozyme or enzyme mixtures is required. The mycolic acid-containing Actinomycetales may be permeabilized by acid hydrolysis with 1 M HCl. When organisms have to be detected

Tab. 1.29 General steps of an in situ hybridization procedure

1.) **Choice and preparation of the probe**
   - Selection of the probe and probe sequence (double- or single-stranded DNA, single-stranded RNA, or oligonucleotide DNA).
   - Preparation of the probe (preparation of DNA fragments, cloning, chemical synthesis of oligonucleotides).
   - Labeling of the probe.

2.) **Preparation of slides**
   - subbing slides with gelatin or poly-lysine.
   - (optional) siliconization of coverslips.

3.) **Fixation of specimens**
   - by coagulation of proteins (e.g., with ethanol).
   - by cross-linkage of proteins (e.g., with formaldehyde).

4.) **(Optional) Pretreatment of the specimen**
   - inactivation of endogenous enzymes that may interfere with the detection system (e.g., alkaline phosphatase).
   - RNase treatment (removal of endogenous RNA to improve the signal-to-noise ratio in DNA-DNA treatment).

5.) **Permeabilization**
   - treatment with alcohols, detergents, or alcohol/detergent mixtures.
   - treatment with diluted acids.
   - treatment with lytic enzymes.

6.) **(Optional) Prehybridization**
   - incubation of the specimen with the hybridization solution minus specific probe, performed under hybridization conditions (especially for complex specimens, prehybridization may be necessary to reduce background staining).

7.) **Denaturation of a chomosomal DNA probe and target**
   - e.g., by simultaneous heating of probe and target to 80 °C.

8.) **Hybridization**
   using hybridization buffers containing:
   - sodium phosphate, EDTA, SDS (sodium dodecyl sulfate) salts in various concentrations.
   - Denhardt's mix (Ficoll, bovine serum albumin, polyvinylpyrrolidone), heterologous nucleic acids.
   - formamide.
   - dextran sulfate.

9.) **Optimization of hybridization conditions (hybridization temperature, pH, formamide, ionic strength, probe concentration, addition of agents for reduction of non-specific binding).**

10.) **Post-hybridization**
   - stringency washes.

11.) **Detection of the indirectly (non fluorescent) labeled probe**
   - blocking,
   - incubation with the secondary marker (secondary antibody, streptavidin), which is directly labeled or requires additional steps for coupling of a fluorescent marker.

12.) **Fluorescence microscopy**

in tissue (sections), the surrounding specimen must also be penetrable. Paraffin sections should be de-waxed and pretreated with proteinase K.

The crucial step of the whole procedure is, of course, hybridization. Conditions for annealing of the probe to the target sequence are mainly determined and "fine-tuned" by buffer composition and hybridization temperature. Formamide decreases the melting temperature of the double strand by weakening the hydrogen bonds. This is important, because the standard temperature for annealing reactions in hybridization experiments (65 °C) has to be reduced to avoid damage of the specimen. Variation of the formamide concentration is used to regulate stringency: the likelihood of non-homologous annealing, which ultimately determines the specificity of the marker. Generally, hybridization experiments are started with low stringency, which favors both specific binding and non-specific background. When a signal is detectable, the background may be reduced by increasing the stringency. Stringency is also influenced by the ionic strengths (adjusted by the concentration of sodium chloride) of the hybridization and washing buffers. Sodium dodecyl sulfate decreases non-specific binding of the probe to surfaces. Other (optional) additions to hybridization and washing buffers are dextran sulfate (increasing the hybridization velocity), non-specific DNA, and bovine serum albumin (both preventing hybridization of the probe with non-specific sequences in the specimen). The temperature used for hybridization is usually between 37 and 55 °C (increased temperature means increased specificity). A whole procedure is shown in the sample protocol (Table 1.30), and optimization of hybridization conditions is summarized in Table 1.31. Oligonucleotide probe sequences for detection of some bacterial groups are listed in Table 1.32.

FISH in tissue sections – an application example using riboprobes

Although the hybridization procedure is, in principle, the same as for microorganisms, the preparation of embedded tissue samples requires additional steps. Detection of nucleotide sequences by FISH has been described for several types of embedding media and methods (e.g., acrylic resins, paraffin, cryoprocedures).

For detection of gene expression in plant tissue, riboprobes are preferably used (i.e., the preparation and detection procedures follow, in principle, a Northern blotting procedure). Generally, FISH is performed with dewaxed paraffin sections. Structural preservation of tissue may, however, be improved by use of synthetic resins. Not all types of resins are suitable for use in conjunction with FISH, due to the masking of target sequences. The glycol methaclate resin Technovit 8100 is one of the more appropriate resins. Low concentrations of a target, however, may not be detected (Ruzin, 1999; Roche Applied Sciences, 2000).

Preparation of the riboprobe requires approximately 1 μg DNA template subcloned into the multiple cloning site of a vector enabling in vitro RNA transcription with T3 and T7 RNA polymerase. The multiple cloning sites of the pBlueskript II plasmids are flanked by T3 and T7 promoters and are generally used for this purpose. After linearization of the vector, antisense RNA (as the specific riboprobe) and sense RNA (if desired; used as negative control for the hybridization experiment) are generated. Normally, UTP is added as the labeled mononucleotide. The marker is then applied to the hybridization procedure. Table 1.34 presents a sample protocol for this procedure.

Tab. 1.30 Preparation of bacterial microorganisms for FISH

Fixation

1.) Fix a cell suspension in 3–4 % formaldehyde solution for 1/3 h at 4 °C. Be aware that rapidly growing cells have highest rRNA contents (i.e., are easiest to detect).
2.) Remove the fixing solution by centrifugation, wash the specimen in an appropriate buffer (e.g., phosphate buffered saline), resuspended in an equal volume of ethanol (95 %, v/v); the specimen may be stored at –20 °C for several months.
3.) Put a small drop on a subbed slide. If necessary, dilute the suspension.
4.) Allow slides to air-dry at room temperature.
5.) Dehydrate and permeabilize the specimen by incubation in an ethanol series (50 %, 80 %, 95 % ethanol, v/v) for three minutes per step.

Hybridization with digoxigenin (DIG) labeled/fluorescent marker-labeled oligonucleotide

1.) Use a non-transparent plastic container with a sealing lid as hybridization chamber. Place paper towels soaked with hybridization buffer (20 mM Tris-HCl, 0.01 % SDS, 0.9 M NaCl, pH 7.2) in the chamber to produce a humidified atmosphere.
2.) Prepare a dilution of the probe that corresponds to 1/5 ng/μl in hybridization buffer (see Step 1). Apply an aliquot (e.g., 25 μl) to each dried specimen drop.
3.) Place a coverslip on the specimen; avoid air bubbles.
4.) Place the slides in the hybridization chamber, seal, and incubate for 4 h or overnight. The temperature used for hybridization is usually between 37 and 55 °C. Use 37 °C for a first experiment.

Post-hybridization washing steps and mounting

Avoid exposition of the specimens to strong light

1.) Fill two staining jars with wash buffer and allow to equilibrate to hybridization temperature in a shaking water bath.
2.) Remove coverslips by dipping the slide with the hybridized specimen in a jar filled with hybridization buffer. The coverslips should drop off after gentle agitation.
3.) Place slides in the jar with washing buffer I (20 mM Tris-HCl, 0.1 % SDS, 0.5 M NaCl, 5 mM EDTA, pH 7.2) under continued agitation for 20 min with one buffer change after 10 min.
4.) Allow slides to air dry at room temperature in the dark.
5.) As an optional step, apply counter-staining here. For counter-staining of nucleic acids, apply 20 μl DAPI (see Tab. 1.22, but use a 0.1 μg/ml solution) to each dried specimen drop. Incubate for 5 min and rinse briefly with distilled water.
6.) Add antifade agent (e.g., Prolong antifade agent; Molecular Probes, Inc; Citifluor AF1 Citifluor, Ltd). Cover with a coverslip, and view under the epifluorescence microscope.

Treatment of a DIG-labeled nucleotide probe

1.) Dilute an Anti DIG antibody or Fab fragment labeled with a fluorescent marker (e.g., fluorescein, rhodamine) in blocking buffer (5 % (w/v), BSA in 0.6 M NaCl, 60 mM sodium citrate, pH 7.0) to 5–10 μg/ml final concentration (other markers may require other dilutions).
2.) Apply an aliquot (e.g., 25 μl) to each specimen drop and incubate in the dark at room temperature for 1 h in a humid container (equilibrated with the blocking buffer).
3.) Place slides in jars filled with washing buffer II (150 mM NaCl, 100 mM Tris-HCl, 0.01 % (w/v) SDS, pH 7.4) and incubate with continued agitation for 20 min with one buffer change after 10 min.
4.) Rinse the slides briefly with sterile water, allow to air-dry at room temperature in the dark.
5.) Add antifade agent (e.g., Prolong antifade agent; Molecular Probes, Inc; Citifluor AF1 Citifluor, Ltd). Cover with a coverslip, and view under the epifluorescence microscope.

Tab. 1.31 Optimization procedures for hybridization and posthybridization treatment

For binding of the oligonucleotide probe:

- Vary formamide concentration in the hybridization buffer. Each percentage of formamide decreases the melting temperature of the probe by 0.4 °C.
- Vary hybridization and wash temperatures between 37 and 55 °C (increased temperature means increased specificity).
- Vary NaCl concentrations (40 mM–900 mM) in washing buffer I (decreased salt concentration means increased specificity).

For binding of the antibody:

- Wash slides in 0.6 M NaCl, 0.03 M sodium citrate buffer (pH 7.0, containing 0.2 % (v/v) Tween 20) (reduction of non-specific binding).
- Vary incubation time and length of washing steps after binding.

Tab. 1.32 Commonly used oligonucleotide probes for microorganisms (Amann et al., 1995)

Phylogenetic group	Sequence (5'-3')
Universal	acgggcggtgtgtrc gwattaccgcggckgctg
Archaea	gtgctcccccgccaatcct
Bacteria (EUB338)	gctgcctcccgtaggagt
Eukarya	gggcatcacagacctg tagaaagggcaggga
High G+C Gram+	ccygatatctgcgca
Low G+C Gram+	tgtagcccargtcata
Flavobacteria	tcagtrccagtgtggggg
Proteobacteria, alpha-subgroup	cgragttagccggggc
Proteobacteria, beta-subgroup	tcactgctacacgyg
Proteobacteria, gamma-subgroup	cttttgcarcccact
Proteobacteria, delta-subgroup	gcttkcawgmagagg
Sulfate-reducing bacteria	cgygcgccrctytact
Pseudomonas sp.	ccttcctcccaactt
Green Sulfur bacteria	cttttarggggatttcctctg
Archaea-Euryarchaeota	cttgcccrgccctt gacctaccgtngyccr
Archaea, Crenarchaeota	ccagrcttgccccccgct

Large transcripts would not penetrate the fixed cytoplasm, whereas transcript lengths lower than 100 bp hybridize unspecifically. Thus, generation of transcript lengths between 150 and 300 bp is recommended. Larger transcripts should be hydrolyzed to this length. This is a critical step.

The hybridization procedure itself should be optimized in the same way as DNA-DNA hybridization. In analogy to the Southern blotting, a Northern blot will prove the specifity of the probe.

The rehydrated paraffin sections (see Table 1.32) are then subjected to hybridization as described in Table 1.36. Technovit 8100 sections of lower thickness (around 5 μm), mounted on a subbed glass slide, may be directly used for the hybridization procedure (Takechi et al., 1999).

Essential points to consider for troubleshooting in FISH

As indicated above, numerous variations of these essential protocols are applied, due to the fact that every specimen and target sequence presents a huge number of variable parameters that influence stringency and specificity of binding. Essentially, it is recommended to optimize the procedure by taking account of the following main points:

Weak or no signals. If the experiment does not show a signal at all, process optimization is very difficult. First of all, the quality of the probe should be routinely checked by Southern/Northern blotting. The pH of all solutions should be checked. Aged solutions may lose buffering capacity or may be contaminated by microorganisms. Weak signals may occur due to insufficient permeabilization. If the enzyme has lost is activity and/or the other agents used for pretreatment of the specimen do not sufficiently solubilize the specimen components, hybridization will not occur. When DNA is the target sequence, denaturation of DNA may be insufficient. The target sequences may also be blocked by proteins. Excessive denaturation, how-

Tab. 1.33 Pre-treatment of tissue sections for hybridization

1.) Remove paraffin by repeated (2×) incubation in Histo-clear (or xylene) for 10 min per step. Use acetone for butyl/methyl methacrylase resins.
2.) Rehydrate sections as follows (for detection of RNA, pretreatment to exclude RNases is necessary):
   - Incubate repeatedly (3×) in ethanol (100 %) for 2 min per step.
   - Incubate in a descending ethanol series (95 % (v/v), 85 %, 70 %, 50 %, 30 % ethanol in aqueous solution containing 0.9 % (w/v) NaCl).
   - Rinse for 2 min in 0.9 % NaCl solution and in phosphate-buffered saline (PBS, 0.9 % NaCl in 50 mM phosphate buffer, pH 7.5).
3.) Treat with proteinase K as follows:
   - Incubate for 45 min at 37 °C in 2 μg/ml proteinase K (in 20 mM Tris-HCl, 2 mM $CaCl_2$, pH 8.0)
   - Rinse at room temperature in 0.2 % (w/v) glycine in PBS to block residual protease activity.
   - Rinse for 2 min in PBS.
4.) Incubate in a 4 % formalin solution in PBS for 10 min, rinse for 2 min in phosphate-buffered saline and then twice in 0.9 % NaCl solution.
5.) Dehydrate in an ascending ethanol series (see step 2.), beginning with 30 % and ending with 100 % ethanol, two 100 % treatments.
6.) Allow to air-dry. The specimen may be now used for hybridization.

Tab. 1.34 Preparation of a riboprobe for in situ hybridization

Prerequisites: the target gene is sub-cloned in an appropriate plasmid vector, the plasmid is linearized at a restriction site downstream of the cloned insert (5' overhangs recommended), DNA is purified by phenol/chloroform extraction and ethanol precipitation.

Pretreatment to exclude RNases from all liquids and labware is necessary: use RNase-free chemicals wherever possible, add 0.1 % diethyl pyrocarbonate (DEP) to solutions, incubate several hours or overnight, autoclave.

The following steps refer to the DIG-RNA labeling kit (purchasable from Roche Diagnostics).

1.) Probe synthesis:

Add to 1 μg of the purified DNA in 13 μl RNase-free, double-distilled water:
- 2 μl of 10-fold NTP labeling mixture,
- 2 μl 10-fold transcription buffer,
- 1 μl RNase inhibitor,
- 2 μl of RNA polymerase T 7.

Mix gently and centrifuge briefly; incubate for 2 h at 37 °C,
- add 2 μl of RNase-free DNase,
- incubate at 37 °C for 30 min.

2.) Run a control minigel. Instead of the DNA template band, a thick RNA band should appear.
3.) Stop the reaction by adding 2 μl of 0.2 M EDTA.
4.) Precipitate by addition of 2 μl EDTA (0.2 M, pH 8.0), 2.5 μl NaOAc (2 M), and 75 μl ethanol (100 %) and subsequent incubation at –80 °C for at least 20 min.
5.) Centrifuge at 15,000×g for 10 min, remove supernatant, and wash pellet twice in 70 % ethanol.
6.) Dry the pellet in a Speedvac. Dissolve the resulting pellet in 50 μl distilled water.

ever, causes loss of DNA. If a secondary marker system is used, the age and quality of the marker system should be checked. As recommended above, the hybridization experiment should be performed with low stringency, since the strength of a signal relative to the background may be estimated under these conditions. The stringency should then be increased.

Background staining may not only be reduced by increasing stringency. If especially strongly expressed DNA sequences are the target, decomposition of RNA prior to hybridization is most important. Disturbing proteins may be dissolved with proteases other than Proteinase K (such as Pepsin). The post-hybridization washing steps are of similar importance to hybridization itself. By increasing the number and duration of washing steps, or the addition of other detergents (e.g., Triton X-100), background signals may be reduced.

Tab. 1.35: Hydrolysis of large riboprobes

1.) Resuspend 50 μl RNA in an equal volume of 0.1 M $NaHCO_3$, pH 10.2 and 10 mM DTT, incubate at 60 °C. Estimate incubation time according to the following formula:
$t = (L_0 - L_f) / 0.11\, L_0 L_F$, with t = time in minutes, L_0 = initial length in kilobases, L_F = desired length in kilobases.
2.) Precipitate with 4 M LiCl and 60 μl ethanol.
3.) Centrifuge at 15000×g for 2 min.
4.) Dissolve the resulting pellet in 50 μl distilled water.

Tab. 1.36: Hybridization procedure for riboprobes

1.) Use a non-transparent plastic container with a tight-sealing lid as hybridization chamber. To create a humid atmosphere during hybridization, use paper towels soaked with hybridization buffer (10 mM Tris-HCl, 10 mM Na_2HPO_4, 5 mM EDTA, 40 % (v/v) formamide, 10 % (w/v) dextran sulfate, 1 µg/ml – 1 mg/ml yeast tRNA; in Denhardt's solution*); pH 6.8) to keep the chamber humidified.
2.) (Optional). Prehybridize each section for 2 h at room temperature by placing an aliquot (e.g., 25 µl) of the hybridization buffer onto the glass slide (especially for complex specimens, prehybridization may be necessary to reduce background staining).
3.) Prepare a dilution of the probe corresponding to 1 – 5 ng/µl in hybridization buffer. Spread hybridization solution evenly over the sections.
4.) Place a coverslip on the specimen; avoid air bubbles.
5.) Place the slides in the hybridization chamber, seal, and incubate overnight at 50 °C.
6.) Remove coverslips by dipping the slide with the hybridized specimen in a jar filled with prewarmed (50 °C) SSC buffer (150 mM NaCl, 15 mM sodium citrate, pH 7.0). The coverslips should drop off after gentle agitation.
7.) Place slides in the jar with SSC buffer under continued gentle agitation for 30 min to 1 h. Repeat this step once.
8.) Rinse slide twice with NTE (500 mM NaCl, 10 mM Tris-HCl, 1 mM EDTA, 10 mM dithiothreitol; pH 8.0) buffer. Incubate slides at 37 °C for 30 min NTE buffer containing RNase A (5 µg/ml), reduce RNase concentration if the signal is too low.
9.) Rinse slide twice with NTE buffer.
10.) Allow slides to air-dry at room temperature in the dark.

Treatment of a DIG-labeled nucleotide probe

1.) Dilute an Anti DIG antibody or Fab fragment labeled with a fluorescent marker (e.g., fluorescein, rhodamine) in BSA solution (1 % (w/v) BSA in 0.15 M NaCl, 100 mM Tris-HCl, pH 7.5) to 5 – 10 µg/ml final concentration (other markers may require other dilutions).
2.) Apply an aliquot (e.g., 25 µl) to each specimen drop and incubate in the dark at room temperature for 1 h in a humid container (equilibrated with the blocking buffer).
3.) Place slides in jars filled with washing buffer (150 mM NaCl, 100 mM Tris-HCl, pH 7.4) and incubate under continued agitation for 20 min with one buffer change after 10 min.
4.) Rinse the slides briefly with sterile water, allow to dry at room temperature in the dark.
5.) Add antifade agent (e.g., Prolong antifade agent; Molecular Probes, Inc; Citifluor AF1 Citifluor, Ltd). Cover with a coverslip, and view under the epifluorescence microscope.

*) Denhardt's solution (50 ×): 10 g Ficoll (Type 400), 10 g polyvinylpyrrolidone, 10 g bovine serum albumin (fraction V) in 1000 ml distilled water

It may be that the specific type of the secondary marker interacts with the specimen. A streptavidin-labeled detection system, for instance, may also react with endogenous biotin. Here, another marker system should be tested.

Controls. Northern or Southern blots should generally be performed to ensure that (in principle) the probes are working. In addition, other controls should be performed to demonstrate the specificity of the specimen. For detection of microorganisms, a combination of oligonucleotide probes with specificities for different phylogenetic levels is generally used. Cells known to lack the sequence(es) of interest, should of course show no signal. Hybridization with an irrelevant probe may also be performed. To ensure that the secondary marker does not bind to the specimen, hybridization in the absence of the probe should be routinely performed.

References

1 Abramowitz, M. (**1994**). Optics: A Primer. Olympus Corp. Publ., New York.

2 Amann R. I., Ludwig W., Schleifer K. H. (**1995**). Phylogenetic identification and in situ detection of individual microbial cells without cultivation. *Microbiol. Rev.* 59, 143–169.

3 Amann, R. and Ludwig, W. (**2000**). Ribosomal RNA-targeted nucleic acid probes for studies in microbial ecology. *FEMS Microbiol. Rev.* 24, 555–565.

4 Amann, R., Fuchs, B. M., Behrens, S. (**2001**). The identification of microorganisms by fluorescence in situ hybridisation. *Curr. Opin. Biotechnol.* 12, 231–236.

5 Andreev, M., Pinkel, D. (**1999**). Introduction to Fluorescence In Situ Hybridization: Principles and Clinical Applications. John Wiley, New York.

6 Atkins, D., Izant, J. G. (**1995**). Expression and analysis of the green fluorescent protein gene in the fission yeast *Schizosaccharomyces pombe. Curr. Genet.* 28, 585–588.

7 Baker, J. R. (**1958**). Principles of Biological Microtechnique. Methuen Co., Ltd.

8 Bancroft, J. D., Gamble, M. (**2002**). Theory and Practice of Histological Techniques. 5th ed., W. B. Saunders Co., London.

9 Barnard, T. (**1987**). Rapid freezing techniques and cryoprotection of biomedical specimens. *Scanning Microsc.* 1, 1217–1224.

10 Baschong, W., Suetterlin, R., Laeng, H. (**2001**). Control of autofluorescence of archival formaldehyde-fixed, paraffin-embedded tissue in confocal laser scanning microscopy (CLSM). *J. Histochem. Cytochem.* 49, 1565–1571.

11 Bast, E. (**2001**). Mikrobiologische Methoden. Eine Einführung in grundlegende Arbeitstechniken. Spektrum Akademischer Verlag, Heidelberg.

12 Berlyn, G. P., Mischke, J. P. (**1976**). Botanical Microtechnique and cytochemisty. Iowa State University Press, Ames, IA.

13 Bradbury, S., Bracegirdle, B. (**1998**). Introduction to Light Microscopy (Microscopy Handbooks, No 42). 2nd ed., Springer Publ., New York.

14 Brand, A. (**1995**). GFP in *Drosophila. Trends Genet.* 11, 324–325.

15 Brelje, T. C., Wessendorf, M. W., Sorenson, R. L. (**1993**). Multicolor laser scanning confocal immunofluorescence microscopy: practical application and limitations. *Meth. Cell Biol.* 38, 97–181.

16 Brouns, I., Van Nassauw, L., Van Genechten, J., Majewski, M., Scheuermann, D.W., Timmermans, J. P., Adriaensen D. J. (**2002**). Triple immunofluorescence staining with antibodies raised in the same species to study the complex innervation pattern of intrapulmonary chemoreceptors. *J. Histochem. Cytochem.* 50, 575–582.

17 Chalfie, M., Kain, S. (**1998**). Green Fluorescent Protein: Properties, Applications, and Protocols. Wiley-Liss, New York.

18 Chalfie, M., Tu, Y., Euskirchen, G., Ward, W.W., Prasher, D.C. (**1994**). Green fluorescent protein as a marker for gene expression. *Science* 263, 802–805.

19 Conn, M. P. (**1999**). Green Fluorescent Protein. *Methods in Enzymology*, Vol. 302. Academic Press, New York.

20 Cubitt, A. B., Heim, R., Adams, S. R., Boyd, A. E., Gross, L. A., Tsien, R. Y. (**1995**). Understanding, improving and using green fluorescent proteins. *Trends Biochem.* 20, 448–455.

21 Czarnik, A.W. (**1993**). Fluorescent Chemosensors for Ion and Molecule Recognition (ACS Symp. Ser. 538). American Chemical Society, Columbus, OH.

22 Falk, M. M., Lauf, U. (**2001**). High resolution, fluorescence deconvolution microscopy and tagging with the autofluorescent tracers CFP, GFP, and YFP to study the structural composition of gap junctions in living cells. *Microsc. Res. Techn.* 52, 251–262.

23 Gage, D. J., Bobo, T., Long, S. R. (**1996**). Use of green fluorescent protein to visualize the early events of symbiosis between *Rhizobium meliloti* and alfalfa (*Medicago sativa*). *J. Bacteriol.* 178, 7159–7166.

24 Gerdes, H. H., Kaether C. (**1996**). Green fluorescent protein: Applications in cell biology. *FEBS Lett.* 389, 44–47.

25 Gray, P. (**1964**). Handbook of Basic Microtechnique. 3rd ed., McGraw-Hill, New York.

26 Gray, P. (**1973**). The encyclopedia of microscopy and microtechnique. Van Nostrand Reinhold, New York.

27 Habs, H., Seeliger, H. P. R. (**1967**). Bakteriologisches Taschenbuch. 38th ed., Johann Ambrosius Barth Verlag, Leipzig.

28 Haseloff, J., Amos, B. (**1995**). GFP in plants. *Trends Genet.* 11, 328–329.

29 Haugland, R. P. (**2002**). Handbook of fluorescent probes and Research Products. 9th ed., Molecular Probes Inc. Eugene, OR.

30 Haugland, R. P. (**2000**). Antibody conjugates for cell biology. In: *Current Protocols in Cell Biology* (eds. J. S. Bonifacino, M. Dasso, J. Lippincott-Schwartz, J. B. Harford, K. M. Yamada), pp. 16.5.1–16.5.22. John Wiley, New York.

31 Hermanson, G. T. (**1996**). Bioconjugate Techniques. Academic Press, New York.

32 Hilger, A. (**1986**). Aberrations of Optical Systems. IOP Publishing Ltd., Philadelphia.

33 Hodgkinson, S. (**1995**). GFP in *Dictyostelium*. *Trends Genet.* 11, 327–328.

34 Hovius, R., Vallotton, P., Wohland, T., Vogel, H. (**2000**). Fluorescence techniques: shedding light on ligand-receptor interactions. *Trends Pharmacol. Sci.* 21, 266–273.

35 Hugenholtz, P., Tyson, G. W., Blackall, L. L. (**2002**). Design and evaluation of 16S rRNA-targeted oligonucleotide probes for fluorescence in situ hybridization. *Methods Mol. Biol.* 179, 29–42.

36 Ishii T, Kasama K, Kondo M. (**1990**). Improvement of the quality of frozen sections from formalin fixed tissue. *Stain Technol.* 65, 43–44.

37 Jefferson, R. A., Kavanaugh, T. A., Bevan, M. W. (**1987**). GUS fusions: β-glucuronidase as a sensitive and versatile gene fusion marker for higher plants. *EMBO J.* 6, 3901–3907.

38 Johnson, I. D. (**1998**). Fluorescent probes for living cells. *Histochem. J.* 30, 123–140.

39 Kasten, F. H. (**1993**). Introduction to fluorescent probes: properties, history and applications. In: *Fluorescent and Luminescent Probes for Biological Activity* (ed. W. T. Mason), pp. 12–33. Academic Press, New York.

40 Klonis, N., Rug, M., Harper, I., Wickham, M., Cowman, A., Tilley, L. (**2002**). Fluorescence photobleaching analysis for the study of cellular dynamics. *Eur. Biophys. J.* 31, 36–51.

41 Lanier, L. L., Recktenwald, D. J. (**1991**). Multicolor Immunofluorescence and Flow Cytometry. *Methods* 2, 192–205.

42 Lillie, R. D. (**1977**). H. J. Conn's Biological Stains. The Williams & Wilkins Comp., Baltimore.

43 Lloyd, R. V. (**2001**). Morphology Methods: Cell and Molecular Biology Techniques. Humana Press, Totowa, WA.

44 Ludwig, W. and Klenk, H. P. (**2001**). Overview: a phylogenetic backbone and taxonomic frame-work for prokaryotic systematics. In: *Bergey's Manual of Systematic Bacteriology* (eds. D. R. Bone, R. W. Castenholz), pp. 49–65. Springer Publ., New York.

45 Maeda, M. (**1984**). Laser dyes: Properties of Organic Compounds for Dye Lasers. Academic Press, Tokyo.

46 Mantis, J., Tague, B. W. (**2000**). Comparing the utility of β-glucuronidase and Green Fluorescent Protein for detection of weak promoter activity in *Arabidopsis thaliana*. *Plant Mol. Biol. Reporter* 18, 319–330.

47 Mason, W. T. (**1999**). Fluorescent and luminescent probes for biological activity. Academic Press, New York.

48 McCrone, W. (**1978**). Polarized Light Microscopy. Butterworth-Heinemann, Oxford.

49 McManus, J. F. A., Mowry, R. W. (1964) Staining methods: histological and histochemical. Harper/Row, New York.

50 Meyer, J. R. (**1995**). Introduction to Classical and Modern Optics. Prentice-Hall Inc., New Jersey.

51 Misteli, T., Spector, D. L. (**1997**). Applications of the green fluorescent protein in cell biology and biotechnology. *Nat. Biotechnol.* 15, 961–964.

52 Moter, A., Göbel, U. B. (**2000**). Fluorescence in situ hybridization (FISH) for direct visualization of microorganisms. *J. Microbiol. Meth.* 41, 85–112.

53 Padawer, J. (**1967**). The Normarski interference-contrast microscope. An experimental basis for image interpretation. *J. Roy. Microsc. Soc.* 88, 305–349.

54 Patterson, G. H., Knobel, S. M., Sharif, W. D., Kain, S. R., Piston, D. W. (**1997**). Use of the green fluorescent protein and its mutants in quantitative fluorescence microscopy. *Biophys. J.* 73, 2782–2790.

55 Pawley, J. B. (**1995**). Handbook of Biological Confocal Microscopy. 2nd ed., Plenum Press, New York.

56 Pearse, A. G. E. (**1980**). Histochemistry, Theoretical and Applied: Preparative and optical technology, Vol. 1. 4th ed., Churchill Livingston, Oxford.

57 Pearse, A. G. E. (**1985**). Histochemistry, Theoretical and Applied: Analytical technology, Vol. 2. 4th ed., Churchill Livingston, Oxford.

58 Pines, J. (**1995**). GFP in mammalian cells. *Trends Genet.* 11, 326–327.

59 Pluta, M. (**1969**). A phase contrast device with positive and negative image contrast. *J. Microsc.* 89, 205.

60 Prento, P., Lyon, H. (**1997**). Commercial formalin substitutes for histopathology. *Biotechnic. Histochem.* 72, 273–282.

61 Rasmussen, D. H. (**1982**). Ice formation in aqueous systems. *J. Microsc.* 128, 167–174.

62 Ray, S. F. (**1994**). Photographic Imaging and Electronic Photography. Focal Press, Oxford.

63 Roche Applied Science (no authors listed) (**2000**). Nonradioactive in situ hybridisation application manual. Roche Diagnostics Corp., Indianapolis.

64 Rost, F. W. D. (**1995**). Fluorescence Microscopy, Vol 2. Cambridge University Press, Cambridge.

65 Russ, J. C. (**1999**). The Image Processing Handbook. 3rd ed. CRC Press, Boca Raton, Fl.

66 Ruzin, S. E. (**1999**). Plant Microtechnique and Microscopy. Oxford University Press, Oxford.

67 Sanderson J. B. (**1994**). Biological Microtechnique. RMS Handbook 28. Springer, New York.

68 Schubert, W. (**1991**) Triple immunofluorescence confocal laser scanning microscopy: spatial correlation of novel cellular differentiation markers in human muscle biopsies. *Eur. J. Cell Biol.* 55, 272–285.

69 Sheppard, C. J. R., Shotton, D. M. (**1997**). Confocal Laser Scanning Microscopy. Springer Publ., New York.

70 Slavik, J. (**1998**). Fluorescence Microscopy and Fluorescent Probes. Plenum Press, New York.

71 Spencer, M. (**1982**). Fundamentals of Light Microscopy. Cambridge University Press, Cambridge.

72 Stoward, P. J. (**1973**). Fixation in Histochemistry. Chapman & Hall, London.

73 Stoward, P. J., Pearse, A. G. E. (**1992**). Histochemistry: Theoretical & Applied. 4th ed., Churchill Livingstone, Edinburgh.

74 Takechi, K., Sakamoto, W., Katsuhara, M., Murata, M., Motoyoshi, F. (**1999**). *In situ* RNA hybridization using Technovit resin in *Arabidopsis thaliana. Plant Mol. Biol. Reporter* 17, 43–51.

75 Tanke, H. J., Herman, B. (**1998**). Fluorescence Microscopy (Microscopy Handbooks, 40). 2nd ed. Springer. Publ., New York.

76 Unge, A., Tombolini, R., Molbak, L., Jansson, J. K. (**1999**). Simultaneous monitoring of cell number and metabolic activity of specific bacterial populations with a dual gfp-luxAB marker system. *Appl. Environ. Microbiol.* 65, 813–821.

77 Wilkinson, M. H. F., Schut, F. (**1998**). Digital Image Analysis of Microbes: Imaging, Morphometry, Fluorometry and Motility Techniques and Applications. John Wiley, New York.

2
Electron Microscopy

2.1
Applications and Limitations in Transmission Electron Microscopy

Transmission electron microscopes are now as easy to use as light microscopes. The costs of standard instruments are in the same range as for such equipment as confocal microscopes or ultracentrifuges. Like the other microscopes, they belong among the standard instruments of larger research facilities.

Transmission electron microscopes used for applications in life sciences generally provide resolutions down to 0.3 nm. This allows, by application of appropriate specimen preparation techniques, imaging of clusters of heavy metal atoms (often used as probes), protein sub-domains, DNA molecules, or the smallest lipid aggregates. At the cellular level, determination of the fine structure of compartments and localization of proteins inside compartments is possible.

The essential information concerning the structures of plastids, mitochondria, the endoplasmic reticulum, and the nucleus has been collected by TEM. Bacterial cell structures such as cellular appendages, the cell wall, and smaller inclusion bodies are only visible by TEM.

In contrast to SEM, a specimen must be electron translucent (i.e., of a limited mass thickness) for visualization in TEM. Specimens therefore either have to be small enough for imaging as a whole unit (protein molecule, phage particle, cellular appendage, etc.) or a larger specimen has to be cleaved or sectioned. Sectioning requires pretreatment similar to that described for light microscopy. The specimen must either be chemically fixed, dehydrated, and embedded in resin, or it must be frozen and finally cut into ultra-thin sections. As in light microscopy, artifacts may be induced by the pretreatment procedures. Since visualization in TEM has to be performed under high vacuum conditions ("environmental chambers" are only suitable for a limited application range), the sample is dry or frozen, and so inactive. In other words, direct monitoring of dynamic processes in samples is not possible. This feature is still the biggest disadvantage of electron microscopy in life sciences.

Though some important techniques for cytological preparations in TEM (embedding and ultra-thin sectioning) require preparation times of several days and additional equipment, there are also techniques that permit sample examination some minutes after sampling. This allows quick examination of a sample taken from a continuous process. Inspection of some cellular inclusions, detection of virus, monitoring of the purity of protein preparations, control of liposome size, etc. are possible after a quick staining procedure.

High-resolution mapping techniques allow detection of cellular compounds (including elements) in a resolution range down to some nanometers. As in scanning electron microscopy (see chapter 2.3), the interaction between beam electrons and specimen atoms produces specific signals (inelastically scattered electrons, X-rays, etc.) that allow determination of element composition and other specimen features. Reconstruction techniques, though quite elaborate, can provide three-dimensional models of subcellular components down to the size of single protein molecules. In principle, resolution allowing reconstruction of coordinates down to atomic scales may be achieved.

2.2
Imaging of Cellular and Large Subcellular Structures

2.2.1
Bacterial Cells

2.2.1.1 Outside the cytoplasmic membrane

Though essential subcellular components may be localized by fluorescence microscopy, as described above, elucidation of their structures is only possible with high-resolution techniques. The complexity of bacterial wall structures is not detectable by light microscopy. In Gram-negative bacteria, the periplasmic space, together with the cytoplasm, represent the only compartments entirely enclosed by membranes (the cytoplasmic membrane and the outer membrane). The periplasm is essentially filled with a viscous peptidoglycan matrix (Hobot et al., 1984).

In Gram-positive bacteria, a thick peptidoglycan (murein) layer represents the borderline with the exterior. In most Gram-positives, no periplasmic space can be detected by electron microscopy (see Fig. 2.1 a). In archaea, a single proteinaceous or glycoproteinaceous S-layer covers the cytoplasmic membrane, leaving no visible space. Although well defined "structures" (a free space between two borderlines, representing the periplasmic space) are absent, a periplasm may very well be present, as indicated in Gram-positive bacteria by "periplasmic functions", such as protein transport, cell wall synthesis, and the presence of periplasmic proteases. A similar "functional" periplasm is postulated for archaea (Beveridge, 1995).

Most data concerning the importance of the periplasm as a compartment have, of course, been collected for Gram-negative bacteria (especially *Escherichia coli*; see also Fig. 2.1 b). A wide variety of hydrolytic enzymes, binding proteins, and chemoreceptors are located in the periplasm. The periplasm serves as compartment for processing of pilus proteins and is involved in the general secretory pathway of proteins across the cytoplasmic membrane and outer membrane: protein folding, protein degradation, DNA cleavage by restriction endonucleases (Fig 2.1 c), and extrusion of noxious compounds such as antibiotics. Though usually visible as a tiny compartment, the peri-

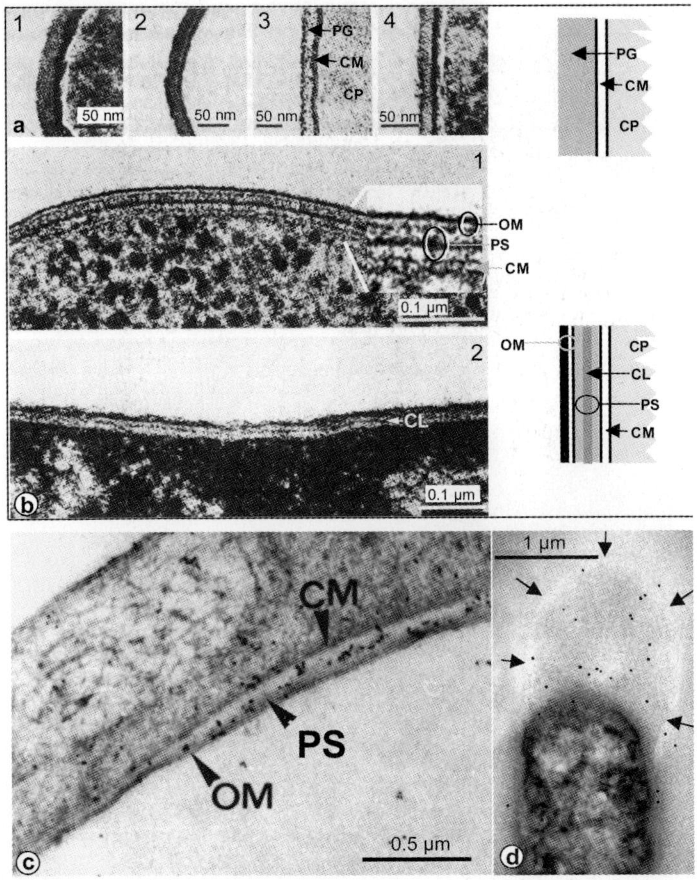

Fig. 2.1 Cell envelopes and periplasm.
(a) Ultrastructure of the cell envelope of *Lactobacillus* after application of different staining procedures to ultra-thin sections (resin embedded): 1, lead citrate; 2, phosphotungstic acid; 3, Thiéry stain (carbohydrate-specific); 4, Ruthenium Red. In spite of the variations in contrast, an outer thick layer and an underlying electron dense double layer can be distinguished, representing the peptidoglycan cell wall (PG) and the cytoplasmic membrane (CM). This is also illustrated by the upper right schematic drawing. (Rastogi et al., 1984, Mayer, 1986).
(b) Ultrastructure of the cell envelope of E. coli. 1, Freeze-substituted *Escherichia coli* cell, ultra-thin section. The cell envelope has a constant, uniform width and is divided into the outer membrane (OM), periplasmic space (PS), and cytoplasmic membrane (CM). The periplasmic space is not empty: it is electron-opaque and contains an additional central layer. Section stained with uranyl acetate and lead citrate; 2, Glutardialdehyde-fixed E. coli B with low-temperature dehydration (–35 °C) and embedded in Lowicryl K4M. The envelope appears to be uniform between the outer membrane and the cytoplasmic boundary. Sometimes a stained central layer (CL) is visible. Section stained with uranyl acetate and lead citrate. This is also illustrated by the lower right schematic drawing. (Hobot et al., 1984; Mayer, 1986).
(c) Localization of the restriction endonuclease EcoRI in *E. coli* by immunogold-labeling. Ultra-thin section of a cell embedded in low-temperature embedding resin. Dark dots represent colloidal gold particles, bound to the enzyme by specific antibodies. CM, cytoplasmic membrane, OM, outer membrane, PS, periplasmic space. (Kohring and Mayer, 1987).
(d) Localization of a pullulanase inside the toga of *Thermotoga maritima* cryosection. The toga represents an extremely enlarged periplasmic space. The toga membrane is marked by arrows

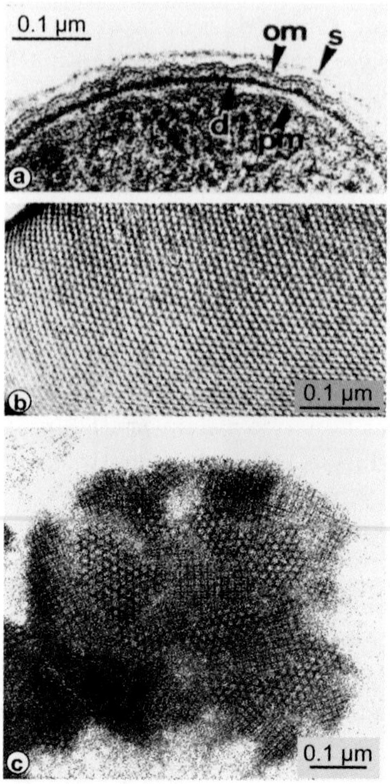

Fig. 2.2 Bacterial surface layers.
(a) Ultra-thin section of the envelope of the Gram-negative bacterium *Acinetobacter* sp. strain *MJT/F5/5* (resin-embedded sample). om, outer membrane; pm, cytoplasmic membrane; s, surface layer (S-layer); d, dense layer. (Sleytr, 1978).
(b) Surface layer in a freeze-etched preparation of *Clostridium thermohydrosulfuricum* (Sleytr and Messner, 1983).
(c) Negatively stained preparation of a fragment of the cell wall from *Thermoanaerobacterium thermosaccharolyticum* showing heterologous reattachment of S-layer subunits isolated from the same organism and from *T. thermohydrosulfuricum* (Sleytr, 1978; Mayer, 1986)

plasmic space may be considerably expanded in several species, such as *Thermotoga maritima* (Fig. 2.1 d). Here, the periplasm also contains metabolic enzymes such as pullulanase (Fig. 2.1 d). Recombinant proteins may also then be deposited in an artificially widened periplasmic space, as depicted in Fig. 2.4 a.

Surface layers (S-layers) are widely distributed in most groups of bacteria and archaea. Usually they represent the outermost envelope, attached as a monolayer to the outer surface of the cytoplasmic membrane (in some archaea), on the peptidoglycan layer of Gram-positive bacteria, or the outer membrane of Gram-negative bacteria (Fig. 2.2). S-layers are regarded as very simple biological membranes, because they are usually composed of multiple copies of a single protein or glycoprotein species and assemble into a regular two-dimensional crystalline array through an entropy-driven process. Figure 2.2 c depicts two different types of S-layer proteins that have been reassembled on the peptidoglycan layer from the cell wall of an appropriate bacterial cell. The morphological unit cells of S-layers consist of two, four, or six monomers and form channels with diameters of approximately 2 – 3.5 nm. S-Layers help to maintain cell shape and envelope rigidity. They may act as molecular sieves, preventing influxes or effluxes of macromolecules exceeding molecular weights of 15 kDa. The anisotropic distribution of surface charges implies that the S-layers may be involved in ion-exchange mechanisms: they may trap useful minerals, help to maintain local pH, and promote specific surface interaction. S-Layers also play an interesting role in the organization of extracellular enzymes (see below). The self-assembly of S-layers into large two-dimensional sheets, their molecular sieve properties, and their intrinsic stability make S-layer proteins building blocks of a biomembrane with application potentials in various fields. S-Layers are currently in development for goals such as the enhancement of the stability of liposomes in drug delivery systems, as immobilization matrices for enzymes in biosensors, and as ultrafiltation membranes (Pum and Sleytr, 1999).

2.2.1.2 **Compartmentation of the bacterial nucleoid**

The bacterial nucleoid has turned out to be of highest complexity (Fig. 2.3). Its overall structure is subject to radical artificial rearrangement of components during chemical fixation procedures, so electron microscopic studies of nucleoid architecture turned out in the past to be very difficult and unreliable. Techniques working without the need for chemical fixation – mainly cryotechniques – allow more reliable conclusions to be drawn from electron microscopic observations in situ. In conjunction with interpretation of images obtained by fluorescence light microscopy, a coralline shape appears to be the most likely overall nucleoid structure (Fig. 2.3 a; Kellenberger and Kellenberger-Van-der-Kamp, 1994). DNA-binding proteins bring about compaction of the nucleoid, in which striking similarities were observed between eukaryotic DNA-compacting histone proteins and archaeal proteins. The eukaryotic nucleosome contains about 200 DNA bp, organized by an octamer of histones, forming the core of the complex. The DNA is wound around the particle surface. The nucleosome structure represents the first organization level (of three) of eukaryal chromatin. Complexes structurally similar to nucleosomes have been detected in several archaea (see Fig. 2.3 b).

The most prominent protein components of the eubacterial nucleoid are the histone-like proteins HU and H-NS. Immunolocalization of HU revealed that the protein is not associated with the bulk DNA in the nucleoid, but is instead located in areas of the cell where metabolically active DNA is associated with ribosomes and where single-stranded DNA, RNA polymerase, and DNA topoisomerase I are also present. Initially described as a transcription factor, the H-NS protein was later shown to be involved in DNA compaction. Like HU, H-NS is known to be involved in global regulation of gene expression. H-NS binds in almost all cases to naturally bent DNA. Both proteins, like nucleosomes, take part in compaction of DNA at the first organization level, but are at the same time deeply involved in regulation of the dynamics of the nucleoid (i.e., global regulation of gene expression).

At a higher organization level, the bacterial nucleoid is separated into independently supercoiled domains, about 100 per genome. The existence of separate domains could permit different degrees of supercoiling to be maintained in different regions of the genome. The domains must be organized into some kind of more compact fiber in order to be in accordance with the overall structural model as derived from electron microscopy (see Fig 2.3 c, d; Mayer and Friedrich, 1986).

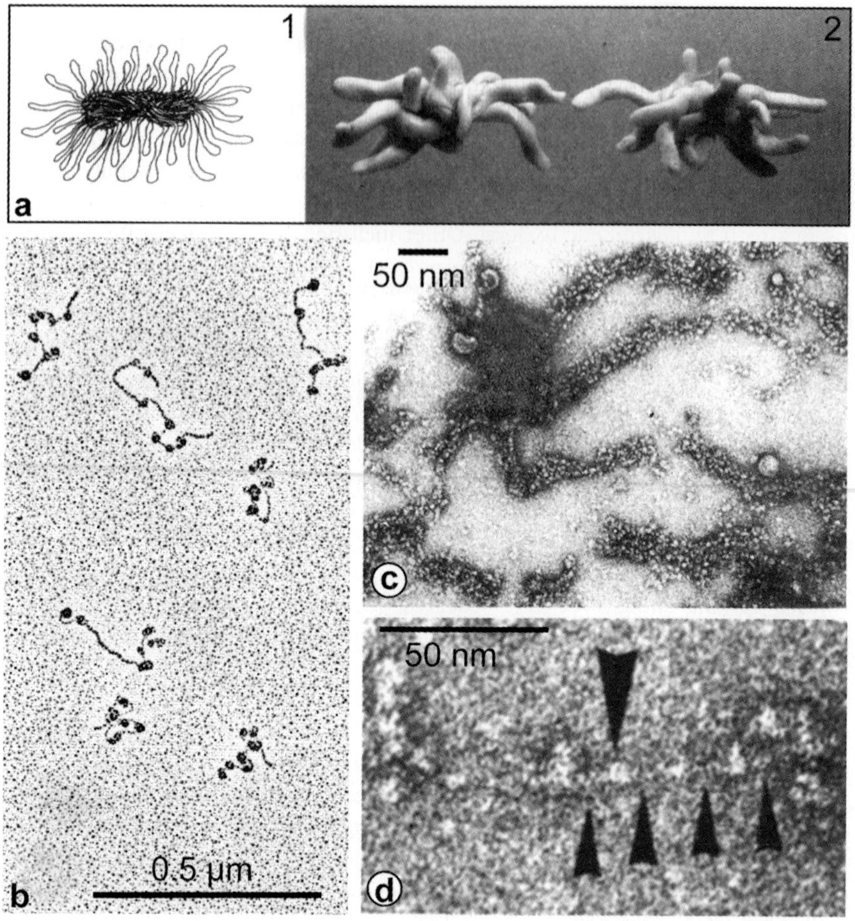

Fig. 2.3 Overall structure of the bacterial nucleoid.
(a) Models of bacterial nucleoids in metabolically active states. 1, Model A (Ryter and Chang, 1975) is deduced indirectly from the autoradiographically determined sites of mRNA synthesis. 2, Model B (Robinow and Kellenberger, 1994) is based on the observation of ribosome-free spaces in the cytoplasm, which were found by immunolocalization to contain DNA.
(b) Sample of HMf from *Methanothermus fervidus*, bound to linearized pUC 18 DNA (Sandman et al., 1990): isolated DNA is spread on a surface of a water drop, transferred to a support film and metal shadowed.
(c) Negatively stained sample from extruded protein-DNA complexes of *Thiosphaera pantotropha* (Mayer and Friedrich, 1986).
(d) Detailed view of a preparation similar to (c), depicting a thin DNA fiber (arrowheads) complexed with particles (large arrowhead) (Mayer and Friedrich, 1986)

2.2.1.3 Inclusion bodies

Numerous types of cytosolic inclusion bodies may be detected in the cytosol of the bacterial cell. Although some of the large inclusions are visible by light microscopy, elucidation of their structures requires an electron microscopic approach. The occurrence and frequency of inclusions are species-specific and depend on nutrient conditions. Cyanobacteria in particular contain a variety of intracellular inclusions. Inclusions of highest complexity, chlorosomes, bear the photosynthetic apparatus in Green photosynthetic bacteria. Other inclusion bodies contain the enzyme ribulosebisphosphate carboxylase (carboxysomes), gas (gas vacuoles), or magnetite crystals (magnetosomes in magnetotactic bacteria). In recombinant bacterial strains, overexpressed proteins and metabolites are deposited as dense granules as depicted in Fig. 2.4 a–d. Inclusion bodies of intracellular proteins are located in the cytosol, but proteins secreted into the periplasm may also accumulate in this compartment. Penicillin G acylase is located in the periplasm of *Escherichia coli*, overexpressed protein accumulating in inclusion bodies in the cytosol as well as in the periplasm. The enzyme is used for production of 6-aminopenicillanic acid in industrial processes. Insoluble metabolites are also deposited in the cytosol, as depicted in Fig. 2.4 d for accumulated indigo in a recombinant *E. coli* strain.

Storage granules containing complexed carbon as sources both of carbon and of energy, especially granules consisting of polyhydroxyalkanoates (PHAs), are widely distributed among all bacterial groups. The most common type of PHA is PHB (poly-β-hydroxybutyrate). PHAs are used as biodegradable plastic materials in various applications. They can be either thermoplastic or elastomeric materials, with melting points ranging from 40 to 180 °C). PHB has properties similar to those of polypropylene, but is stiffer and more brittle. In recombinant stains, the synthesis of PHAs may also be modified in various ways, and recombinant PHA-producing crop plants are in cultivation (see below). Whereas the inclusion bodies of recombinant proteins and indigo may be viewed as intracellular precipitates, without any further defined separation from the cytosol, the PHA granule is a well defined compartment. About 2 % of the granule mass is contributed by proteins that cover the surface of the granule in a highly ordered two-dimensional array). Of these proteins, PHA synthase and structural proteins, "phasins", have so far been characterized in greater detail (Steinbüchel et al., 1995). PHA synthases and phasins are two interesting examples of the specific association of proteins with storage granules at the borderline between the hydrophilic cytosol and a hydrophobic compartment (Fig. 2.4 e–g). PHA synthases appear to form the initial core of the growing PHA inclusion body by a process of emulsion polymerization, more and more of them becoming bound to the PHA granule upon onset of PHA biosynthesis. The enzymes catalyze the polymerization of the PHA monomers (e.g., hydroxybutyryl-CoA) to the respective polymer (e.g., polyhydroxybutyric acid), consisting of approximately 35,000 monomeric units (Steinbüchel et al., 1995).

Depiction of the activity of biogenic PHA depolymerization by SEM has already been shown above. Whereas the extracellular degradation of PHA, as exogenous substrate, is well studied, very little is known about depolymeration of intracellular (native) PHA from granules.

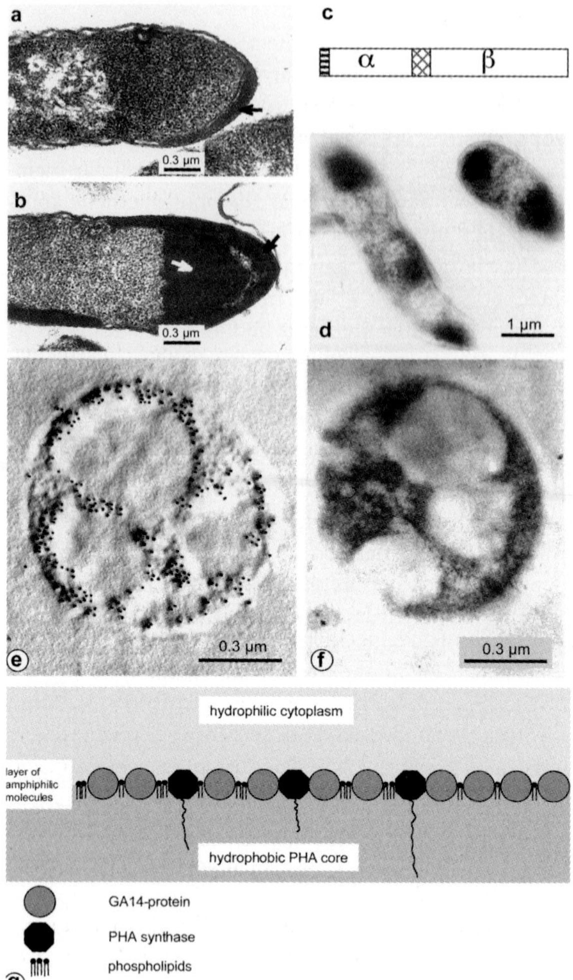

Fig. 2.4 Large bacterial inclusion bodies.
(a, b, c) Inclusions of Penicillin G acylase in *E. coli*. a, periplasmic inclusion body (arrow); b, cytoplasmic and periplasmic inclusion bodies (white and black arrow, respectively), visualized in ultra-thin sections of embedded samples; c, schematic structure of penicillin G acylase preproenzyme. The molecule is divided into four parts as indicated. The preproenzyme is processed by proteolytic cleavage of the signal peptide (yielding the proenzyme) that determines export of the enzyme through the cytoplasmic membrane into the periplasm. In a second step, the spacer peptide is cleaved, yielding the α and β subunits of the mature enzyme (Sriubolmas et al., 1997).
(d) Dark inclusion bodies of indigo in a recombinant strain of *E. coli* predominantly located at the cell poles (Drewlo et al., 2001).
(e) Immunolocalization of a phasin (GA14-protein), bound to the surface of polyhydroxyalkanoic acid (PHA) granules in *Rhodococcus ruber*. Immunogold-labeled ultra-thin section of *Rhodococcus ruber* (wild type) cells, metal-shadowed to give a three-dimensional impression of the section surface (Pieper-Fürst et al., 1994).
(f) Ultra-thin section of *Rhodococcus ruber* containing PHA granules, appearing as bright areas inside the cell (Pieper-Fürst et al., 1994).
(g) Model of the structure and composition of PHA granule surface in *R. ruber* (Pieper-Fürst et al., 1994)

2.2.2
Plant Cells

Visualization of ultra-thin sectioned cells can reveal the structure and complexity of the cell wall (if present) and intracellular membrane compartments. Figure 2.5 shows a view of a meristematic cell from the root tip of garden cress (*Lepidium sativum*). The cytoplasmic membrane can clearly be distinguished from the cell wall, the fibrillar texture of a plant cell wall becoming visible. The cytosol displays a characteristic granulation, similar to its appearance in bacteria, made up of ribosome particles. These are either bound to cisternae of the endoplasmic reticulum ("rough" endoplasmic reticulum) or are distributed in the cytoplasm. A typical feature of plant cells is the interconnection between adjacent cells through plasmodesmata: holes in the cell wall lined by membrane tubes. Cisternae of the endoplasmic reticulum pass through the plasmodesmata, linking the endomembrane systems of the neighboring cells. Typically for meristematic cells, the vacuoles are small. Mitochondria and plastids (here as proplastids) show their highly differentiated fine structures. Other membraneous compartments (microbodies and the Golgi apparatus) are also present. The nuclear envelope is surrounded by a double membrane. The nucleus is differentiated into a highly condensed region (nucleolus) and less condensed chromatin regions. The fine structure of these compartments may be resolved by various preparation techniques. Freeze-fracturing, together with freeze-etching of the specimen, allows visualization of the membrane surface: Fig. 2.6 a shows a view of the membrane surface of the nucleus, covered with nuclear pores, prepared by freeze-etching of a freeze-fractured specimen. The structure of nuclear pore complexes has been deduced from preparations as depicted in Fig. 2.6 b, a negatively stained sample from an enriched nuclear pore complex fraction. The organization of chromatin itself is depicted in Fig 2.6 c. DNA is bound to nucleosomes, proteins that appear like globules, arranged in a regular distance on a DNA molecule. There is an apparent overall similarity to the archaeal protein-DNA complexes depicted in Fig. 2.3 b.

The smaller membraneous compartments can be divided into the endoplasmic reticulum (ER), the Golgi apparatus (GA), and vesicles of various kinds such as peroxisomes, glyoxisomes, lysosomes, and secretory vesicles. The ER consists of interconnected cisternae and both tubular and reticular structures. Rough and smooth ER (rER and sER) is distinguished by the attachment or absence of ribosomes at the plasmatic side of the membranes. Figure 2.7 a shows extended rER cisternae in an (unusual) regular arrangement. Whereas the rER is the compartment for processing of nascent proteins synthesized by the attached ribosomes, the sER is the site of synthesis of membrane lipids, storage lipids, and other lipophilic components such as steroids.

The membrane flow interconnecting the ER, GA, and membrane vesicles has been reconstructed from observations of these compartments in cells with secretory activity (Fig. 2.7 b, Fig. 2.8). A continuous membrane flow occurs from the ER (and the nuclear envelope as part of the ER) to the GA and, finally, to the vacuole or the cellular exterior. This is mediated by vesicular transport between ER and GA and between GA and the outside or the vacuole. In this way, (glycosylated) proteins or cell wall polymers are transferred to the outside. Immunolocalization illustrates the directed transport of

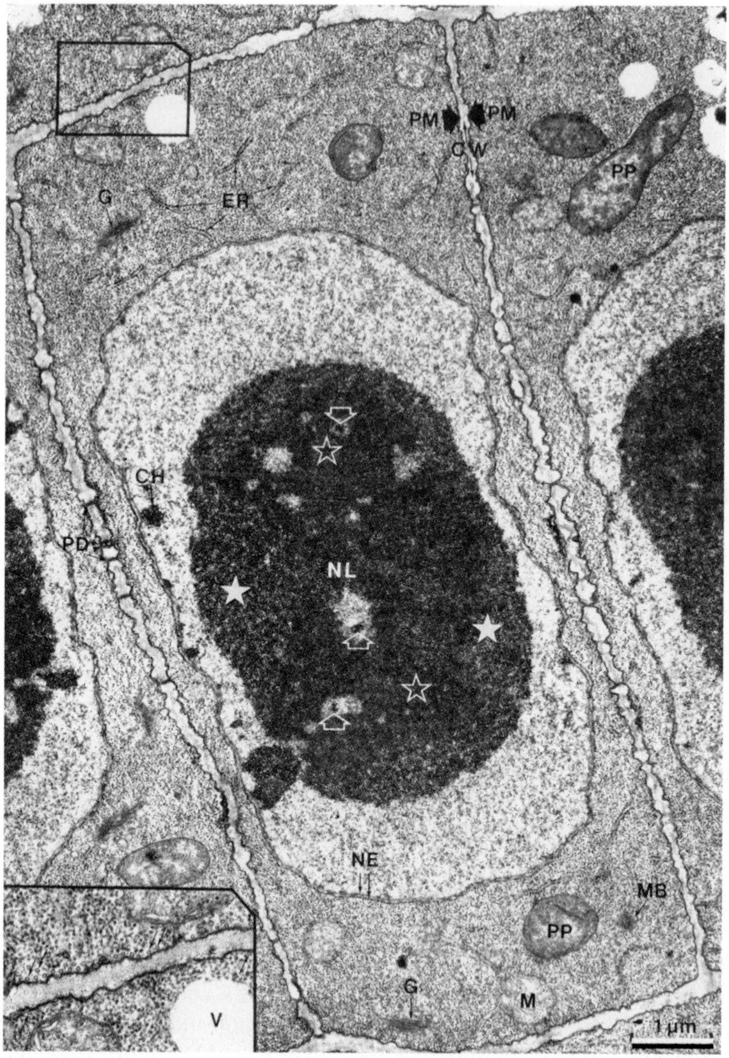

Fig. 2.5 Ultra-thin section of a meristematic cell from the root tip of *Lepidium sativum*. The electron microscopic view allows assignment of nearly all compartments by their morphological features. PM, cytoplasmic membrane; CW, cell wall; PD, plasmodesma; V, vacuoles; ER, endoplasmic reticulum; M, mitochondria; PP, proplastids; G, dictyosomes; MB, microbodies; NE, nuclear envelope; NL, nucleus; CH, chromatin; asterisks, granular areas; asterisks (outlines), areas consisting of thin fibrils (Gunning and Steer, 1996)

proteins and other compounds. This is documented for vicilin, a storage protein transferred to the vacuole. Detection of vicilin in the respective compartments implies that the protein is synthesized in the lumen of the rER, modified after transfer to the dictyosomes, and then deposited in the vacuole as depicted in Fig. 2.7 c. Vicilin is only detectable at the *cis* (directed towards the endoplasmic reticulum) side of the Golgi stack. The sorting of vicilin and the presence of vicilin in one type of Golgi vesicle (dense vesicles, storage protein-containing vesicles) has been only demonstrated in specimens processed by various cryotechniques (Hillmer et al., 2001).

The plant vacuoles are functionally diverse compartments, but often appear morphologically similar, differentiation between mature vacuoles and prevacuolar compartments in particular remaining difficult. In freeze-etched specimens, the membrane of a mature vacuole (tonoplast) shows a defined structure. Membrane pro-

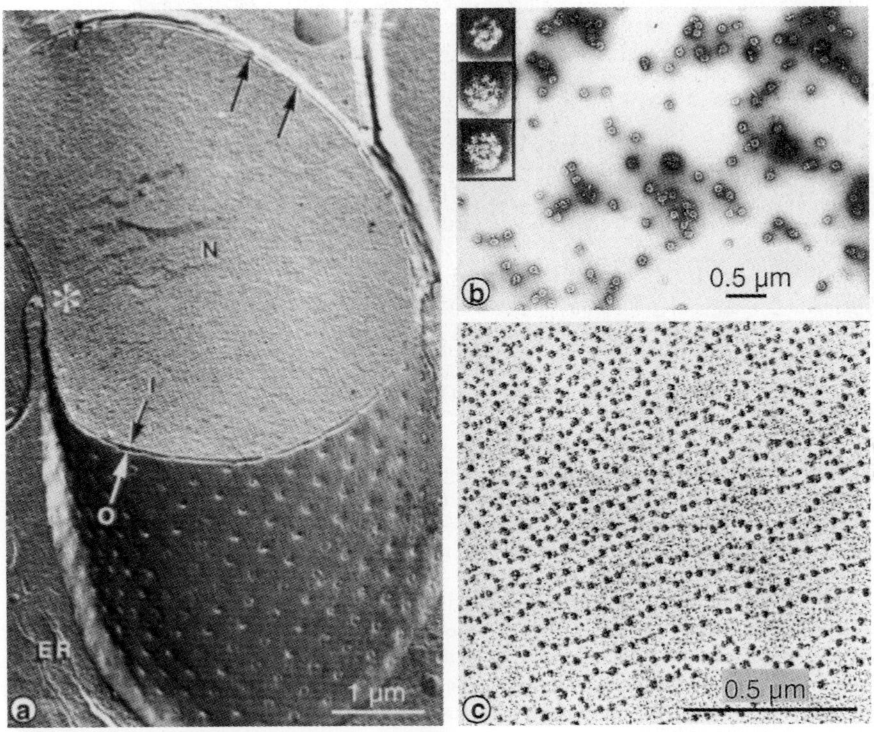

Fig. 2.6 The plant nucleus.
(a) Freeze-etched preparation of the nucleus from *Selaginella kraussiana*. The lower half of the image shows the surface of the nucleus with nuclear pores. O, outer nuclear membrane; I, inner nuclear membrane; N, nucleoplasma; arrows, side-view-on nuclear pores; asterisk, junction between nuclear envelope and ER (Gunning and Steer, 1996).
(b) Electron micrograph of negatively stained particles from a highly enriched, detergent-solubilized yeast nuclear pore complex fraction. Inset: three examples at high magnification (diameter of each particle is 90 nm) (Rout and Blobel, 1993).
(c) Rotation-shadowed sample of DNA spread on the surface of a water drop. The DNA is folded to nucleosomes in distinct distances by histones (Gunning and Steer, 1996)

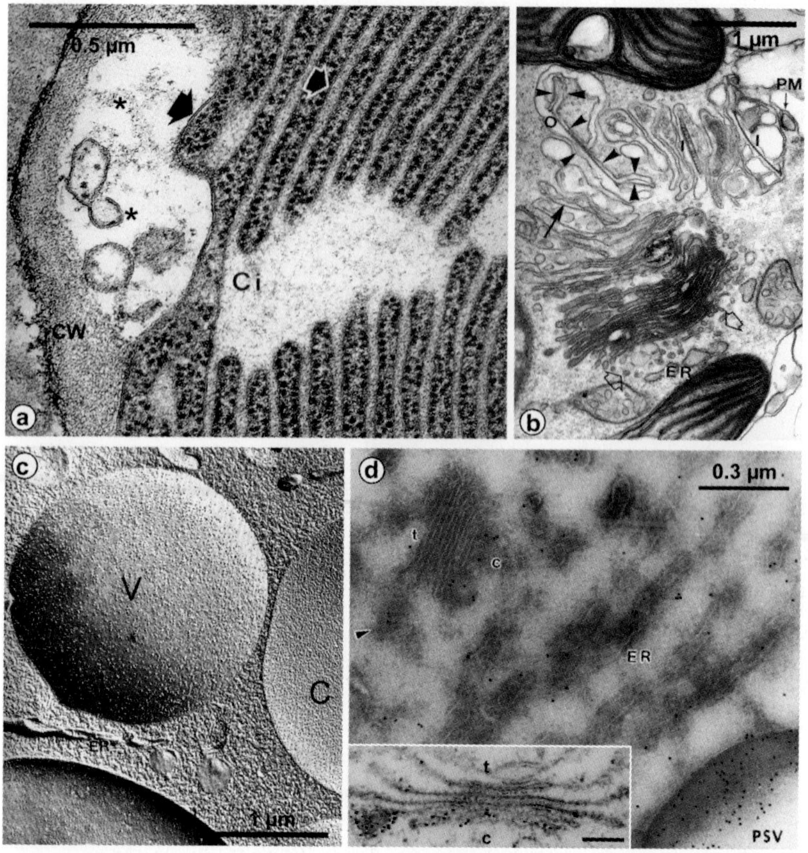

Fig. 2.7 Endoplasmic reticulum, Golgi apparatus, vacuoles, and traffic between the compartments.
(a) Cisternae from the endoplasmic reticulum (ER) trichome cells from *Coleus blumei*. The membranes of the ER (black arrow, edged in white) show a unit membrane structure as well as the cytoplasmic membrane (black arrow). Dark dots represent ribosomes at the outside of the ER. Ci, lumen of the ER cisternae; CW, cell wall; asterisks mark a lacuna of unknown function between the cytoplasmic membrane and the cell wall (Gunning and Steer, 1996).
(b) Development of cell wall scales in a dictyosome (a stack of the Golgi apparatus) from *Chrysochromulina chiton*. From the *cis* to the *trans* side of the stack, various stages of maturity are noticeable. The arrow depicts an earlier stage, whereas the mature scales are on the *trans* side and close to the cytoplasmic membrane (arrowheads). ER and transitional vesicles are only on the *cis* side of the dictyosome (arrow outlines). Two types of scales can be discerned: O, outer, I, inner scales; PM, cytoplasmic membrane; ER, endoplasmic reticulum (Gunning and Steer, 1996).
(c) Freeze-etched preparation of a vacuole. The fracture plane runs along the inner face of the tonoplast (which is a unit membrane). The convex and concave surfaces are equivalent to the vacuolar (V) cytoplasmatic (C) half of the tonoplast. Numerous small particles at the fracture plane represent transmembrane proteins (Gunning and Steer, 1996).
(d) Localization of vicilin in cryosections (large image) and high-pressure frozen specimens (inset). The endoplasmic reticulum (ER), dense vesicles (arrowhead) in close vicinity to the Golgi stacks, the Golgi stack itself (c, *cis* side; t, *trans* side) and in the vacuole (V) of a pea (*Pisum sativum*) cotyledon is labeled with immunogold markers. Especially in high-pressure frozen specimens, only the *cis* side of a Golgi stack, together with a dense vesicle, is heavily labeled, indicating the presence of vicilin. Directed trafficking of the protein from ER to vacuole via dictyosome and dense vesicles may be inferred (Hillmer et al., 2001)

teins, mainly involved in the transport of ions, are visible on a fraction plane transmitting the inner leaflet of the tonoplast (Fig. 2.7 c).

Cytoskeletal elements appear as networks made up of tubular or filamentous structures (microtubuli, microfilaments; Fig. 2.9). Microfilaments, occurring as single filaments or bundles, are abundant in cells with intensive intracellular transport. The section depicted in Fig. 2.9 is approximately tangential to the cylindrical cell. The cortical (adjacent to the cytoplasmic membrane) microtubuli appear as short hollow tubules, orientated approximately parallel to each other.

A view of the wall surface, prepared by freeze-fracturing and freeze-etching of the sample, is depicted in Fig. 2.10 a. The primary wall shows a parallel arrangement of cellulose fibrils. During growth the regular arrangement is disturbed, as depicted in the metal-shadowed specimen (Fig. 2.10 b). The principal model of primary cell wall genesis has mainly been reconstructed from electron microscopic observations (Emons and Mulder, 2000). A schematic drawing (Fig. 2.9 b) summarizes the essential data. Cortical microtubuli are arranged at the plasmic surface of the cytoplasmic membrane, and influence the direction of the newly synthesized cellulose fibrils in the wall (see Fig. 2.10 c). Cellulose fibrils appear to be synthesized by rosette complexes, as depicted in Fig. 2.10 d.

The fine structures of the large organelles are depicted in Fig. 2.11. Mitochondria and chloroplasts exhibit two envelope membranes of the unit-membrane type. The inner membrane of a mitochondrion is invaginated, forming membrane cristae (Fig. 2.11 a). Two sections show – in side view and face-on view – chloroplasts of *Zea mays* and *Avena ventricosa*, respectively (Fig. 2.11 b, c). Thylakoids appear either

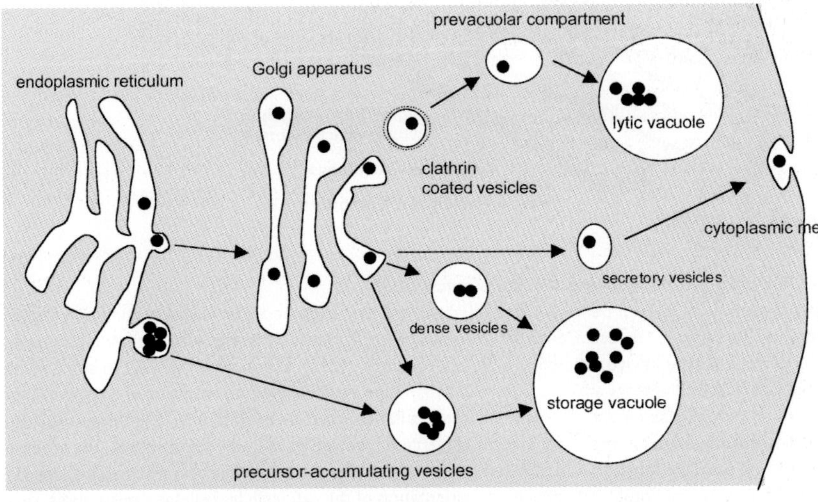

Fig. 2.8 Simplified view of some routes of soluble proteins (black dots) to vacuoles and the cellular exterior, which may be deduced from electron microscopic investigations as presented in Fig. 2.7. Transport between the large compartments is mainly achieved by small vesicles of different types (as indicated) allowing directed targeting to the respective compartment (after Hawes et al., 1999; Vitale and Raikhel, 1999, with modifications)

as disc-shaped structures in face-on view and as stacks in side view. After differentiation to amyloplasts, the plastid lumen is filled with large starch grains (Fig. 2.11 d). A developing chromoplast (from a tomato fruit) is depicted in Fig. 2.11 e. The chromoplast shows various inclusions. Whereas plastoglobuli (dark-stained, lipid-containing inclusions) are also present in other plastid types, the pigment crystals (lycopene in the

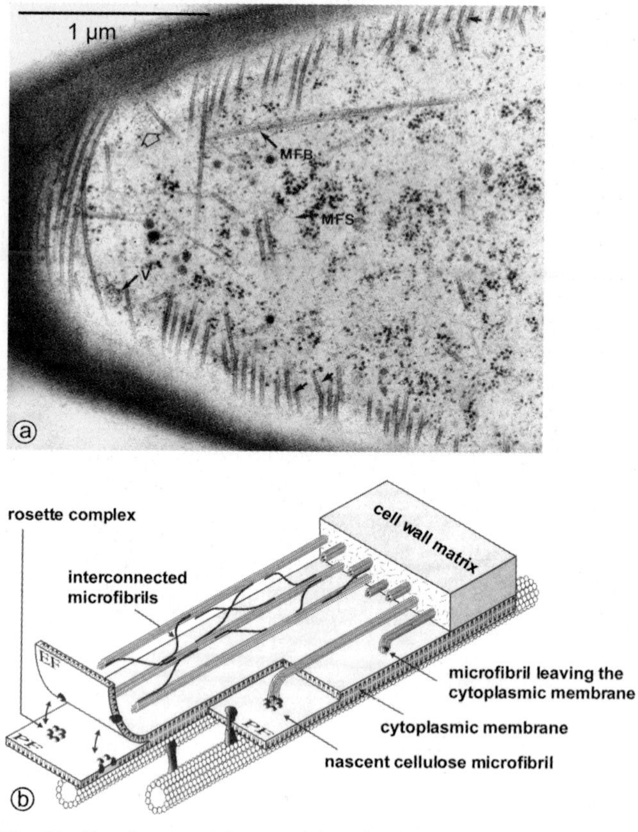

Fig. 2.9 The plant cytoskeleton and the role of microtubuli.
(a) Ultrathin section of the cell periphery from a wheat leaf hair, prepared after deep-freezing and freeze-substitution of the sample. The occurrence of numerous microfilaments correlates with the fast cytoplasmic streaming of this cell type in vivo. Microfilaments appear singly (MFS) or in bundles (MFB). In some areas microtubuli-associated filaments are visible (arrowheads). One vesicle shows an envelope of clathrin (V). In the top view on the cytoplasmic membrane there is also a lattice structure of clathrin visible (arrow outline).
(b) Cortical microtubuli are an important structural element involved in cell wall biosynthesis. The scheme summarizes one possible functional model that has also been deduced from electron microscopy (see (a) and Fig. 2.10). Cortical microtubuli determine the orientation of the extracellular cellulose microfibrils, perhaps by influencing the positioning of the transmembrane rosette complexes (cellulose synthase). These complexes are visible after freeze-fracturing at the plasmic face (PF) of the cytoplasmic membrane. At the plasmic face, globules obviously interconnected with the rosette complex in the intact membrane are visible. EF, extracellular face.
(Gunning and Steer 1996; see also Emons and Mulder, 2000)

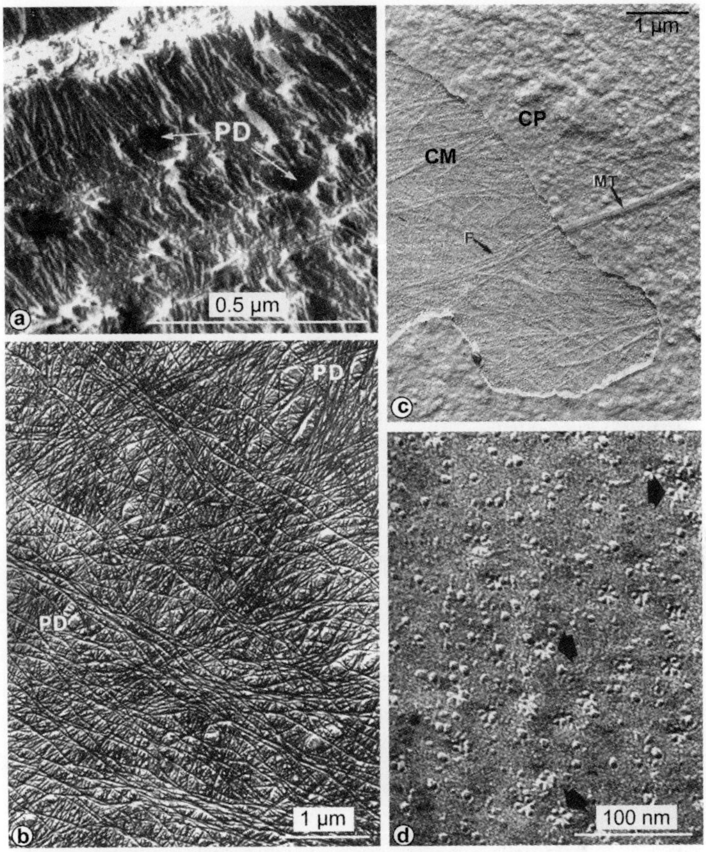

Fig. 2.10 Structure of the primary wall.
(a) Freeze-etched preparation, top view of parts of the cell wall from a leaf cell of *Asparagus*. Layers of microfibrils are visible. PD, plasmodesma.
(b) Metal-shadowed preparation of the primary cell wall from a growing cell in the ground tissue (pith cell) in *Ricinus*. The matrix material was extracted, so the cellulose microfibrils are clearly visible. The recently deposited fibrils forming the uppermost layer still exhibit an orientation parallel to the long axis of the cell, while fibrils deposited earlier are deflected from this orientation, caused by cell growth. PD, plasmodesma.
(c) Freeze-etched preparation of a root cell from *Trianea bogotensis (Limnobium stoloniferum*, frogbit). The fracture shows different layers of the cell envelope with a view from the inside of the cell. Cortical microtubuli (MT) are embedded in a thin cytoplasmic layer (CP). In the area marked with CM, the fracture has exposed the extracellular face of the cytoplasmic membrane. Imprints of numerous microfibrils (F), located outside the cytoplasmic membrane, are visible. The microfibrils are oriented parallel to the microtubulus.
(d) Freeze-fracture preparation of the root from *Lepidium sativum* (cress). The fracture runs between the two phospholipid layers of the cytoplasmic membrane, the view is from the outside of the cell. Membrane particles are visible as single globules or in irregular groups of two or three, as well as rosettes of six particles (arrows). These complexes represent the cellulose synthase.
(Gunning and Steer, 1996)

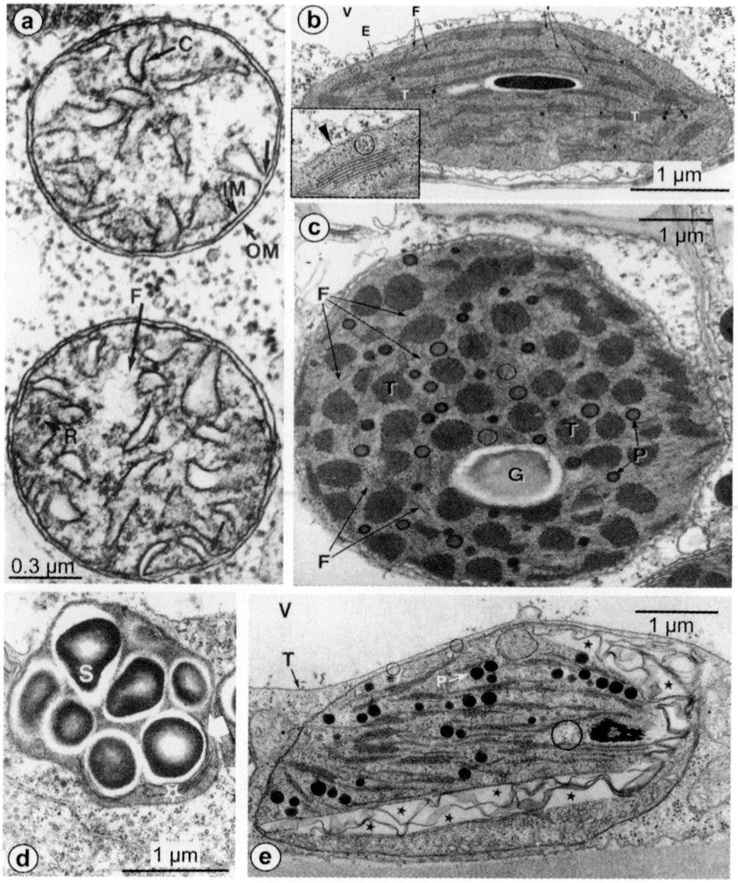

Fig. 2.11 Structure of mitochondria and plastids.
(a) Mitochondria in a parenchymatic cell from *Vicia faba* (faba bean). Each mitochondrion is surrounded by an outer (OM) and inner (IM) membrane. The inner membrane is invaginated (arrow) and forms cristae (C). In the matrix of some mitochondria the nucleoid can be recognized as a bright area containing fibrillar material (F). The mitochondrial ribosomes (R) are located in the darker stained areas of the matrix.
(b, c) Side view and face-on view of chloroplasts (a, *Avena ventricosa*; b, *Zea mays*). The double membrane of the chloroplast envelope is visible (in b: E and marked by an arrowhead in the inset); thylakoid grana stacks (T) are visible in side view (b, inset) and face-on view (c); dark stained ribosomes in the stroma (circles); starch granules (G); stroma thylakoids ("frets" F), connect the grana; plastoglubuli (P).
(d) Amyloplast from a root cell from the soybean (*Glycine*) with starch granules (S). A few stacks of thylakoids (asterisk, arrow) are oriented around the starch granules.
(e) Developing chromoplast from a tomato fruit. During the course of chromoplast development, the grana will disappear, leaving plastoglobuli and lycopine crystals (precursor of α-carotene). The crystals are extracted during the embedding procedures, leaving empty membraneous cavities with angular contours (asterisks); artificial membrane rearrangement may also take place. The stroma contains ribosomes and a nucleoid region (big circle) V, vacuole; T, tonoplast.
(Gunning and Steer, 1996)

tomato fruit) are characteristic for chromoplasts, but difficult to prepare by conventional electron microscopy. The crystals have been dissolved during the dehydration procedure used on the specimen; their location, size, and outer shape are marked by empty cavities (indicated by asterisks). The structural organization of mitochondria and plastid membranes has mainly been elucidated by electron microscopy, as documented here in some examples for the structure of the thylakoid membrane and membrane-bound enzyme complexes. Figure 2.12 depicts the overall distribution of large membrane-bound complexes in a freeze-fracture of thylakoid membranes. The structures of the large complexes themselves, down to the resolution of single domains, have been obtained by averaging techniques with single particles and ordered crystals of the complexes, as depicted here for plastidal ATPase (Fig. 2.12 b) and the photosystem II complex (Fig. 2.12 c–e). The organization of thylakoid membranes as shown in the schematic drawing summarizes the essential data relating to distribution and domain structure of the particles that may be obtained by electron microscopy.

The structures of the narrow contact regions between cells are only accessible at high resolution with transmission electron microscopes. The close connection between a sieve tube of the phloem and a companion cell is depicted in Fig. 2.13 a. The sieve tube itself appears to be nearly empty, since neither ribosomes nor nucleus are present, though specially adapted plastids and mitochondria are in evidence (not shown). Plasmodesmata are viewed as partially branched tubes between both cells. The amorphous polysaccharide deposit of callose may be clearly distinguished from darker stained, fibrillar network of the cellulose wall. A close-up view of a plasmodesma reveals the continuity of the endoplasmic reticulum between two adjacent cells, known as desmotubulus (2.13 b, e). A sieve plate connecting the sieve vessels in longitudinal direction is depicted in Fig 2.13 c. The plasmodesmata (sieve pores) are difficult to visualize in their native state, because their structures change during wound reaction in milliseconds. A unique filamentous protein (P-protein) is released from its anchoring sites after injury to the phloem, and accumulates at sieve pores because of the hydrostatic pressure of the sieve tube sap, blocking the pores and preventing assimilate loss at the injury site. Such aggregations of P-protein, sometimes also called "slime bodies" are usually formed during the processing of phloem for microscopy. Since plasmodesmata are the only gaps in a rigid plant cell wall, they are also the channels for spreading of plant viruses. The channeling is enabled by viral movement proteins that interact with the cytoskeleton and with the ER of the plant cell and direct the nucleic acid or the virus particle through the desmotubulus. Movement proteins induce an increase in the plasmodesma size exclusion limit and allow the spread of the infective agent from cell to cell. Whereas the overall mechanism is still unknown, the movement proteins are localizable in the plasmodesma by immunolocalization (Fig. 2.13 d).

The figures illustrate the crucial function of plasmodesmata. On the one hand, they provide continuity of the cytoplasm (the symplast) in the plant and allow rapid transport of nutrients through sieve vessels, but on the other hand they are also potential gates for the spreading of infective virus particles. Quick sealing of plasmodesmata is mainly achieved through the action of heavy mechanical or chemical sensors, preventing – for instance – severe infections upon rupture of a plant organ, but spreading of

Fig. 2.12 Large protein complexes of the thylakoid membrane ▶
(a) Freeze-fracture preparation of thylakoids from Pisum. Particles with variations in size and density are visible on the thylakoid fracture faces. PFG , EFG, membrane faces of grana, PFS, of stroma regions (see Section 2.4.5.7, Fig. 2.41 for freeze-fracture nomenclature) Grana stack membranes exhibit large particles (EFG) on one face and imprints of these particles as well as smaller particle types at the complementary PFG face. Stroma thylakoids are decorated with smaller particles (Kleinig and Mayer, 1999).
(b) Side view of the FOF1 ATPase from chloroplasts exhibiting distinct subunit structure, such as the "orange-shaped" head (11.4×12 nm), representing the F1 complex, which is composed of two types of elongated subunits (α and β), and two stalks, interconnecting the head with the membrane-integrated FO complex of the molecule (Böttcher and Gräber, 2000).
(c) Photosystem II-enriched thylakoids membrane preparation with densely-packed particles. Large arrowheads mark contact regions between ordered and non-ordered parts of the membrane. The ordered array consists of two separate crystalline areas (borderline marked by arrowheads) (Stoylova et al., 1998).
(d, e) Three-dimensional surface-rendered representations of PSII complexes, obtained essentially by analysis of two-dimensional protein crystals as depicted in c), but from a frozen-hydrated preparation. Positions of the main domains are marked I – IV, a central cavity is marked by an arrow; d) face-on, e) side view. The putative position of the lipid bilayer is marked by two lines (Stoylova et al., 1998).
f) Schematic representation of thylakoid membranes. The distribution and shapes of the molecules have essentially been estimated after results obtained by electron microscopic specimens as depicted in Figs. 2.11 and 2.12 a–d (Allen and Forsberg, 2001)

virus particles or minute changes in phloem sap pressure (caused by the action of sucking insects) are not sensed.

2.2.3 Animal Cells

Animal cells contain the same compartments – with some variations – as plant cells. Plastids are missing, and vacuoles are smaller and usually occur as lysosomes. The lack of a cell wall causes differentiation of the cell surface, rare or absent in plant cells. Cytoskeletal elements are more involved in maintenance of the overall cell structure. Their structures, as resolvable by electron microscopy, are thus described here in more detail. The fine structures of the elements have mainly been revealed by electron microscopy. Figure 2.14 a and b depicts cross- and longitudinal sections of microvilli from mouse small intestine epithelium. In cross-sections, filament bundles of actin are visible as small dots inside each microvillus. Longitudinal sections show filaments arranged parallel in each microvillus. The bundles reach into the cellular lumen, where they interconnect with other cytoskeletal elements. After extraction of all other cellular components, the cytoskeletal network becomes directly visible. Figure 2.14 c depicts a small area of a detergent-extracted cell after critical point drying in a high voltage TEM. Microtubuli, intermediate filaments, actin filaments, and smaller filaments are shown. Intermediate (10 nm) filaments may be further separated into different groups, specific for certain cell types (some groups are not found in plants, fungi, and protists). For further ultrastructural studies, the isolated filamentous structures may be visualized by application of negative staining techniques. Figure 2.14 d shows actin filaments isolated from a skeletal muscle. Together with myosin, actin

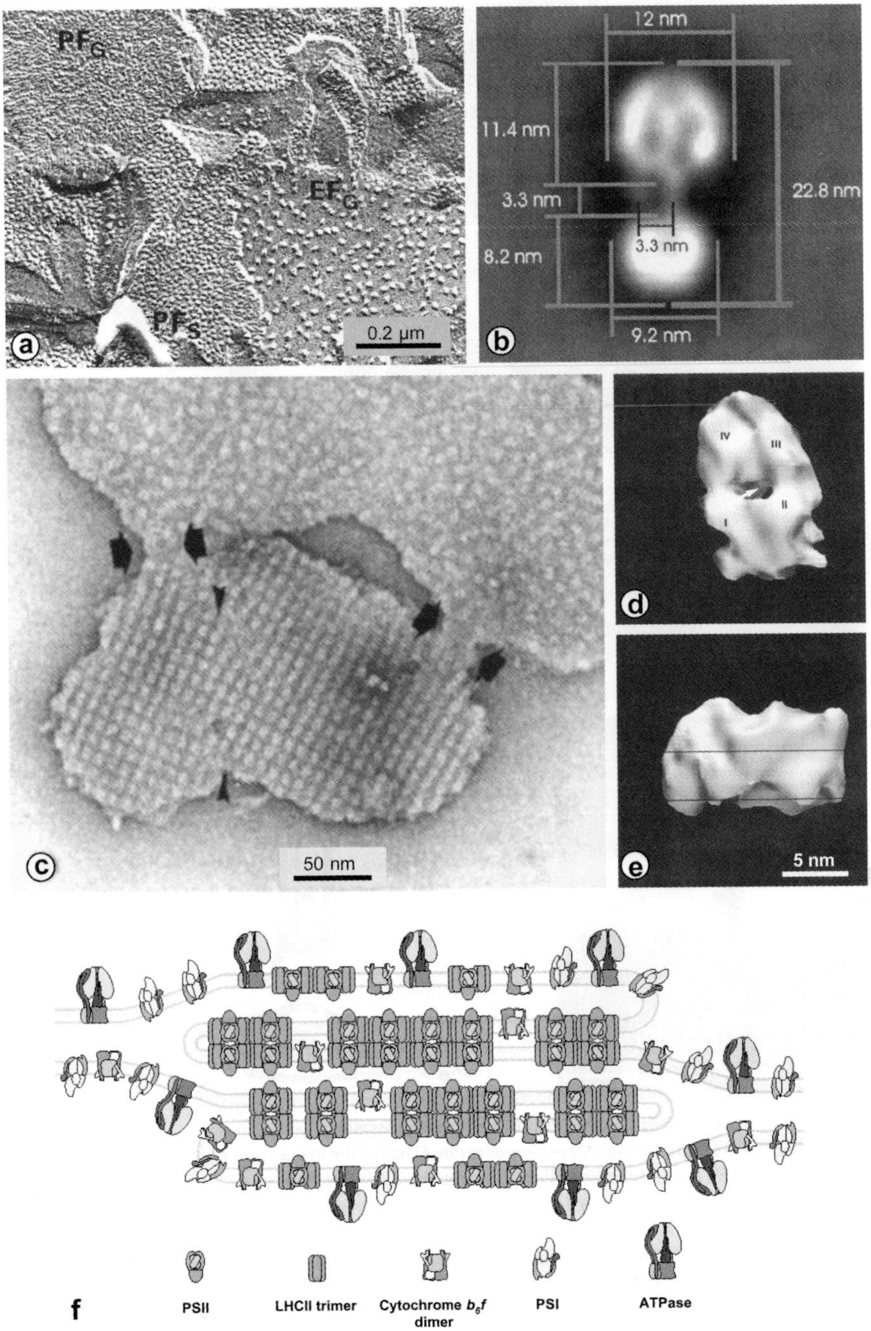

PF$_G$
EF$_G$
PF$_S$
0.2 µm
a
12 nm
11.4 nm
3.3 nm
3.3 nm
22.8 nm
8.2 nm
9.2 nm
b
c
50 nm
IV
III
II
I
d
e
5 nm
f
PSII
LHCII trimer
Cytochrome b_6f dimer
PSI
ATPase

progress regarding the fine structure has been made by image analysis of the filamentous elements, as illustrated by Fig. 2.14 e (inset) and Fig. 2.14 f. The relevant techniques, which are explained in the Methods Section (see chapter 2.4.9.4), allow modeling of the interaction between the filament complexes and accessory proteins, which is not possible by other techniques such as X-ray structural analysis.

2.2.3.1 Structural aspects of the endocytic and exocytotic pathways

The lytic vacuoles of differentiated plant cells usually exist in animal cells in the form of (smaller) lysosomes. As in plant cells, the lysosomal compartments are components of the endocytic pathway. The pathway has been intensively investigated, especially in animal (and human) cell lines, due to its importance for the uptake and processing of viruses, bacteria, pharmaceuticals, toxic compounds, and all processes of cellular communication. Here, to illustrate the application potential of electron microscopy, only some features are summarized. Since various stages of lysosomal activity are often visible – especially in phagocytic cells – their function can be deduced from interpretation of electron microscopic data. Large cell organelles, once internalized in vesicular structures known as endosomes or autophagic vacuoles, can be completely degraded by fusion of the endosomes with lysosomes. Phagosomes engulf large particles (such as bacteria) from the cellular exterior, small molecules are ingested by (receptor-mediated) endocytosis, after which micropinocytotic vesicles and multivesicular bodies are formed. These transient vesicles and lysosomes develop into secondary lysosomes, the central "digestive vacuole" in animal cells. Of the various types of lysosomal compartments, here a secondary lysosome is depicted. Lysosomal morphology varies with the state of the cell and its degree of degradative activity. Pieces of membranes, granules, and parts of mitochondria inside the lysosome are detectable by TEM (Fig. 2.15 g). A schematic view of lysosome development is depicted in Fig. 2.15 l. The events depicted in this "classical" drawing may all be deduced directly from ultra-thin sectioned cells, although the detailed mechanisms of internalization, recycling of the mediating receptor, the fusion events, and the presence of hydrolytic enzymes during the stages of lysosome development have been elucidated by application of various localization techniques at the TEM level. Some prominent types of vesicles may be identified without specific markers. Figure 2.15 a depicts a membrane invagination covered with a rim consisting of clathrin at the cytoplasmic side (arrowheads). The coated pit (and the closed coated vesicles, Fig. 2.15 b) are the well known vehicles of receptor-mediated endocytosis. Identification of specific vesicular contents, however, requires the application of specific markers. Figure 2.15 c depicts the fate of liposomes after invagination by an endothelial cell of a blood vessel. The liposomes themselves may not be identified in a section, but they are tagged with small lipid-gold particles (nanogold particles coupled to lipid) as marker systems. Their targeting to vesicles of the endocytic pathway may be followed with the aid of these markers. After invagination, the liposomes are enclosed in a multivesicular body (see also Fig. 2.15). The liposomes are tested for potential targeting of therapeutic agents to angiogenic blood vessels in tumors (Thurston et al., 1998). The pathways of specific receptors during endocytosis may also be documented with marker techniques. Figure

2.15 d–f depict the initial stages of epidermal growth factor receptor (EGFR) uptake from the cytoplasmic membrane by coated vesicles. EGFR is a signal transducer for a wide range of cellular signaling pathways (cell proliferation, cell migration, apoptosis, and others). EGFR has served as a model for study of ligand-induced receptor-mediated endocytosis. Binding of the ligand (epidermal growth factor, EGF) results in down-regulation of EGFR due to rapid internalization of the receptor through the clathrin-coated pits. After uptake, the receptor will be either recycled to the cytoplasmic membrane or targeted to lysosomes. The internalization proccss induced by the action of EGF is depicted in Figs. 2.15 d and e. In Fig. 2.15 f, a double labeling experiment reveals the co-localization of a ligand (eps15) with EGFR observed on a pit at the cell surface (Torrisi et al., 1999).

In analogy with investigation of endocytosis, immunoelectron microscopic techniques have revealed the participation of important key processes in exocytosis pathways. Only selected experiments exemplifying the localization experiments are presented here. Thanks to their high relevance in biomedical research, compartmentation of insulin-regulated enzymes and transport proteins underlying the action of insulin have been investigated in some detail. Here, some aspects of the compartmentation of the glucose transporter GLUT4 are summarized. This molecule is predominantly found in adipose (fat) cells, as well as in skeletal and heart muscle cells. In insulin-induced cells, GLUT4 becomes redistributed from intracellular compartments (Fig. 2.15 h) to the cytoplasmic membrane (Fig. 2.15 i). Immunolocalization, as depicted in Fig. 2.15 h, was applied to mechanically disrupted cells in which the exposed cellular exterior was held in place due to the preparation conditions. The technique maintains the three-dimensional structure of vesicles (Ramm et al., 2000). Various tubular and vesicular compartments contain GLUT4. The compartments originate from the Golgi apparatus or are endosomes involved in recycling of the protein. The vesicle-associated membrane protein VAMP2 co-localizes in these compartments (Fig. 2. 15 h). VAMP2 ("v-SNARE" partner) participates in the formation of a fusion complex with the cytoplasmic membrane targets ("t-SNARE" partner). The localization of GLUT4 at the cytoplasmic membrane in insulin-induced cells is depicted here in stimulated muscle fibers. Stimulation (i.e., induced contraction) enhances the intensity of labeling at the cytoplasmic membrane (due to the need for glucose inside the cell). Immunolocalization was performed with silver-enhanced nanogold particles linked to antibody Fab fragments prior to embedding of the cell in resin. This results in an amplification of marker deposition, compared with other techniques (Ploug et al., 1998).

2.2.4
Interaction between Cells in Animal Tissue

In animal cells, intercellular contact is mediated by desmosomes, tight junctions, and gap junctions. As indicated in Fig 2.16 a, the contact zones are of some structural complexity. The contact zones of these cells from mouse small intestine epithelium consist of a desmosome, the zonula adhaerans, and a tight junction. As deducible from

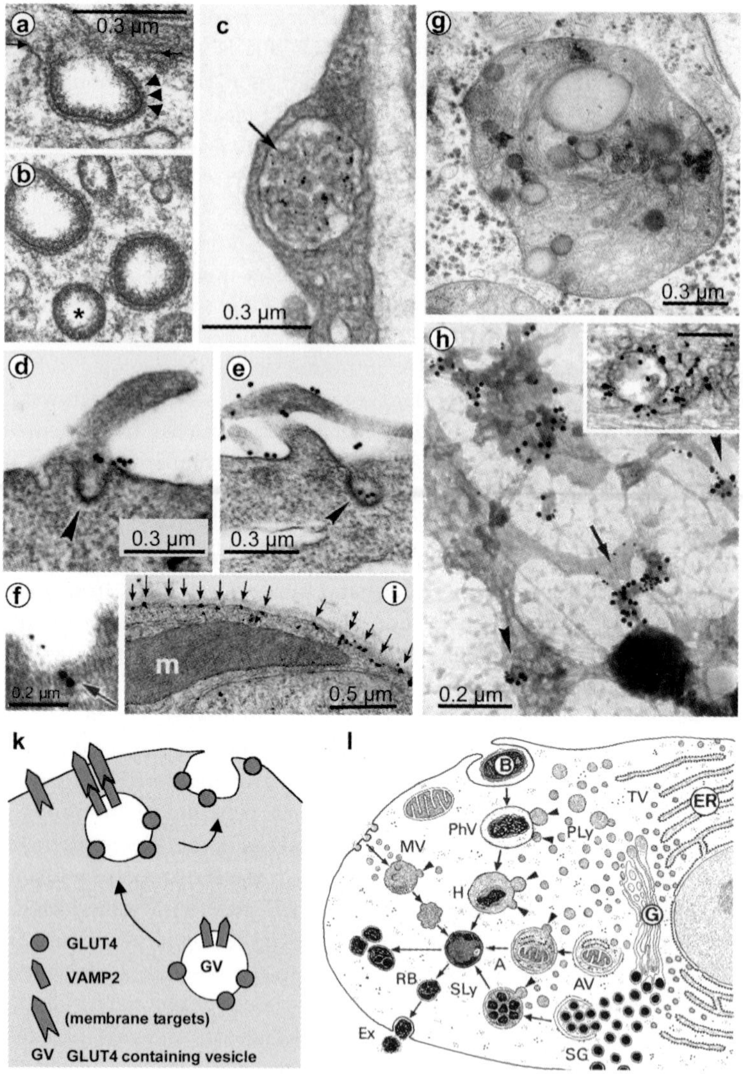

electron micrographs, there are tight interconnections between these structures and cytoskeletal elements (actin, intermediate filaments). Zonula adhaerans and desmosomes are mainly responsible for mechanical interconnection of cells, whereas tight junctions act as a seal against the cellular exterior. Freeze-etching and negative staining of gap junctions and gap junction protein (connexon) preparations show arrays of hexagonally arranged particles (Fig. 2.16 c, e). Electron crystallographic analysis has revealed that gap junction proteins are transmembrane proteins with central channels allowing molecules with a molecular weight of up to 1,500 Da to pass from cell to cell (Fig. 2.16 f). A variety of gap junction types are present in vertebrates, all involved in

Fig. 2.15 Structural aspects of endocytosis and exocytosis in ultra-thin sectioned cells
(a, b) Two stages of coated vesicle development. A coated pit (a) at the cytoplasmic membrane (small arrows). Coated pits/vesicles originating from the cytoplasmic membrane are covered with a clathrin coat (arrowheads). Two coated vesicles and a uncoated one (asterisk; after loss of the clathrin coat) in the cytoplasm. Cells from a frog kidney tubulus (Bucher and Wartenberg, 1989).
(c) Multivesicular body (arrow), containing small dark dots of a specific marker system. The gold-lipid conjugate used as marker indicates the presence of liposomes (or their remnants) in the compartment. For detection, silver enhancement (deposition of silver grains around gold particles) was applied (Thurston et al., 1998).
(d, e, f) Localization of the epidermal growth factor receptor (EPSR) in epidermal growth factor (eps15) treated cells before (d) and after (e) warming of the cells from 4 °C to 37 °C. Exclusion from coated pits (d) and engulfment (e) may be clearly discerned. Coated pits are marked by arrowheads. (f) Co-localization of EPSR (large gold particles, arrow) and eps15 (small gold particle) in a pit (Torrisi et al., 1999).
(g) Secondary lysosome from a neurosecretory cell of *Periplaneta americana* (American cockroach). The secondary lysosome contains numerous inclusions subjected to decomposition (Ude and Koch, 1994).
(h) Immunolocalization of GLUT4 compartments in a whole-mount preparation of a rat adipocyte (fat cell). Large gold particles are markers for GLUT4, small particles for VAMP2. both markers are found in various types (vesicular: arrowheads, tubular, arrows) of compartments. Inset: localization of Glut4 in an endosome of a skeletal muscle cell by application of a specific Antibody Fab-fragment conjugated to 1.4 nm gold particles prior to embedding and sectioning (Ramm et al., 2000; inset: Ploug et al., 1998).
(i) Immunolocalization of GLUT4 in insulin-stimulated muscle cells after induced contraction (technique as described for (h), inset). The specific marker is deposited at the cytoplasmic membrane as indicated by arrows (Ploug et al., 1998).
(k) Simplified representation of GLUT4 redistribution upon insulin induction of cells, illustrating the immunolocalization data (h, i). GLUT4 vesicles fuse with the cytoplasmic membrane. VAMP2 participates in the formation of a fusion complex with the cytoplasmic membrane t-SNARE.
(l) Structural aspects of lysosome development. ER, endoplasmic reticulum; TV, transport vesicle; G, Golgi apparatus; PLy, primary lysosome; B, bacterial cell; PhV, phagocytic vacuole; H, heterophagosome; SG, secretory granula; AV, autophagic vacuole; A, autophagosome; MV, multivesicular compartments/endosomes; RB, residual body; Ex, exocytosis (Kristi, 1984)

cell-cell interaction, from exchange of electrochemical signals to metabolite transfer. Consequently, various diseases (such as skin disease, hearing loss, cataract formation) are connected with the impairment of mutated gap junction proteins.

2.2.5
Enzyme Complexes In Vivo and In Vitro

Transmission electron microscopy is a versatile tool for imaging of subcellular components down to the size of small proteins. Though enzymes may be indirectly detected by localization techniques (see Fig. 2.17 a) or sometimes also directly visualized in situ (Fig. 2.17 d, e), in vitro studies are necessary for elucidation of their fine structure (symmetry, subunit composition, domain structure, etc.). Cryotechniques allow examination of small samples in a near-native state, embedded in a layer of amorphous (i.e., non-crystalline) ice. After fractionation of cells, all subcellular compartments may be further characterized in TEM preparations. Non-cellular small particles such as viruses may also be directly visualized after concentration of an environmental sam-

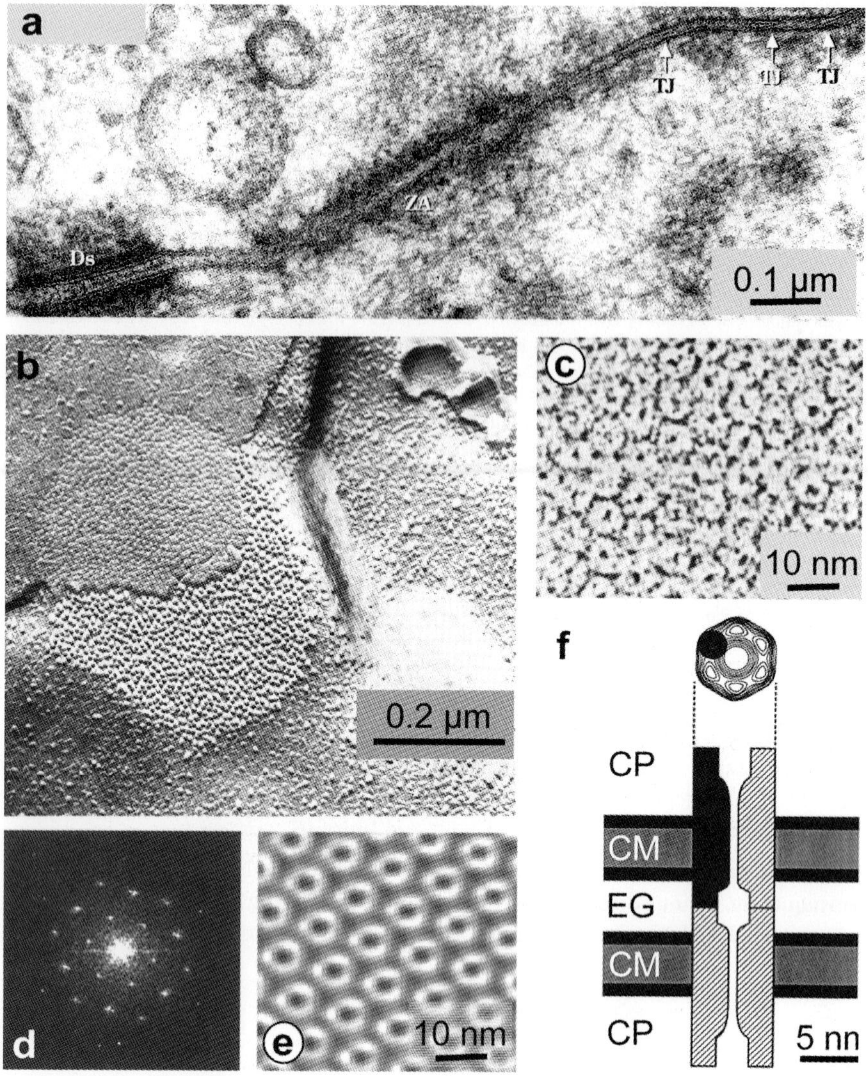

Fig. 2.16 Cell-cell contact in animal cells.
(a) Tight junctions (TJ), zonula adhaerens (ZA), and desmosome (DS) between epithelial cells of a mouse small intestine (Ude and Koch, 1994).
(b) Gap junction patch from small intestine epithelial cells of a frog, freeze-fractured preparation. Small, densely packed vesicles are visible at the P-face of the cell (Kleinig and Sitte, 1986).
(c, d, e, f) Isolated gap junction form a hexagonal lattice (c), consisting of identical particles that may be further characterized by image averaging by diffraction analysis (d) and subsequent filtering, resulting in a processed image (e), and finally a representation as depicted in f (arrowhead): a hexagonal complex with a central pore. The hexagonal complex represents the face-on view of the gap junction as indicated in the schematic drawing. CP, cytoplasm; CM cytoplasmic membrane, EG extracellular gap (c: Kleinig and Sitte, 1986; d, e: Sosinsky and Perkins, 2000; f: Yeager and Nicholson, 1996)

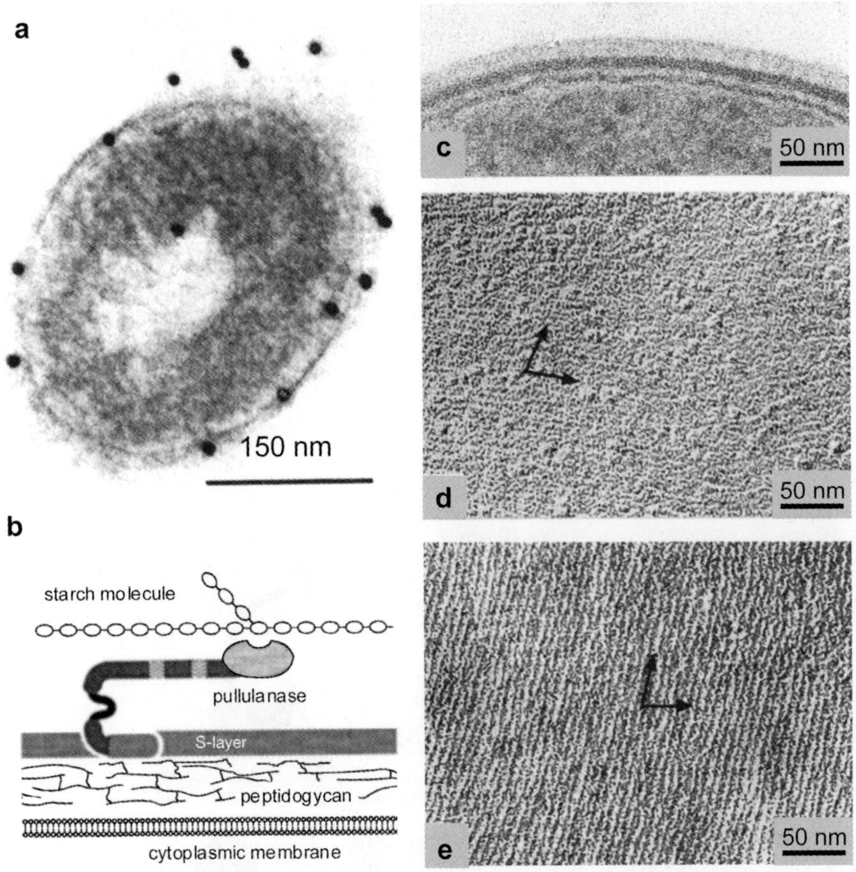

Fig. 2.17 Surface layer association of extracellular enzymes.
(a) Immunolocalization of the pullulanase from *Thermoanaerobacterium thermosulfurigenes* (Specka et al., 1991).
(b) Model for the anchoring of the enzyme to the S-layer lattice. The enzyme essentially consists of a catalytic domain, a connecting part, and an S-layer like domain, responsible for integration of the molecule into the surface layer of the cell envelope (Matuschek et al., 1994).
(c–e) S-layer association of an extracellular amylase. Ultra-thin-sectioned (c) and freeze-etched (d, e) preparations of whole cells from *Bacillus stearothermophilus* DSM 2358. Cells were investigated before (c, d) and after (e) treatment with 2 M guanidinium hydrochloride, leading to extraction of the high-molecular-weight amylase. Ultra-thin sectioned cell exhibit the thick dark peptidoglycan-containing layer completely covered with an outermost additional S-layer (c). Untreated S-layer does not exhibit a clearly visible crystalline structure, indicating the presence of a bound uppermost protein layer (d). In (e) this layer is stripped by the action of the chaotropic agent guanidinium hydrochloride. Arrows indicate the base vectors of the crystal lattice (Egelseer et al., 1995)

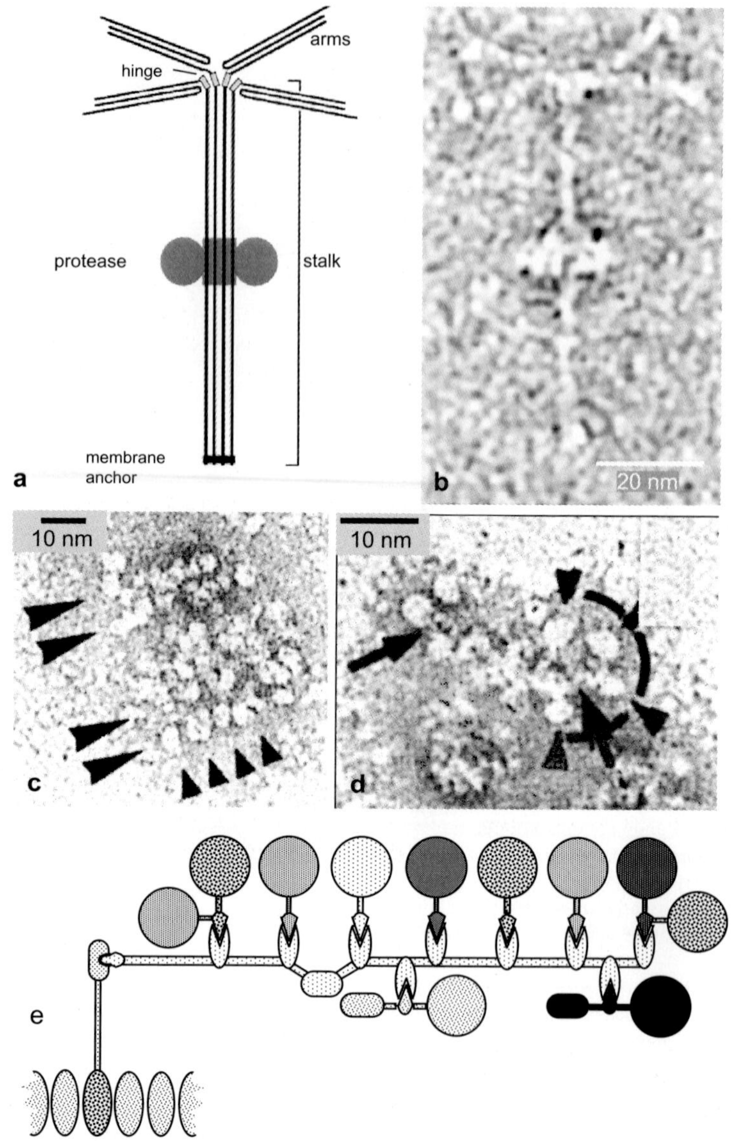

Fig. 2.18 Large polymer-degrading assemblies.
(a, b) Negatively stained "tetrabrachion" protease complex. The respective schematic drawing is based upon electron microscopic and sequence data (simplified after Peters et al., 1996).
(c–e) Cellulosome from *Clostridium thermocellum*. Negatively stained whole, slightly decomposed complex ("open complex"; c) and a respective fragment (d) reveal the composition of globular particles (small arrowheads) arranged in rows (large arrowheads). The aggregates appear to be interconnected by a filamentous scaffolding structure. The complex composition has been deduced from electron microscopic and biochemical and genetic data (e). The catalytic units (large globules) are interconnected by a scaffolding protein without catalytic activity. This protein is anchored, through an additional subunit, in the S-layer membrane (Mayer et al., 1987)

ple. In fermentation processes, contamination by viruses may be detected after preparation of a small culture volume.

In the same way, rapid detection of contaminating proteins during a purification process, checking of the presence of cells and free subcellular compounds after maceration of a sample, and monitoring of the sizes of liposome vesicles is possible with a preparation time similar to those needed for light microscopic observation, but at high resolution. For most research purposes, of course, purified preparations are used. Determination of the size of a whole protein or a virus and of protein subunits is possible. Although X-ray analysis and NMR can provide protein structural data down to atomic resolution, electron microscopy shows the domain structure of large complexes, and also tolerates the presence of high quantities of impurities. Examples of investigations of structure-function relationships, mainly of isolated enzyme complexes, are presented below.

It has already been mentioned that bacterial S-layers may serve as sites for the attachment of these enzymes, as depicted here for the pullulanase of *Thermoanaerobacterium thermosulfurigenes*. The enzyme is located at the cell periphery, as demonstrated by immunolocalization (Fig 2.17 a), and it appears to be attached to the cell surface (Specka et al., 1991; Matuschek et al., 1994). The C-terminus of pullulanase contains three domains displaying close similarities to S-layer proteins and may therefore anchor in the S-layer lattice (Fig. 2.17 b). The catalytic domain (hydrolyzing α-1,4-and α-1,6-glycosidic linkages) of the enzyme is located in the N-terminal half of the molecule. A high-molecular-weight amylase of *Bacillus stearothermophilus* is located on the outer face of the S-layer lattice, too. Obviously, this protein is not integrated into the S-layer lattice as described for the pullulanase, because S-layer-like domains are absent in the amylase. The amylase may be stripped from the cell surface without affecting the structure of the underlying S-layer (Fig. 2. 17 c–e). It has been suggested that the protein binds to domains located either on the surface of the lattice or inside the pores of the S-layer units (Egelseer et al., 1995).

The S-layer of the archaeon *Staphylothermus marinus* serves as a holdfast for a protease. The S-layer itself exhibits unique structural features: it consists of a poorly ordered meshwork of glycoprotein structures resembling dandelion seed-heads (Peters et al., 1996). The molecule ("tetrabrachion") consists of a filamentous stalk 70 nm in length, ending in four filamentous arms of 24 nm in length (Fig. 2.18 a, b). In vivo, these stalks represent the outside of the S-layer lattice and are laterally connected to each other. At half of the length of the long filamentous stalk are attached two globular particles that exhibit protease activity. The highly thermostable enzyme is very likely to be involved in peptide fermentation of the organism. Isolated tetrabrachion molecules assemble into a poorly ordered two-dimensional network or into micelles.

The cellulolytic system of *C. thermocellum* is organized as a large multiprotein complex, known as a cellulosome (Fig. 2.18 c–e; Coughlan and Mayer, 1992). The cellulosome, similarly to the pullulanase, is interconnected with the surface layer of the bacterial cell. The complex mediates adherence of the bacterium to the cellulose and promotes depolymerization of crystalline cellulose. The majority of the 14–18 proteins in the cellulosome function as β-1,4-endoglucanases, hemicellulases, and cellobiohydrolases. A working model for the structural organization of the cellulosome and the

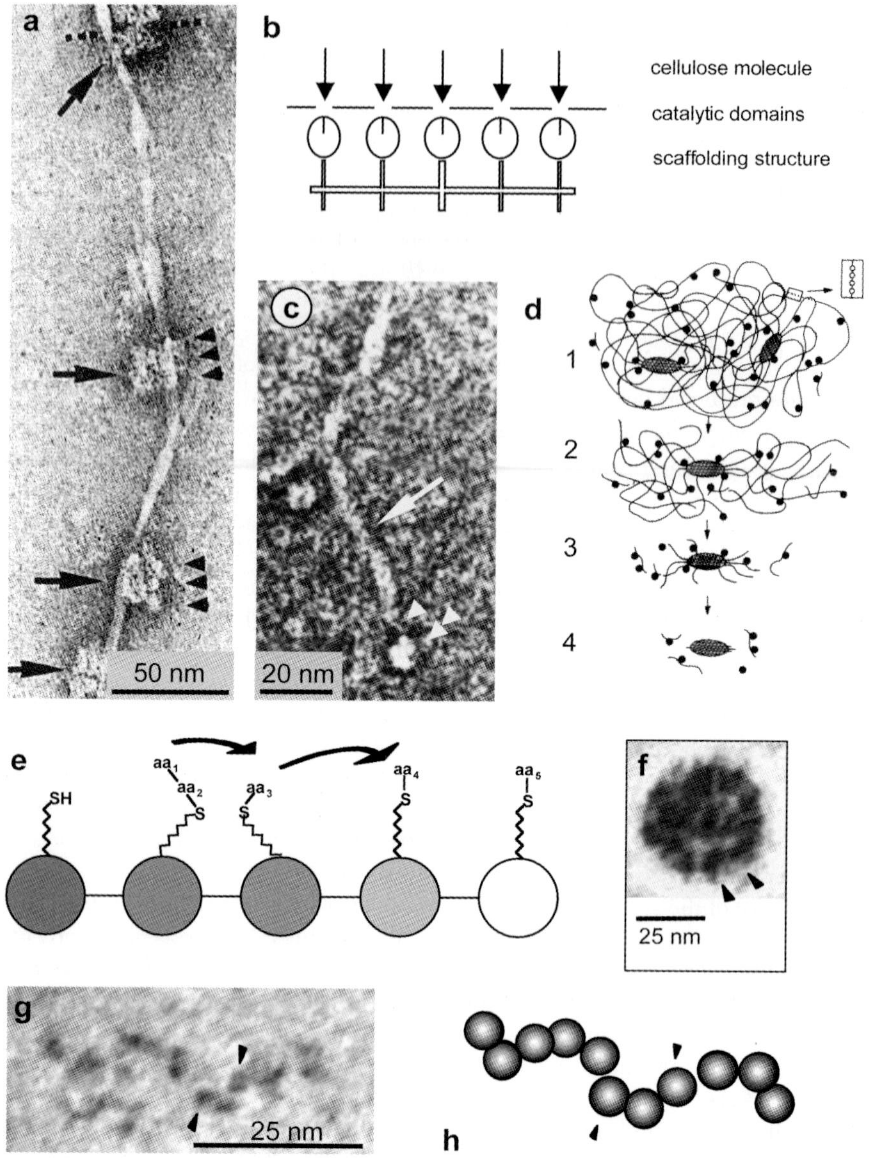

functional implications has been developed from electron microscopic data, later confirmed by biochemical investigations (see Fig. 2.18 e). The enzymes are bound to a scaffolding structure (scaffoldin or CipA). Arranged in rows, the catalytic domains attack the same cellulose chain in a set of quasi-simultaneous cutting events. The scaffoldin is composed of the subunit-binding domains (cohesins) and a cellulose-binding domain. The enzymes are bound to cohesins through 23-residue C-terminal domains (dockerin). The scaffoldin itself is anchored to the bacterial cell surface

Fig. 2.19 Large enzyme complexes and their interaction with substrates (negatively stained specimens). (a) Cellulosomes (arrows) bound to a cellulose fiber; arrowheads mark distinct subunits.
(b) Functional model deduced from the cellulosome structure and interaction with cellulose. Several subunits are arranged by a scaffolding protein (CipA) in such a way that cellulose is decomposed to oligomers in a simultaneous cutting event.
(c, d) Cellulase from a *Bacillus* isolate decomposes amorphous but not crystalline cellulose areas of Avicel. A cellulase molecule (c; double arrowhead) is bound to a thin fiber of amorphous cellulose (arrowhead) emerging from a highly packed crystalline area. The decomposition process is depicted in the schematic drawing (d), cellulose molecules (black dots) decompose amorphous cellulose. Finally, only crystalline areas (spindle-shaped) remain visible in the preparation (Jörn Kricke, Univ. Göttingen).
(e–h) Correlation between a functional model and the overall enzyme structure in peptide synthetases. A functional module (consisting of several domains) catalyzes incorporation of one specific aminoacid residue (aa_n) into the growing peptide chain (e). Overall structure of cyclosporin synthetase as globular (f) and chain-like complexes (g, h). The images are negatively stained and recorded with electron spectroscopic imaging at the energy loss of the staining solution (uranyl acetate), which results in an enhanced (but reverse) contrast. Single particles on a chain represent modules as depicted in (e). The separate functional modules appear as structurally distinct units (e: adapted from Mootz and Marahiel, 1997; f, g: Hoppert et al., 2001)

through proteins (SdbA, OlpA, OlpB) with COOH-terminal S-layer homologous repeats. These domains are probably involved in anchoring the proteins to the bacterial surface layer (see below). The cellulosome itself is packed into polycellulosomal protuberance-like particles that may be released from the cell surface.

The interaction of the polymer-degrading enzyme with its substrate is deductible from electron micrographs, as depicted in Fig. 2.19 a. Cellulosomes (here detached from the S-layer) are bound to a cellulose fiber (composed of multiple cellulose molecules). The catalytic units are arranged in rows, as can be deduced from images such as Fig. 2.18 c, d and as depicted in the schematic drawing 2.19 b, and appear to cut one cellulose molecule at a defined distance. The cut creates oligosaccharides of identical size, which are further decomposed to cellobiose, the building block of one cellulose molecule. Whereas decomposition of crystalline cellulose requires the concerted action of a cellulosome multi-enzyme complexes, amorphous cellulose may be decomposed by single enzyme activities. The molecule, as depicted in Fig. 2.19 c, is much smaller than a cellulosome. Here the enzyme consists of only four identical subunits. Cellulases of this simple composition (here from a *Bacillus* isolate) are common in soil bacteria, but only decompose amorphous cellulose (hydrated molecules that are only loosely associated to each other). The action of the enzyme with carboxymethylcellulose may easily be documented with negatively stained samples taken at time intervals after onset of the enzyme reaction. The molecule decomposes non-crystalline areas but does not attack the crystalline cellulose. The scheme in Fig. 2.19 d depicts the successive decomposition of carboxymethylcellulose. The crystalline parts of the molecule remain unchanged.

The sequential elongation of a peptide chain in non-ribosomal (i.e., enzymatic) synthesis of peptides appears to be reflected by the overall structure of the cyclosporin synthetase, which catalyses synthesis of the undecapeptide cyclosporin. Cyclosporin and numerous other peptides are highly bioactive agents and are used as immunosuppressants, antibiotics, surfactants, etc. The incorporation of an amino acid into a peptide chain (see Fig. 2.19 e) is brought about by a functional unit (module), which is itself separated into intensively interacting domains. The modules represent more or less separate activities in the whole enzyme complex and may even be exchanged for one another, affording peptide products with altered sequences. The functional biosynthetic chain is reflected by the structure of the large cyclosporin synthetase molecule, where the modules – in unfolded enzyme complexes (otherwise they appear as large globules; see Fig. 2.19 f) – appear to be arranged like beads on a chain (Fig 2.19 g), one bead representing a module, or one incorporation step.

Direct interaction between enzymes and their substrates is only visible under the electron microscope as long as the substrate is large enough to be readily detectable. Nevertheless, postulation of functional models behind an observed structure in most cases needs some more information. When antibodies directed against specific epitopes are available, it is possible to determine the position of the epitope in a complex. This is shown in Fig. 2.20 for a large molecule (D-ribulose-1,5-bisphosphate carboxylase/oxygenase, RuBisCO) and a small enzyme (glucoamylase). From the relative positions of the protein complexes it may be deduced that the subunits of the RuBisCO are arranged as depicted in Fig. 2.20 a and that the epitope of glucoamylase is located at an exposed site of the protein (Fig 2.20 b). These techniques are helpful only for obtaining essential data on the topology of subunits or domains in enzymes; no information on domain structure is gained directly. When, on the other hand, structural data from other sources are available, comparison between results from negative staining and epitope mapping with these data may provide a more refined – but hypothetical – structural model. Figure 2.20 c compares three projection forms of the glucoamylase from a bacterial organism with three views of a tertiary structure of the respective enzyme from *Aspergillus awamori*, the structure of which has been determined by X-ray crystallography. Since homologies and similarities are present in the primary sequences of the two enzymes, other analogies between the arrangements of the secondary structure elements are also reasonable, as implied in Fig. 2.20 c. The result, however, is only a hypothetical working model. A reliable model in which atomic coordinates obtained from, say, X-ray data are accurately merged with structural data from the electron microscope is a very laborious task. Only under ideal conditions do the structural data obtained by electron microscopy provide a resolution that also allows reliable fitting with X-ray data (see Section 2.4.9 for further details).

2.2.6 Interaction of Organisms

It is possible to obtain overviews from large samples (such as biofilms or tissues) by TEM, but this often requires time-consuming combination of a series of detail images. Oligonucleotide probes therefore do not have potential for application for detection of whole organisms in ultra-thin sections by electron microscopy, though they are used for subcellular detection of mRNA. However, TEM allows a gap to be closed between the documentation of interacting whole cells and the involved macromolecular structures down to the scale of macromolecules. In suitable preparations, direct visualization of holdfast structures and the biofilm matrix at high resolution is possible. Figure

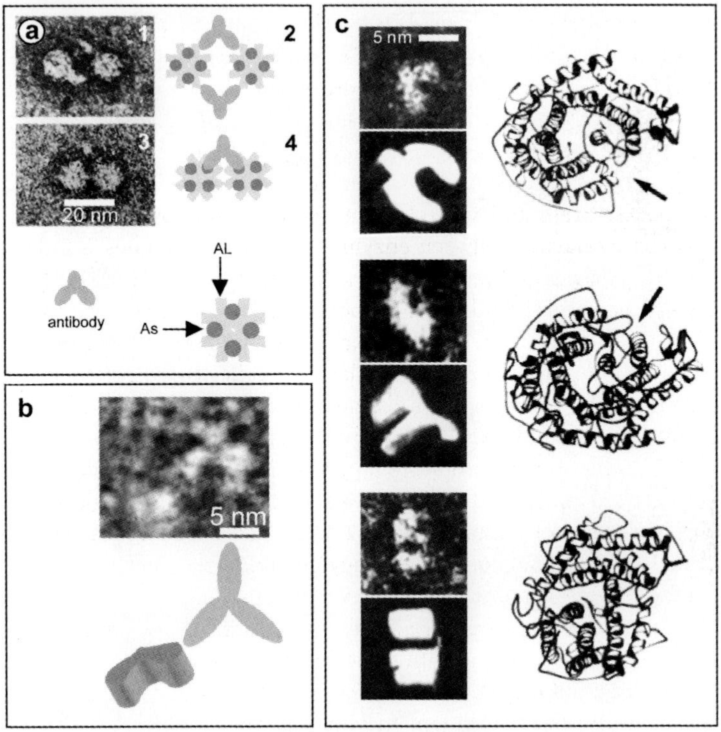

Fig. 2.20 Epitope mapping of enzymes with IgG-antibodies (negatively stained specimens).
(a) Localization of the large and small subunit of the hexadecameric (L8S8) d-ribulose-1,5-bisphosphate carboxylase/oxygenase (RuBisCO) from *Ralstonia eutropha*. Purified enzyme was labeled with polyclonal antibodies directed against the large enzyme subunit (AL, see a1 and a2) and the small subunit (AS, see a3 and a4). Antibody-crosslinked enzyme molecules (a1, a3) show different intermolecular distances depending on the antibody used, leading to a model of the subunit topology of the enzyme as indicated (Holzenburg and Mayer, 1989).
(b) An IgG antibody directed against a specific epitope of glucoamylase from *Thermoanaerobacterium thermosaccharolyticum*, bound to an epitope exposed at one end of the c-shaped molecule (Ducki et al., 1998).
(c) Three projection forms of the glucoamylase and their respective models compared with structural data, obtained from *Aspergillus awamori* glucoamylase (Aleshin et al., 1992; Specka and Mayer, 1993)

2.21 a shows two types of pili emerging from the cell surface of an *Acinetobacter* strain responsible for adhesion and twitching motility. Adhesion brought about by pili is a temporary event. The formation of a thick biofilm is brought about by an extracellular polymer matrix, as mentioned above. The matrix material that surrounds a cell either as a tight capsule or as a loosely, structurally less defined slime appear in ultra-thin sections as loosely fibrillar material (Fig. 2.21 b–d). The fibrillar appearance is an artifact caused by dehydration of the homogeneous gel-like extracellular matrix. Figure

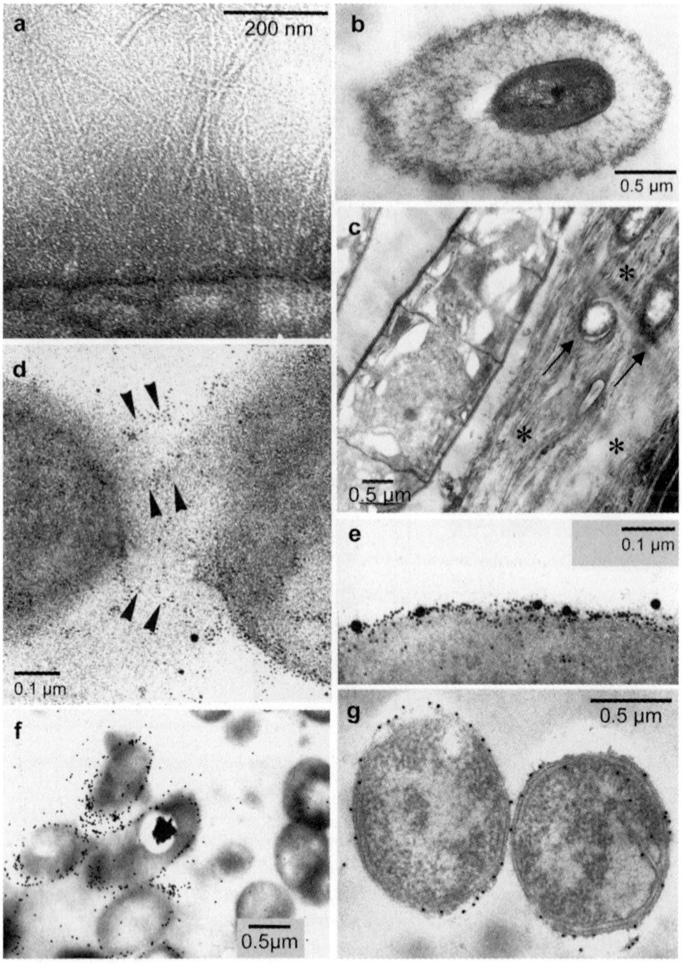

Fig. 2.21 Fine structure of cellular constituents involved in biofilm formation.
(a) Pili emerging from the surface of an *Acinetobacter* strain. The pili play an unknown role for the mechanism of twitching motility, a prerequisite for movement on a solid surface (Melanie Bolte, Univ. Göttingen).
(b) Bacterial strain isolated from a biofilm surrounded by a thick layer of filamentous material (Martin Kämper, Univ. Göttingen).
(c) Filamentous cyanobacterium from a cryptobiotic soil crust surrounded by extracellular polymers, appearing as tightly packed filaments. Typically, non-photosynthetic bacteria (arrows) are embedded in the matrix (asterisks) (Anne Kemmling, Univ. Göttingen).
(d) Extracellular polysaccharides detected by lectin-gold (arrowheads) indicate the action of extracellular polymers as a glue between the microorganisms.
(e) Binding of two types of gold-coupled lectins to a bacterial cell surface in a biofilm preparation obtained form a human urethral catheter (Martin Kämper, Univ. Göttingen).
(f) Differentiation of two biofilm organisms with lectins. Not all bacteria bind the lectin-gold complex (Martin Kämper, Univ. Göttingen).
(g) Localization of levansucrase in a phytopathogenic *Pseudomonas syringae* strain. The extracellular enzyme degrades extracellular polysaccharide and is therefore an important component of biofilm dynamics (Ulrike Hettwer, Univ. Göttingen)

2.21 c shows a typical example of the massive production of extracellular polymers acting as matrix for other organisms as well. A large cyanobacterial filament (taken from a cryptobiotic soil crust) provides rigid matrix material for the biofilm community. The specificity of lectins as oligosaccharide probes allows differentiation between extracellular polysaccharides from various organisms at high resolution. The applications show the detection of two different lectin-binding affinities on the surface of a bacterial cell taken from a biofilm (Fig. 2.21 e). In addition, the role of intercellular contact through the polymer matrix may be visualized with the aid of lectins (Fig. 2.21 d). As in the respective light microscopic techniques, lectin probes also allow

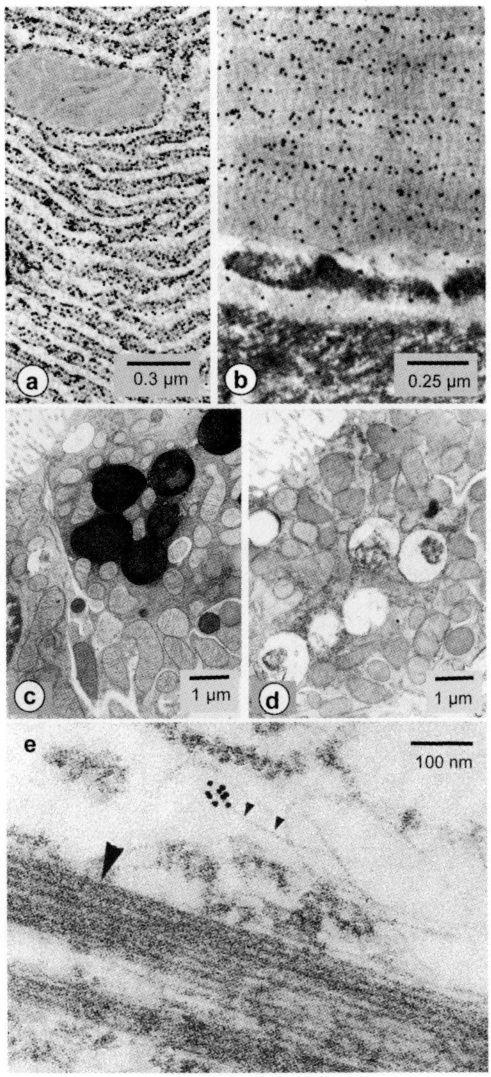

Fig. 2.22 Various cytochemical markers applied to ultra-thin section of specimens.
(a) Localization of RNA in a rat pancreatic acinar cell with RNase, coupled to colloidal gold markers. Labeling of the rough endoplasmic reticulum is restricted to the ribosomal sites on the membrane (Bendayan, 1985).
(b) Localization of collagen in the rat duodenum with collagenase coupled to colloidal gold markers. The striated collagen fibrils appear densely labeled by gold particles (Bendayan, 1985).
(c, d) Acid phosphatase in lysosomes localized with the aid of *p*-nitrophenyl phosphate as enzyme substrate and cerium as the electron-dense marker (c), and untreated control (d).
(e) Localization of myosin heavy-chain mRNA in cultured muscle cells (myotubes) from chicken embryos with a biotinylated DNA probe coupled to anti-biotin-antibody gold marker. Cells have been pretreated by triton-extraction. A small cluster of nine gold particles is associated with a thin filament (small arrowheads), corresponding to the diameter of myosin. A myofiber is marked with a large arrowhead (Silva et al., 1989)

differentiation between organisms by their varying affinities for lectin probes (Fig. 2.21 f). It is reasonable to assume that extracellular enzymes involved in polymer reorganization are also present in the polymer matrix. In fact, as depicted in Fig. 2.21 g, a polymer-degrading levan-sucrase is attached to the surface of a levan-producing *Pseudomonas syringae* strain.

2.2.7
Specific Markers for Intracellular Components

Numerous probes specific either for single proteins or for substance classes exist, allowing localization of DNA, specific proteins, polysaccharides, etc. The most widely used technique remains immunogold staining, with specific antibodies as probes for proteins and colloidal gold as the electron-dense marker. Applications of this method have already been depicted in Figs. 2.1, 2.4, 2.7, 2.13, and 2.17. A big advantage for this detection is, of course, the high spatial resolution of electron microscopy, which normally allows clear assignment of signals to a structure (often a problem at the resolution levels obtained by light microscopy).

Besides the widely used immunogold markers, alternative markers for localization of a target may also be used. The application of lectin markers has already been presented in Fig. 2.2.1 d, e and f. Especially when macromolecules are present in high concentrations, a target based on a substrate-specific enzyme may be applied. Localization with an RNase-gold probe is depicted in Fig. 2.22 a. RNase binds to RNA, indicating the presence of the nucleic acid (rRNA) in the endoplasmic reticulum. Figure 2.22 b shows collagenase-gold labeling in the rat duodenum. Striated collagen fibrils appear on labeling by the enzyme coupled to gold particles, appearing to follow the pattern of striation in many instances. Enzyme-substrate interaction may also be detected with a substrate inducing the precipitation of an electron-dense marker in order to localize an enzyme activity. In Fig. 2.22 c the acid phosphatase in lysosomes from rat kidney cells has been localized with the aid of *p*-nitrophenyl phosphate as

Fig. 2.23 Compartmentation of biochemical pathways. ▶

(a, b) Localization of alanine racemase (a) and cyclosporin synthetase (b) the two key enzymes of the cyclosporin biosynthetic pathway, at the periphery of vacuoles in the ascomycete fungus *Tolypocladium inflatum*.

(c) Localization of the product cyclosporin in vacuoles of *Tolypocladium inflatum*.

(d) Possible compartmentation of the cyclosporin biosynthetic pathway as deduced from the immunolocalization experiments. Alanine racemase (AR) and cyclosporin synthetase (CS) are bound to the vacuolar membrane, and use the cytosolic pool of aminoacids (aa) to synthetize d-alanine and cyclosporin, which is transferred to the vacuolar compartment (Hoppert et al., 2001).

(e) Localization of enzymes involved in glycine fermentation in *Eubacterium acetaminophilum*. Protein P3 (large image) labeled by large gold complexes, Pa by small complexes reveals co-localization of both proteins at the cell periphery (CW, cell wall). The small image gallery shows P1 labeled by large complexes (arrows), P2 by small complexes (arrowheads) reveals co-localization of the enzymes in the cytoplasm.

(f) Model for interactions of the enzymes. P1, pyridoxal phosphate-containing decarboxylase protein; P2, lipoamide containing hydrogen carrier protein; Pa, selenoprotein; GR, glycine reductase; THF, tetrahydrofolate; CH2=THF, methylenetetrahydrofolate; cm, cytoplasmic membrane (Freudenberg et al., 1989)

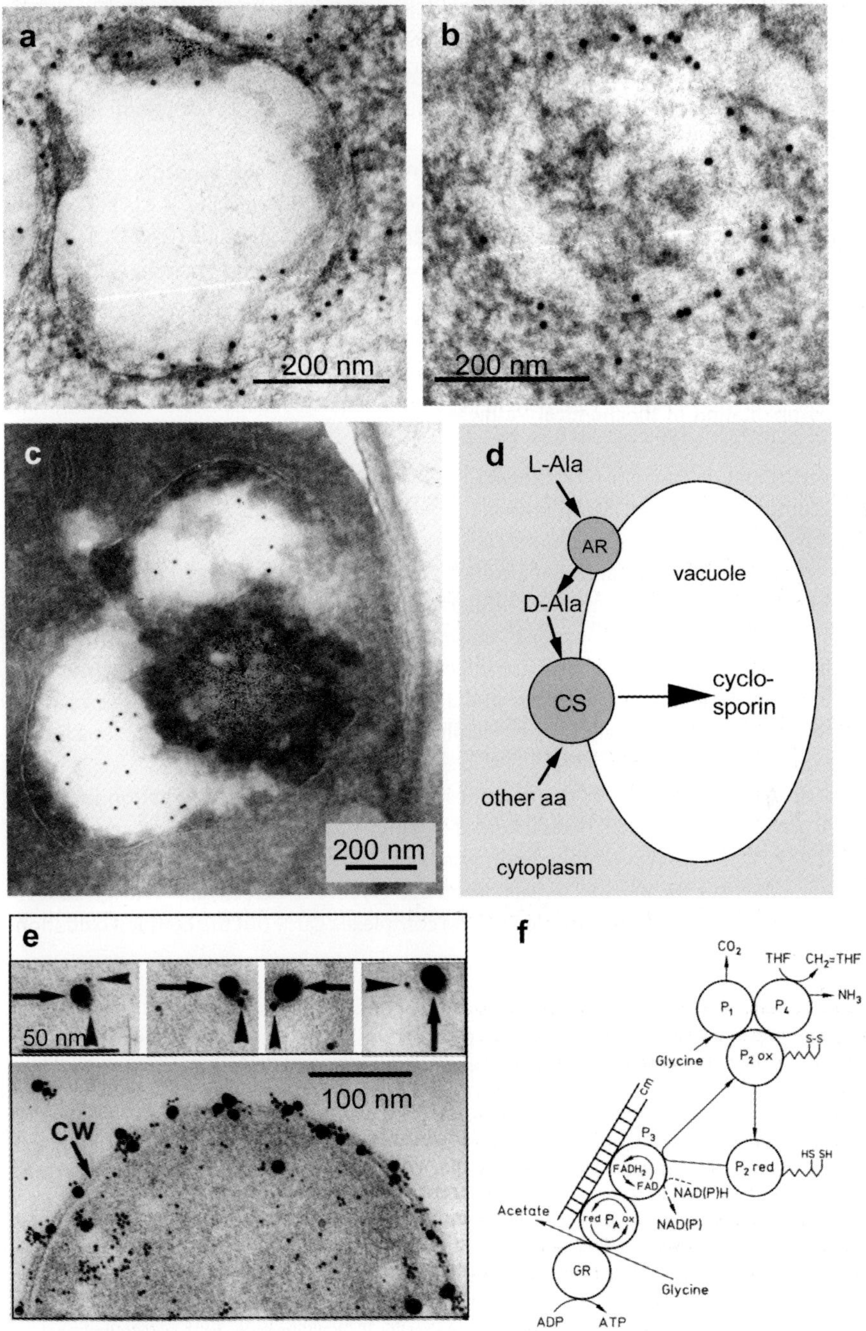
a
200 nm
b
200 nm
c
200 nm
d
L-Ala
AR
vacuole
D-Ala
CS
cyclo-
sporin
other aa
cytoplasm
e
50 nm
100 nm
CW
f
CO2
THF
CH2=THF
NH3
P1
P4
Glycine
P2 ox
S-S
cm
P3
FADH2
FAD
NAD(P)H
P2 red
HS SH
Acetate
red PA ox
NAD(P)
GR
Glycine
ADP
ATP

enzyme substrate and cerium as the electron-dense marker. The method is equivalent to the respective techniques used for histochemical staining in light microscopy and is regularly used for detection of high concentrations of enzyme activities in compartments (lysosomes, for instance). The technique is often used for complementation of immunolocalization procedures.

Equivalent to in situ hybridization in light microscopy, the electron microscopic approach can allow high-resolution mapping of specific mRNA (Fig. 2.22 e). The technique may be combined with specific signal enhancement methods (CARD, see chapter 2.4.8.1) and is suitable for high-sensitivity and high-resolution in situ hybridization in medical diagnostics (Cheung et al., 1999).

2.2.8 Compartmentation of Biochemical Pathways

The functional organization of metabolic pathways may be reflected in their subcellular compartmentation. Simultaneous labeling of several enzymes belonging to a pathway may therefore help understanding of the flow of metabolic products. Knowledge of, say, compartmentation of antibiotic biosynthesis may be helpful for optimization of product yields. Figure 2.23 illustrates compartmentation of cyclosporin biosynthesis. The product cyclosporin is detectable by use of monoclonal antibodies and is located in vacuoles of the ascomycete *Tolypocladium inflatum*. The important key enzymes cyclosporin synthetase and alanine racemase are located at the vacuolar membrane. The results allow the metabolite flow to be inferred, as depicted in Fig. 2.23 d (Hoppert et al., 2001).

The spatial organization of the enzymes involved in glycine fermentation in *Eubacterium acetaminophilum* has been revealed by double immunolocalization experiments (Fig. 2.23 e, f). Here, two readily distinguishable markers are applied simultaneously. The key enzymes for glycine fermentation are glycine decarboxylase and glycine reductase, two multi-enzyme complexes. The complexes carry out the coupled oxidation-

Fig. 2.24 Deposition of polyhydroxyalkanoates (PHA) in various compartments of transgenic plants. ▶
(a) Polyhydroxybutyrate granula (arrrow) in the nucleus of a parenchymatic leaf cell of transgenic *Arabidopsis thaliana*; in the same plant, the granula are also located in the cytoplasm and the vacuole (Poirier et al., 1992).
(b) PHA-granula in peroxisomes from cells in *Arabidopsis thaliana* cotelydones. Bright granules appear to be PHA, dark gold particles mark the PHA-synthase enzyme, located surround the granulum (Mittendorf et al., 1998).
(c–e) PHA-granula inside plastids, either in chloroplasts of parenchymatic leaf cells (d) from *Arabidopsis thaliana* plants (c: wild type control; arrow depicts a starch granule) or leukoplasts of *Brassica napus* oilseeds (e). The cytosol of the cell is filled with numerous lipid bodies (L) (Bohmert et al., 2000; Houmiel et al., 1999).
(f) Schematic representation of the biochemical pathways for PHA synthesis in the three compartments of plant cells based upon transgenic *Arabidopsis* plants. The engineered parts of the pathway are highlighted in bold letters and with solid arrows, the endogenous plant pathways in plain letters and open arrows), the various transgenes expressed in plants are given in parentheses (Poirier, 1999)

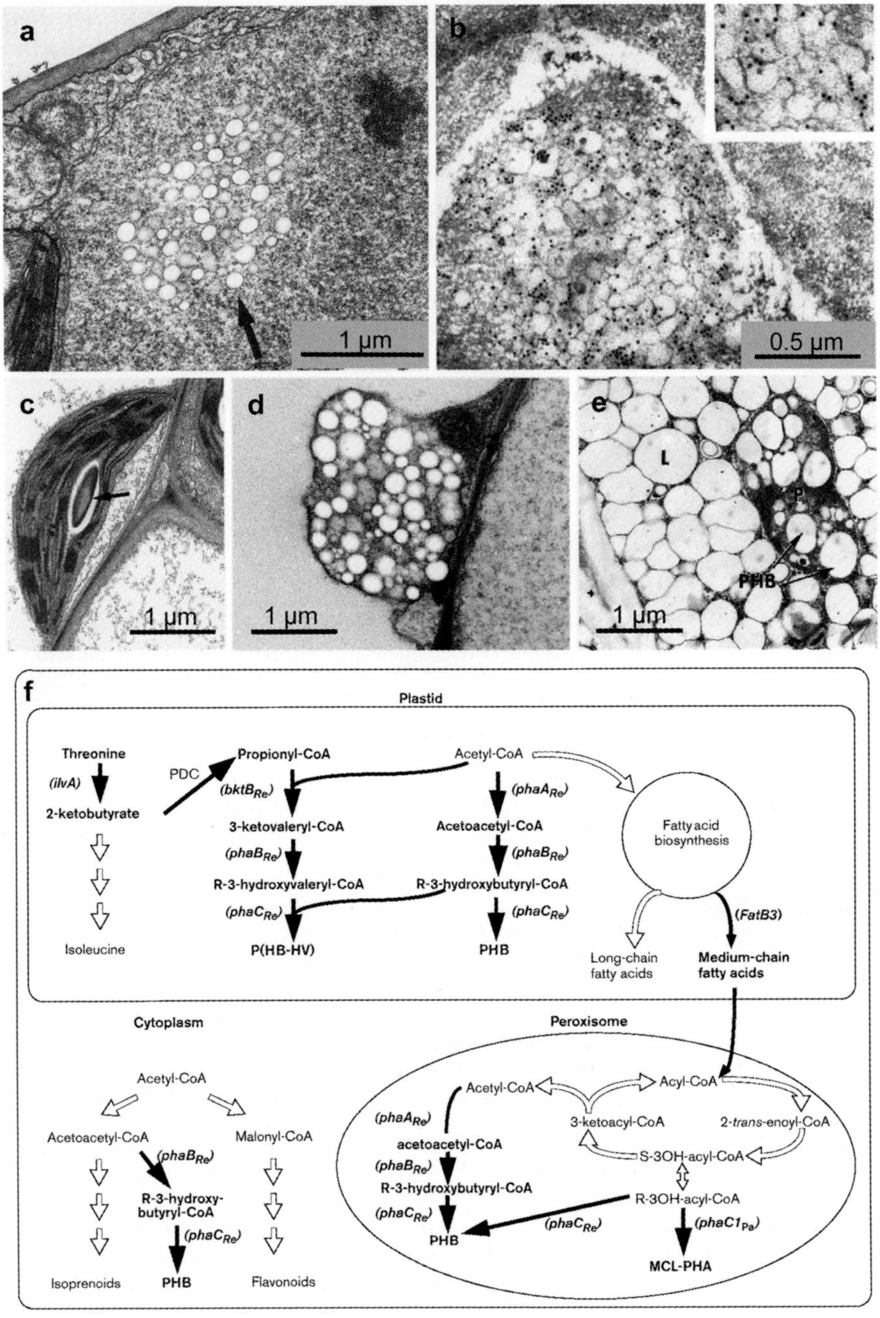
a
1 µm
b
0.5 µm
c
1 µm
d
1 µm
e
L
P
PHB
1 µm
f
Plastid
Threonine
(ilvA)
2-ketobutyrate
Isoleucine
PDC
Propionyl-CoA
(bktB_Re)
3-ketovaleryl-CoA
(phaB_Re)
R-3-hydroxyvaleryl-CoA
(phaC_Re)
P(HB-HV)
Acetyl-CoA
(phaA_Re)
Acetoacetyl-CoA
(phaB_Re)
R-3-hydroxybutyryl-CoA
(phaC_Re)
PHB
Fatty acid biosynthesis
(FatB3)
Long-chain fatty acids
Medium-chain fatty acids
Cytoplasm
Acetyl-CoA
Acetoacetyl-CoA
Malonyl-CoA
(phaB_Re)
R-3-hydroxy-butyryl-CoA
(phaC_Re)
Isoprenoids
PHB
Flavonoids
Peroxisome
Acetyl-CoA
Acyl-CoA
(phaA_Re)
3-ketoacyl-CoA
2-trans-enoyl-CoA
acetoacetyl-CoA
S-3OH-acyl-CoA
(phaB_Re)
R-3-hydroxybutyryl-CoA
R-3OH-acyl-CoA
(phaC_Re)
(phaC_Re)
(phaC1_Pa)
PHB
MCL-PHA

reduction of the amino acid (Stickland reaction). The decarboxylase complex is responsible for glycine oxidation to CO_2, methylene-THF, and NADPH. The electrons are passed over to the glycine reductase, where glycine is reduced to acetyl phosphate and NH_3. The decarboxylase complex is located in the cytoplasm; the reductase complex is membrane-bound. A small protein acts as free diffusible electron carrier between these two complexes. The reductase itself may interact with a hydrogenase and a formate dehydrogenase, both membrane-bound (Freudenberg et al., 1989; Andreesen et al., 1994). The structures of the complexes and their interactions are depicted in Fig. 2.23 f.

The documentation of engineered pathways is illustrated in Fig. 2.24. The localization of PHB in different compartments of *Arabidopsis thaliana* (and other plants) – especially when the respective biosynthetic enzymes are also localized – is necessary for elucidation of the compartmentation of this engineered pathway, and also for optimization of the heterologously expressed product.

2.2.9
Elemental Analysis

Similarly to SEM (see chapter 2.3), the interaction of the electron beam with matter can also be used to analyze the elemental composition and the distribution of mass within cells. For elemental analysis, inelastically scattered beam electrons or X-rays generated during the interaction between electron beam and specimen atoms are used. The methods – electron energy loss spectroscopy (EELS) or electron probe microanalysis – differ with respect to the required instrumentation and the detection limits for specific elements. In principle, both methods are suitable for detection of specific elements in ultra-thin sections of samples.

Both methods are especially suitable for detection of deposits of homogeneous composition (such as salt crystals, dense intracellular inclusions of sulfur or phosphate, elemental probes). Figures 2.25 a and b show the detection of toxic heavy metal complexes in algae. X-ray spectra revealed the presence of heavy metals in plastoglobuli after exposition to lead (Wong et al., 1997).

Since the elemental compositions of most cellular compounds are too homogeneous to give substantial results relating to specimen structure, elemental probes may be introduced for labeling of the cell component of interest. This technique is used in conjunction with direct imaging of the element distribution in the specimen by electron energy loss spectroscopy. This method allows imaging of electrons with a specific energy loss, resulting from interaction with the electron shell of a specific element, and the direct imaging of elements in samples. When elemental probes are specifically marked with antibodies, high-resolution immunolocalization may be achieved. The element should not, of course, be a native component of the specimen. For immunolocalization, peptide-tagged carborane clusters covalently linked to antibody fragments allow mapping of the antigen with high resolution. Figures 2.25 c and d show the distribution of boron used for localization of somatotrophic hormone in porcine pituitary cells (Kessels et al., 1996). The distribution of boron signals is

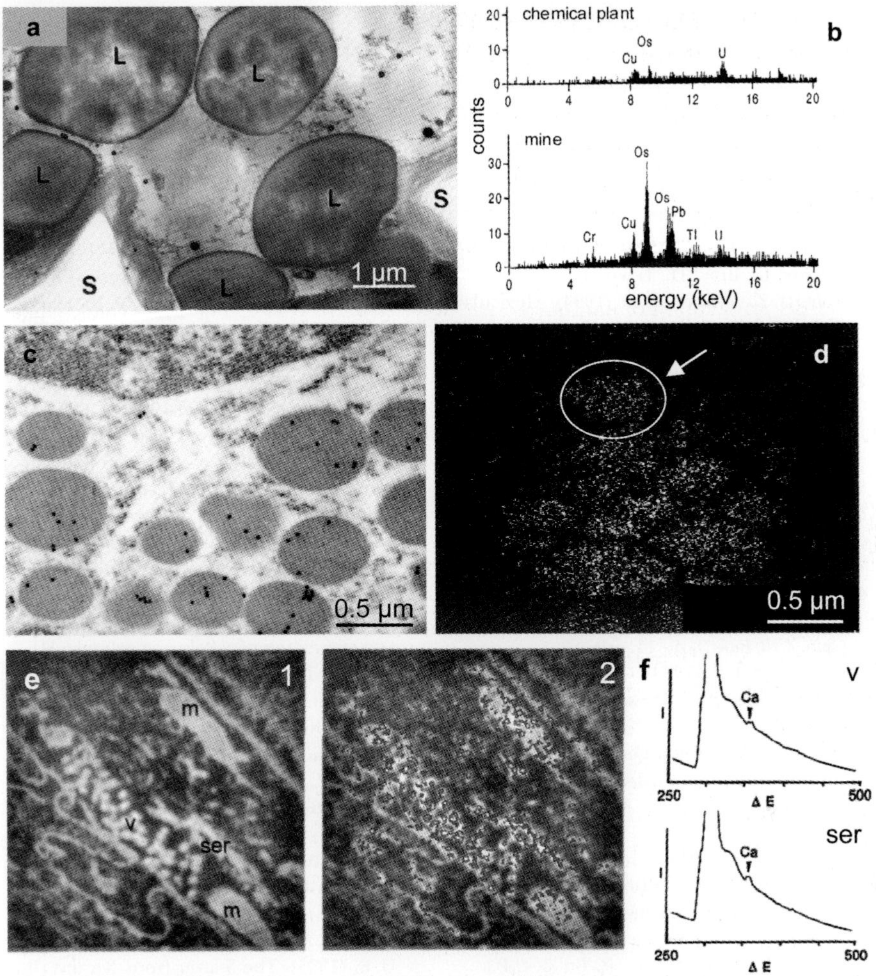

Fig. 2.25 Detection of elements in ultra-thin sections of specimens.
(a, b) Detection of toxic metals in the filamentous alga Cladophora glomerata. A lipid body (L) of a cell exposed to a mine effluent was subjected to X-ray microanalysis (EDX) as depicted in (b), revealing the presence of heavy metals as indicated. The metals are accumulated as organometallic compounds in lipid bodies (Wong et al., 1997).
(c, d) Localization of somatotrophic hormone by immunogold labeling (c) compared with detection by boron-labeled antibody Fab fragments (d). Two similar sections from somatotrophic pituitary cells from pig are shown. The boron has been specifically detected by electron spectroscopic imaging (ESI). Unlike colloidal gold markers, the boron signal fills the secretory vesicles (one vesicle encircled).
(e, f) A motor nerve terminal synaptic vesicle (v) and also the smooth endoplasmic reticulum (ser) exhibit strong calcium signals as revealed by ESI analysis (1,2). In 1, an image taken at an energy loss of 240 eV is shown, resulting in an unusual (reverse) contrast brought about by inelastically scattered electrons. In 2, the element specific image is projected in the original image. The calcium signal is confirmed by electron energy loss spectra (f) (Grohovaz et al., 1996)

compared with that of conventional immunomarkers (colloidal gold). Mapping of naturally occurring elements may also be performed in a similar manner. An example is shown in Fig. 2.25 e and f (Grohovaz et al., 1996), in which the intracellular calcium distribution in nerve-muscle preparations is depicted.

References

1 Alberts, B., Bray, D., Lewis, J., Raff, M., Roberts, K., Watson, J. D. (**1994**). Molecular Biology of the Cell. 3rd ed., Garland Publ., New York.

2 Aleshin, A., Golubev, A., Firsov, L. M., Honzatko, R. B. (**1992**). Crystal structure of glucoamylase from *Aspergillus awamori* var. X100 to 2.2 Å resolution. *J. Biol. Chem.* 267, 19291–19298.

3 Allen, J. F., Forsberg, J. (**2001**). Molecular recognition in thylakoid structure and function. *Trends Plant Sci.* 6, 317–326.

4 Amos, L. A., Hirose, K. (**1997**). The structure of microtubule-motor complexes. *Curr. Opin. Cell Biol.* 9, 4–11.

5 Andreesen, J. R. (**1994**). Glycine metabolism in anaerobes. *Antonie van Leeuwenhoek* 66, 223–237.

6 Beveridge, T. J. (**1995**) The periplasmic space and the periplasm in Gram-positive and Gram-negative bacteria. *ASM News* 61, 125–130.

7 Bendayan, M. (**1985**). The enzyme-gold technique: a new cytochemical approach for the ultrastructural localization of macromolecules. In: *Techniques in Immunocytochemistry*, Vol. 3. (eds. G. R. Bullock, P. Petrusz), pp. 179–201. Academic Press, London.

8 Bohmert, K., Balbo, I., Kopka, J., Mittendorf, V., Nawrath, C., Poirier, Y., Tischendorf, G., Trethewey, R. N., Willmitzer, L. (**2000**). Transgenic *Arabidopsis* plants can accumulate polyhydroxybutyrate to up to 4 % of their fresh weight. *Planta* 211, 841–845.

9 Böttcher, B., Gräber, P. (**2000**). The structure of the H^+-ATP synthase from chloroplasts and its subcomplexes as revealed by electron microscopy. *Biochim. Biophys. Acta* 1458, 404–416.

10 Bucher, O., Wartenberg, H. (**1989**). Cytologie, Histologie und mikroskopische Anatomie des Menschen. 11th ed., Hans Huber, Bern.

11 Cheung, A. L., Graf, A. H., Hauser-Kronberger, C., Dietze, O., Tubbs, R. R., Hacker, G. W. (**1999**). Detection of human papillomavirus in cervical carcinoma: comparison of peroxidase, Nanogold, and catalyzed reporter deposition. *Mod. Pathol.* 12, 689–696.

12 Coughlan, M. P., Mayer, F. (**1992**). The cellulose-decomposing bacteria and their enzyme systems. In: *The Prokaryotes* (eds. A. Balows, H. G. Trüper, M. Dworkin, W. Harder, K.-H. Schleifer), pp. 460–516. Springer, New York.

13 Drewlo, S., Brämer, C. O., Madkour, M., Mayer, F., Steinbüchel, A. (**2001**). Cloning and expression of a *Ralstonia eutropha* HF39 gene mediating indigo formation in *Escherichia coli*. *Appl. Environ. Microbiol.* 67, 1964–1969.

14 Ducki, A., Grundmann, O., Konermann, L., Mayer, F., Hoppert, M. (**1998**). Glucoamylase from *Thermoanaerobacterium thermosaccharolyticum*: sequence studies and analysis of the macromolecular architecture of the enzyme. *J. Gen. Appl. Microbiol.* 98, 327–335.

15 Egelseer, E., Schocher, I., Sára, M., Sleytr, U. B. (**1995**). The S-layer from *Bacillus stearothermophilus* DSM 2358 functions as an adhesion site for a high-molecular-weight amylase. *J. Bacteriol.* 177, 1444–1451.

16 Emons, A. M., Mulder, B. M. (**2000**). How the deposition of cellulose microfibrils builds cell wall architecture. *Trends Plant Sci.* 5, 35–40.

17 Freudenberg, W., Mayer, F., Andreesen, J. R. (**1989**). Immunocytochemical localization of proteins P1, P2, P3 of glycine decarboxylase and of the selenoprotein PA of glycine reductase, all involved in anaerobic glycine metabolism of *Eubacterium acidaminophilum*. *Arch. Microbiol.* 152, 182–188.

18 Grohovaz, F., Bossi, M., Pezzati, R., Meldolesi, J., Tarelli, F. T. (**1996**). High resolution ultrastructural mapping of total cal-

cium: electron spectroscopic imaging/electron energy loss spectroscopy analysis of a physically/chemically processed nerve-muscle preparation. *Proc. Natl. Acad. Sci. USA* 93, 4799–4803.

19 Gunning, B. E. S., Steer, M. W. (**1996**). Bildatlas zur Biologie der Pflanzenzelle. 4th ed., Gustav Fischer Publ., Stuttgart.

20 Hawes, C. R., Brandizzi, F., Andreeva, A. V. (**1999**). Endomembranes and vesicle traffikking. *Curr. Opin. Plant Biol.* 2, 454–461.

21 Hillmer, S., Movafeghi, A., Robinson, D. G., Hinz, G. (**2001**). Vacuolar storage proteins are sorted in the cis-cisternae of the pea cotyledon Golgi apparatus. *J. Cell Biol.* 152, 41–50.

22 Hobot, J. A., Carlemalm, E., Villiger, W., Kellenberger, E. (**1984**). Periplasmic gel: new concept resulting from the reinvestigation of bacterial cell envelope ultrastructure by new methods. *J. Bacteriol.* 160, 143–152.

23 Holzenburg, A. and Mayer, F. (**1989**). D-ribulose-1,5-bisphosphate carboxylase/oxygenase: function-dependent structural changes. *Electron Microsc. Rev.* 2, 139–169.

24 Hoppert, M., Gentzsch, C., Schörgendorfer, K. (**2001**). Structure and localization of cyclosporin synthetase, the key enzyme of cyclosporin biosynthesis in *Tolypocladium inflatum*. *Arch. Microbiol.* 176, 285–293.

25 Houmiel, K. L., Slater, S., Broyles, D., Casagrande, L., Colburn, S., Gonzalez, K., Mitsky, T. A., Reiser, S. E., Shah, D., Taylor, N. B., Tran, M., Valentin, H. E., Gruys, K. J. (**1999**). Poly(beta-hydroxybutyrate) production in oilseed leukoplasts of *Brassica napus*. *Planta* 209, 547–550.

26 Kellenberger, E., Kellenberger-van-der-Kamp, C. (**1994**). Unstained and in vivo fluorescently stained bacterial nucleoids and plasmolysis observed by a new specimen preparation method for high-power light microscopy of metabolically active cells. *J. Microsc.* 176–142.

27 Kessels, M. M., Qualmann, B., Klobasa, F., Sierralta, W. D. (**1996**). Immunocytochemistry by electron spectroscopic imaging using a homogeneously boronated peptide. *Cell Tissue Res.* 284, 239–245.

28 Kleinig, H., Maier, U. (**1999**). Zellbiologie. 4th ed., Gustav Fischer Publ., Stuttgart.

29 Kleinig, H., Sitte, P. (**1986**). Zellbiologie. 2nd ed., Gustav Fischer Publ., Stuttgart.

30 Kohring, G. W, Mayer, F. (**1987**) In situ distribution of EcoRI methylase and restriction endonuclease in cells of *Escherichia coli* Bs 5. *FEBS Lett.* 216, 207–210.

31 Kristić, R. V. (**1984**). Illustrated Encyclopedia of Human Histology. Springer Publ., Berlin.

32 Matuschek, M., Burchhardt, G., Sahm, K., Bahl, H. (**1994**). Pullulanase of *Thermoanaerobacterium thermosulfurogenes* EM 1 (*Clostridium thermosulfurogenes*): molecular analysis of the gene composite structure of the enzyme and a common model for its attachment to the cell surface. *J. Bacteriol.* 176, 3295–3302.

33 Mayer, F. (**1986**). Cytology and morphogenesis of bacteria (Encyclopedia of Plant Anatomy, VI, 2). Borntraeger Publ., Berlin.

34 Mayer, F., Coughlan, M. P., Mori, Y., Ljungdahl, L. G. (**1987**). Macromolecular organization of the cellulolytic enzyme complex of *Clostridium thermocellum* as revealed by electron microscopy. *Appl. Environ. Microbiol.* 53, 2785–2792.

35 Mayer, F., Friedrich, C. (**1986**). Higher orders structural organization of the nucleoid in the thiobacterium *Thiosphaera pantotropha*. *FEMS Microbiol. Lett.* 37, 109–112.

36 Mittendorf, V., Robertson, E. J., Leech, R. M., Krüger, N., Steinbüchel, A., Poirier, Y. (**1998**). Synthesis of medium-chain-length polyhydroxyalkanoates in *Arabidopsis thaliana* using intermediates of peroxisomal fatty acid β-oxidation. *Proc. Natl. Acad. Sci. USA* 95, 13397–13402.

37 Mootz, H. D., Marahiel, M. A. (**1997**). Biosynthetic systems for nonribosomal peptide antibiotic assembly. *Curr. Opin. Chem. Biol.* 1, 543–551.

38 Peters J., Baumeister W., Lupas A. (**1996**). Hyperthermostable surface layer protein tetrabrachion from the archaebacterium *Staphylothermus marinus*: evidence for the presence of a right-handed coiled coil derived from the primary structure. *J. Mol. Biol.* 257, 1031–1041

39 Pieper-Fürst, U., Madkour, M. H., Mayer, F., Steinbüchel, A. (**1994**). Purification and characterisation of a 14-kilodalton protein that is bound to the surface of polyhydroxyalkanoic acid granules in *Rhodococcus ruber*. *J. Bacteriol.* 176, 4328–4337.

40 Ploug, T., van Deurs, B., Ai, H., Cushman, S. W., Ralston, E. (**1998**). Analysis of GLUT4 distribution in whole skeletal muscle fibers: identification of distinct storage compartments that are recruited by insulin and muscle contractions. *J. Cell Biol.* 142, 1429–1446.

41 Poirier, Y. (**1999**). Production of new polymeric compounds in plants. *Curr. Opin. Biotechnol.* 10, 181–185.

42 Poirier, Y., Dennis, D., Klomparens, K., Nawrath, C., Somerville, C. (**1992**). Perspectives on the production of polyhydroxyalkanoates in plants. *FEMS Microbiol. Rev.* 103, 237–246.

43 Pum, D., Sleytr, U. B. (**1999**). The application of bacterial S-layers in molecular nanotechnology. *Trends Biotechnol.* 17, 8–12.

44 Ramm, G., Slot, J. W., James, D. E., Stoorvogel, W. (**2000**). Insulin recruits GLUT4 from specialized VAMP2-carrying vesicles as well as from the dynamic endosomal/*trans*-Golgi network in rat adipocytes. *Mol. Biol. Cell* 11, 4079–4091.

45 Rastogi, N., Frehel, C., David, H. L. (**1984**). Evidence for taxonomic utility of periodic acid-thiocarbohydrazide-silver proteinate cytochemical staining for electron microscopy. *Internatl. J. System. Bacteriol.* 34: 293–299.

46 Robinow, C., Kellenberger, E. (**1994**). The bacterial nucleoid revisited. *Microbiol. Rev.* 58, 211–232.

47 Robinson, D. G. (**1985**). Plant membranes. John Wiley & Sons, New York.

48 Rout, M. P., Blobel, G. (**1993**). Isolation of the yeast nuclear pore complex. *J. Cell Biol.* 123, 771–783.

49 Ryter, A., Chang, A. (**1975**). Localization of transcribing genes in the bacterial cell by means of high resolution autoradiography. *J. Mol. Biol.* 98, 797–810.

50 Sandman, K., Krzycki, J. A., Dobrinski, B., Lurz, R., Reeve, J. N. (**1990**). HMf, a DNA-binding protein isolated from the hyperthermophilic archaeon *Methanothermus fervidus,* is most closely related to histones. *Proc. Natl. Acad. Sci. USA* 87, 5788–5791.

51 Silva, F. G., Lawrence, J. B., Singer, R. H. (**1989**). Progress toward ultrastructural identification of individual mRNAs in thin section: Myosin heavy-chain mRNA in developing myotubes. In: *Techniques in Immunocytochemistry*, Vol. 4 (eds. G. R. Bullock, P. Petrusz), pp. 147–165. Academic Press, London.

52 Sleytr, U. B. (**1978**). Regular arrays of macromolecules on bacterial cell walls: structure, chemistry, assembly, and function. *Internatl. Rev. Cytol.* 53, 1–64.

53 Sleytr, U. B., Messner, P. (**1983**). Crystalline surface layers of bacteria. *Annual Rev. Microbiol.* 37, 311–339.

54 Sosinsky, G. E., Perkins, G. A. (**2000**). Electron crystallographic methods for investigating gap junction structure. *Methods* 20, 140–155.

55 Specka, U., Mayer, F. (**1993**). Cellular location, activity states, and macromolecular organization of glycoamylase in *Clostridium thermosaccharolyticum. Arch. Microbiol.* 160, 284–287.

56 Specka, U., Spreinat, A., Antranikian, G., Mayer, F. (**1991**). Immunocytochemical identification and localization of active and inactive α-amylase and pullulanase in cells of *Clostridium thermosulfurigenes* EM 1. *Appl. Environ. Microbiol.* 57, 1062–1069.

57 Sriubolmas, N., Panbangred, W., Sriurairatana, S., Meevootisom, V. (**1997**). Localization and characterization of inclusion bodies in recombinant *Escherichia coli* cells overproducing penicillin G acylase. *Appl. Microbiol. Biotechnol.* 47, 373–378.

58 Steinbüchel, A., Aerts, K., Babel, W., Föllner, C., Liebergesell, M., Madkour, M., Mayer, F., Pieper-Fürst, U., Pries, A., Valentin, H. E., Wieczorek, R. (**1995**). Considerations on the structure and biochemistry of bacterial polyhydroxyalkanoic acid inclusions. *Can. J. Microbiol.* 41 (Suppl. 1), 94–105.

59 Stoylova, S. S., Flint, T. D., Kitmitto, A., Ford, R. C., Holzenburg, A. (**1998**). Comparison of photosystem II 3-D structure as determined by crystallography of frozen-hydrated and negatively stained specimens. *Micron* 29, 341–348.

60 Thurston, G., McLean, J. W., Rizen, M., Baluk, P., Haskell, A., Murphy, T. J., Hanahan, D., McDonald, D. M. (**1998**). Cationic liposomes target angiogenic endothelial cells in tumors and chronic inflammation in mice. *J. Clin. Invest.* 101, 1401–1413.

61 Torrisi, M. R., Lotti, L. V., Belleudi, F., Gradini, R., Salcini, A. E., Confalonieri, S., Pelicci, P. G., Di Fiore, P. P. (**1999**). Eps15 is recruited to the plasma membrane upon epidermal growth factor receptor activation and localizes to components of the endocytic pathway during receptor internalization. *Mol. Biol. Cell* 10, 417–434

62 Ude, J., Koch, M. (**1994**). Die Zelle – Atlas der Ultrastruktur. 2nd ed., Gustav Fischer, Jena.

63 Vitale, A., Raikhel, N. V. (**1999**). What do proteins need to reach different vacuoles? *Trends Plant Sci.* 4, 149–155.

64 Wong, S. L., Nakamoto, L., Wainwright, J. F. (**1997**). Detection of toxic organometallic complexes in wastewaters using algal assays. *Arch. Environ. Contam. Toxicol.* 32, 358–366.

65 Yeager, M., Nicholson, B. J. (**1996**). Structure of gap junction intercellular channels. *Curr. Opin. Struct. Biol.* 6, 183–192.

66 Yoshikawa, N., Oogake, S., Terada, M., Miyabayashi, S., Ikeda, Y., Takahashi, T., Ogawa, K. (**1999**). Apple chlorotic leaf spot virus 50 kDa protein is targeted to plasmodesmata and accumulates in sieve elements in transgenic plant leaves. *Arch. Virol.* 144, 2475–2483.

2.3 Applications and Limitations in Scanning Electron Microscopy

Scanning electron microscopy (SEM) is the technique for visualization of surfaces at the cellular and subcellular level. The depth of focus allows imaging of wide areas and large, irregular-shaped objects. SEM produces images of surfaces that have a strong three-dimensional plasticity. Scanning electron microscopes routinely operate between the magnification range of stereo light microscopes up to 100,000 fold (which is a comfortably large range), with a resolution limit of 2.5 nm.

The SEM employs a finely focused electron beam to scan across the specimen surface. The diameter of the focused beam on the specimen surface determines the resolution of the image. The structures of interest must be stable after exposure to the electron beam, and also vacuum stable (usually high vacuum-stable, but environmental SEMs allow a working pressure of 0.19 Pa). For visualization in standard instruments, the specimen must remain stable after exposure to the electron beam, must reflect sufficient back-scattered electrons and emit sufficient secondary electrons – and should develop as few surface charge as possible (i.e., must be conductive). Some biological structures, such as hard exoskeletons, mineralized structures (teeth, bone), and rigid plant structures (pollen, wood, etc.) fulfil these prerequisites directly after metal coating. Most biological specimens, though, have to be fixed and dehydrated or, alternatively, frozen and covered with a conductive layer to meet the requirements for SEM imaging. Environmental scanning electron microscopy, low-dose electron beams, and cryo techniques allow the chemical and thermal stress applied to the specimen to be reduced to a minimum, lessening the danger of artificial alterations. The interaction of beam electrons with the specimen enables more than structural information to be gained. X-ray microanalysis allows monitoring of chemical elements on the specimen surface. The distribution of elements can be visualized either optically as an image of the specimen or graphically as emission spectra. In cathodoluminescence, electrons interacting with an object cause luminescence (in principle, the same kind of excitation response as in optical fluorescence microscopy). The use of fluorescence is limited due to rapid "bleaching" of the specimen and low intensities of the emitted light, but the "autofluorescence" capability of biological material can successfully be exploited to visualize, for example, chalk deposits in tissues or sclerified walls in blood vessels. Treatment with conventional fluorochromes leads to additional excitation.

The main draw back-compared with light microscopy – is that prior to examination under the electron microscope the object has to be dehydrated or frozen. For most complex samples this requires chemical fixation and treatment with organic solvents or rapid freezing. Therefore, unlike in light microscopy, the biological specimen is inactive or dead; "on-line" observation of the specimen is impossible. Although low-temperature preparation techniques and environmental SEM allow more gentle specimen pretreatment, most specimens are routinely processed by chemical fixation and dehydration techniques, with a certain risk of inducing artifacts.

2.3.1 Visualization of Surface and Internal Cellular Structures

SEM is a versatile method for collection of data on surface structures, with magnifications ranging from that of a magnifying glass scale down to resolutions of several nanometers, and at high depth of field. The method thus allows high-resolution views ranging from growth patterns of cells on natural and artificial carriers to macroscopically visible organisms. Views of internal structures are only possible if the surface has been artificially removed. Localization techniques, with antibodies as specific probes and colloidal gold as detectable marker, for example, allow detection of cellular components, but are rather rarely applied due to problems in visualization of internal structures.

Though most filamentous appendages of the bacterial cell and surface layers are better visualized by techniques that allow higher resolution (such as transmission electron microscopy), large bacterial surface structures are very readily visible. Figure 2.26 compares whole cells of a cellulose-degrading *Clostridium* strain with a non-cellulose degrading mutant. The surface of the mutant appears smooth, whereas the wild-type strain depicts numerous protuberances. Although structural details of these protuberances (cellulosomes) can only be unraveled by other techniques, SEM gives a good view of the whole surface.

Figure 2.27 a depicts the interactions between the surface of a pollen grain and papillae on the pistil of *Arabidopsis thaliana*. The low magnification does not in principle provide more information than the light microscopic image, but the high depth of field allows the spatial arrangement of all structures to be overviewed and understood. Nevertheless, the imaging technique does not allow the creation of a transparent sample, and the pollen tube (arrow) inside the papillum is invisible.

Visualization of cell components after cleavage of a specimen is possible. In Fig. 2.27 b, some cells appear to be nearly empty because they are highly vacuolated. The predominant cell components are globular bodies, the starch-containing amyloplasts, and a filamentous network of cytoplasmic strands. It is difficult to differ-

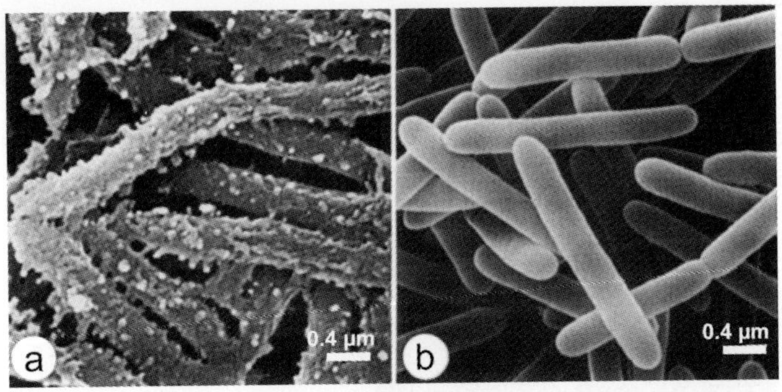

Fig. 2.26 (a) Cells of a *Clostridium thermocellum* wild-type strain with numerous protuberances on their cell surfaces. The protuberances are composed of cellulosomes, the cellulase-degrading complexes (Mayer, 1988).
(b) A mutant strain deficient in cellulase biosynthesis shows a smooth cell surface (Mayer, 1988)

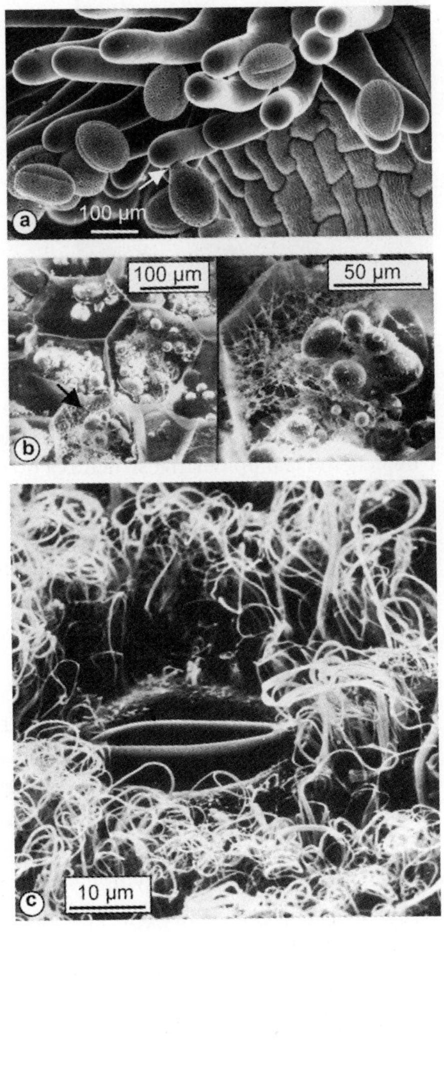

Fig. 2.27 (a) Pollen grains sticking to papillae on the stigma of *Arabidopsis thaliana*. The pollen tubes grow into the papillae when the grains germinate (arrow) (Gunning and Steer, 1996).
(b) REM images of storage cells in a critical-point dried sample of a potato. The right image shows one cell from the left (arrow) in a higher resolution (Gunning and Steer, 1996).
(c) Deposit of wax on the bottom of a banana leaf. Wax filaments (25–50 μm long cover) the epidermis, apart from the stomata (Gunning and Steer, 1996)

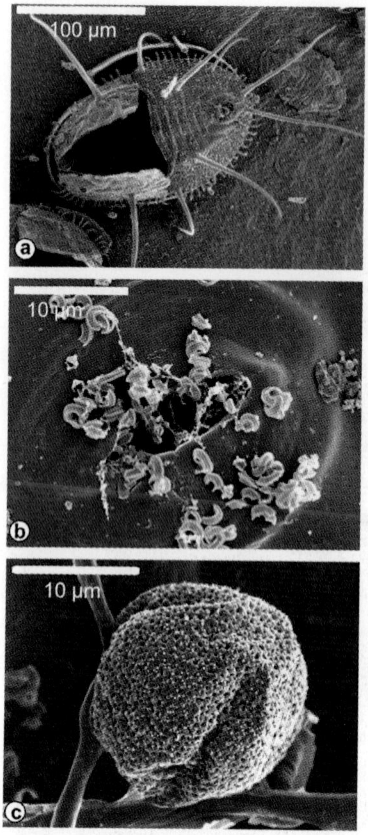

Fig. 2.28 Surfaces of whitefly-infested *Fuchsia* leaves (Marina S. Chow, Smith College, Northampton, Mass).
(a) An empty pupa (final whitefly nymph stage) on a leaf surface.
(b) Blocked stoma on the leaf lower side.
(c) Fungal sporangium and mycelium growing on the surface of a diseased leaf

entiate the other cellular contents. Though SEM is not the method of choice for depiction of cellular organelles, the sculpturing of rigid cell walls is easily visible. Figure 2.27 c depicts the leaf epidermal structures (lower leaf surface) from banana. The stoma is surrounded by epidermal cells with a highly sculptured cuticle: wax filaments of 25–50 μm in length.

2.3.2
Interactions between Organisms

At lower magnification ranges, light microscopy, confocal microscopy (in extended focus mode), and SEM may be of equal value for obtaining the relevant image information. Figure 2.28 illustrates the interaction between the plant host and an insect parasite, a whitefly, in the light microscope magnification range. The large depth of focus allows visualization of all structures in an area of about one mm^2. The images illustrate the adhesion of a nymph stage of the insect on the leaf surface (Fig. 2.28 a). Tiny wax particles secreted by the insect attach to the leaf surface and may be one reason for the plant-pathogenic effect of the organisms (Fig. 2.28 b). Whiteflies secrete honeydew at all stages in their lives, promoting the growth of fungal mycelia on the leaf epidermis (Fig. 2.28 c).

Interaction between microorganisms and surfaces requires higher magnification and resolution. Light microscopic techniques and specific labeling allow visualization of overall distribution of microorganisms, but do not directly reveal the structural details. Figure 2.29 illustrates the importance of a large magnification range for mon-

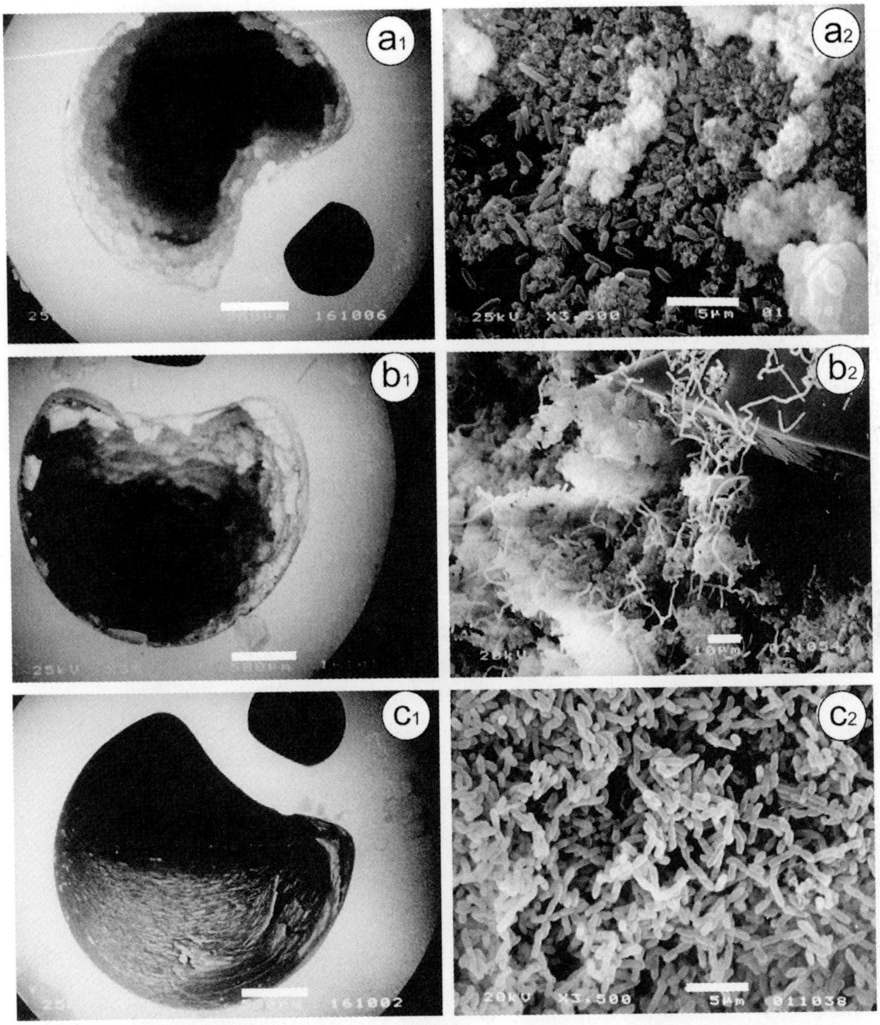

Fig. 2.29 Colonization of urethral catheters (Stickler et al., 1998).
(a) *Proteus vulgaris* NM19,
(b) *Proteus mirabilis* NM60, and
(c) *Pseudomonas aeruginosa*. NM33 At higher magnifications, crystal formation could sometimes be observed (c2). Bars represent 500 μm in a1 – c1, 5 μm in a2 and c2, 10 μm in b2)

itoring of biofilms. The example depicts a common biofilm on a urethral catheter, grown in an environment modeling conditions in the urinary bladder. The overall view shows the differences in thickness and texture of the biofilms, developed within 24 hours after inoculation with various bacterial strains. At higher magnification, the organisms and crystalline deposits are visible (Fig. 2.29 b2). These deposits are formed by urease-positive organisms and lead to a rapid increase in biofilm thickness. These biofilms apparently cause rapid clogging of the catheters.

Surfaces of thin biofilms are depicted in Fig. 2.30 a and b. SEM was used in this application to give an overall view of a blocked membrane before and after a cleaning procedure with a membrane-cleaning detergent. In spite of the obvious reduction in structural heterogeneity, the residual biofilm still reduces the performance of the filtration process. Figures 2.30 c and d show a layer of non-cellular debris on a membrane surface.

Figure 2.31 illustrates decomposition of a polymeric surface, the biodegradable polymer BIOPOL. After 7 days incubation with *Pseudomonas fluorescens*, the bacteria have formed numerous erosion pits on the surface of a poly-hydroxyoctanoate derivative. A freeze-fracture of the specimen reveals that the erosion has also taken place in deeper layers of the material (Fig. 2.31 d). A close-up view shows bacteria associated with an erosion pit (Fig. 2.31 e).

Figure 2.32 shows a SEM preparation obtained simply by cracking a small piece of limestone, colonized by algal and fungal cells distributed inside the stone to a depth of several millimeters under the surface. These endolithic biofilms may be involved

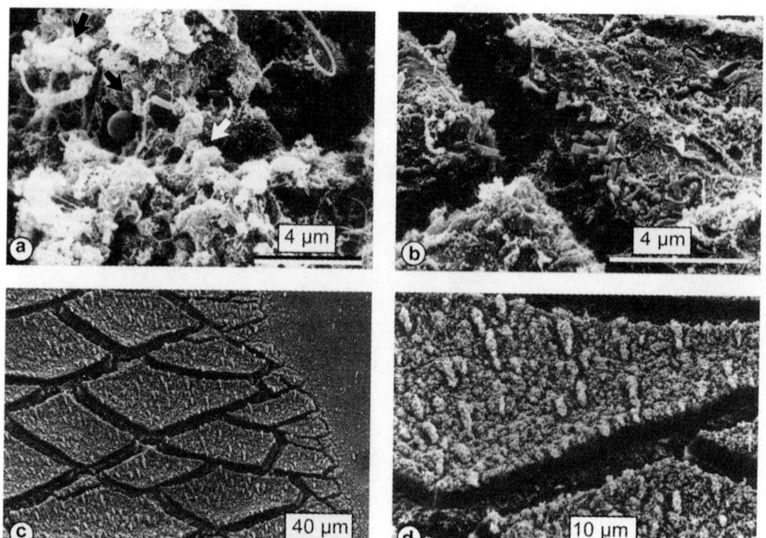

Fig. 2.30 (a, b) Membrane, blocked by biofouling. a, before; b, after 48 h treatment with cleaning agent. 80 % of the coating was removed (Flemming, 1995).
(c, d) Abiotic coating, developed in microbially enriched seeping water. b, detail from a, higher resolution (Flemming, 1995)

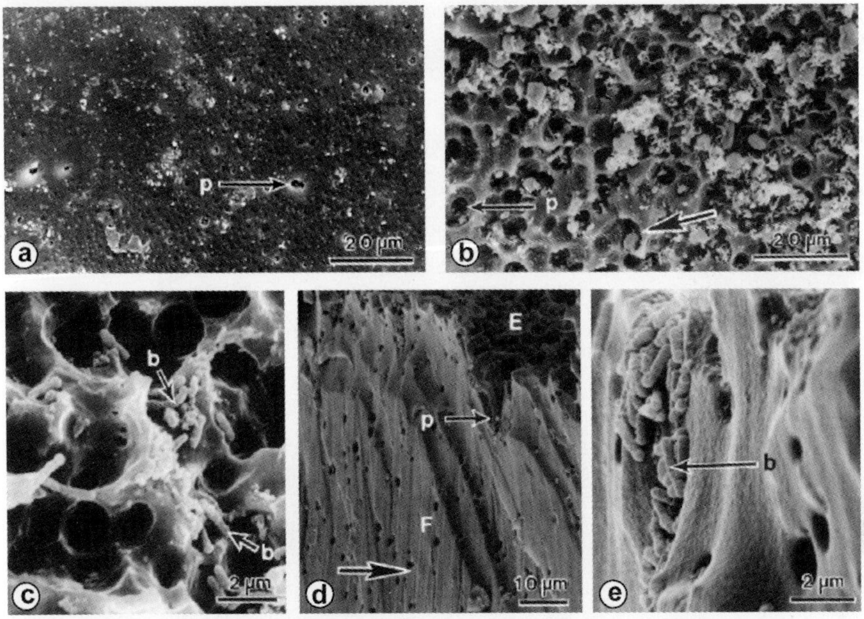

Fig. 2.31 Poly(3-hydroxyoctanoate) (PHO) and cross-linked derivatives (Molitoris et al., 1996).
(a), PHO-100CL, surface of sterile-incubated control showing discrete irregular pits (p).
(b, c) PHO-100CL, *Pseudomonas fluorescens*, 7 days incubation. The exposed surface is covered with discrete erosion pits (p) separated by less eroded areas (arrows). Note bacteria (b) within the pits.
(d, e) PHO-100CL, freeze-fractured surface, *Pseudomonas fluorescens*, 7 days incubation (d), The pitted exposed surface (E) and fractured surface (F) shows holes of 1 – 3 μm diameter (arrows) and 10 to 20 μm deep erosion pits (p) originating from the exposed surface. (e) An erosion pit with associated bacteria (b)

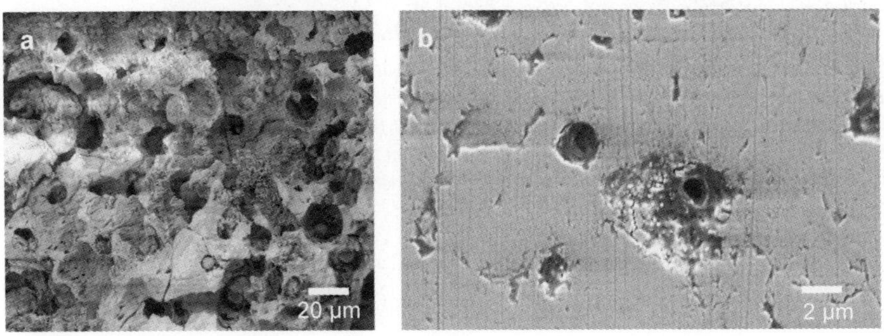

Fig. 2.32 Limestone colonized by algal and fungal cells. Scanning electron micrographs of sections perpendicular to the limestone surface. (W. Pohl, Göttingen University).
a) Layer of algal cells. The small cavities are created by active substrate dissolution of the algae.
b) Detail of hyphae dissolving the limestone

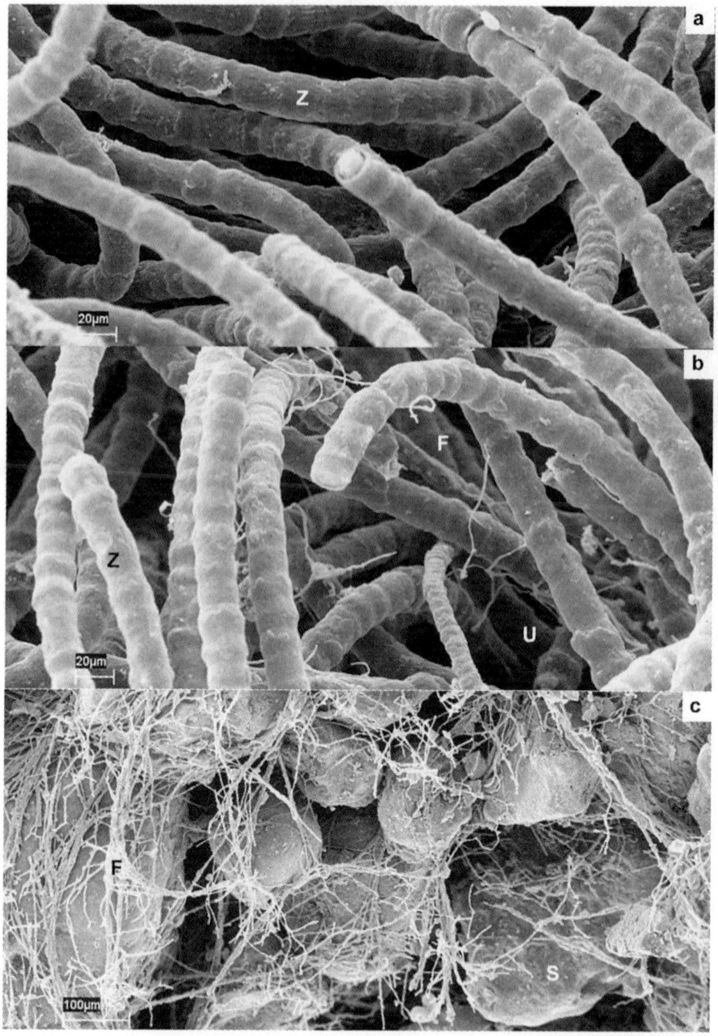

Fig. 2.33 Samples taken from a microbial soil crust (R. Reimer, Göttingen University).
(a) The upper crust layer contains exclusively algal filaments (*Zygogonium ericetorum*).
(b) Filaments in a lower crust layer are infested with fungi (*Fusarium sp.*).
(c) The bottom layer of the crust is dominated by *Fusarium sp.* mycelium.
Z, *Z. ericetorum*; U, *Ulothrix sp.*; F, *Fusarium sp*; S, sand grains

Fig. 2.34 Analysis of wind-borne particulates in a lichen thallus (Lee, 1999).
(a) Angular mineral fragments derived from granite (the rock substratum) and spherical particles (arrows), also concentrated in a cluster (rectangle).
(b) Higher magnification image of the cluster; the spheres have an average diameter of 0.96 μm and the smallest is 0.35 μm.
(c) Energy-disperse X-ray spectrum of one of the fly ash spheres reveals the composition in the elements Al, Si, K, and O

in successive decomposition of stone material. Though organisms inhabiting rock and building stone are ubiquitous, the organisms in deeper stone layers are invisible before cleavage of the material, because they are entirely surrounded by their rigid substratum. Small cracks and fissures may allow diffusion of gas or mineral nutrients.

In addition, when loose mineral grains are integrated in a biofilm network, a light microscopic view of such a biofilm sample is difficult and requires the preparation of thin transmittable slices (Fig. 1.30). Fig. 2.33 shows a view of a soil crust biofilm – perpendicular to the surface – obtained simply by breaking a small piece of crust.

SEM imaging is not restricted to monitoring of sample morphology. Element analysis allows high resolution mapping of elements inside samples. An analysis of the X-ray emission of a sample area is depicted in Fig. 2.34. Here, X-ray spectra of small particles inside an endolithic lichen thallus reveal their compositions in the elements Al, Si, K, and O. The element composition and the spheroidal shape allow these particles to be identified as fly ash with a composition corresponding to alkali feldspar.

2.3.3 Selective Detection of Surface Structures

Although fluorescent light microscopy and transmission electron microscopy are commonly used for localization of specific targets, specific probes are also detectable under the scanning electron microscope. Detection of specific sites is, of course, restricted to the specimen surface. SEM is therefore an interesting tool for detection of surface antigens, allowing mapping of antigen distribution in complex three-dimensional objects.

Cellular surface antigens – such as sugar residues – are detectable with marker-conjugated antibodies or lectins. In Fig. 2.35 a, the smooth membrane (arrow) is covered with microvili. Figure 2.35 b shows the membrane labeled with the lectin Con A. While direct imaging of Con A is not possible, a large or electron-dense marker particle (here the large protein molecule hemocyanin) allows detection of the lectin attached to the cell surface. Small hemocyanin particles are distributed over the whole membrane surface, indicating the presence of Con A binding sites.

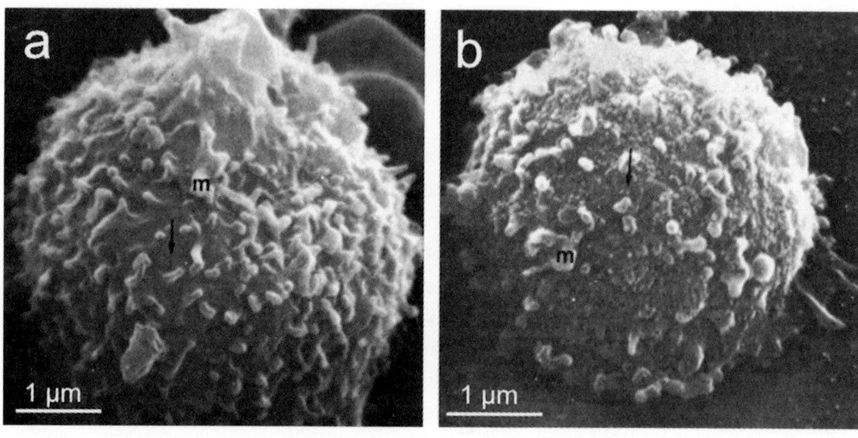

Fig. 2.35 Localization studies with SEM specimens (Hayat, 1978).
(a) Normal rat lymphocyte. Although the surface is covered with irregular protrusions (microvilli, m), the membrane itself is smooth (arrow).
(b) Rat lymphocyte exposed to Con A and hemocyanin (marker) at 20 °C for 30 min. Bound hemocyanin molecules (arrow) identify Con A-binding sites. The complexes are distributed over the entire surface of the cell, except for the periphery near the substrate

References

1 Flemming, H. (**1995**). Biofouling bei Membranprozessen. Springer, Berlin.
2 Gunning, B. E. S., Steer, M. W. (**1996**). Bildatlas zur Biologie der Pflanzenzelle. 4th ed. Gustav Fischer Verlag, Stuttgart.
3 Hayat, M. A. (**1978**). Introduction to biological scanning electron microscopy. University Park Press, Baltimore.
4 Lee, M. R. (**1999**). Organic-mineral interactions studied by controlled pressure SEM. *Microscopy and Analysis* 58, 23–25.
5 Mayer, F. (**1988**). Cellulolysis: Ultrastructural aspects of bacterial systems. *Electron Microsc.* Rev. 1, 69–85.
6 Molitoris, H. P., Moss, S. T., DeKoning, G. J. M., Jendrossek, D. (**1996**). Scanning electron microscopy of polyhydroxyalkanoate degradation by bacteria. *Appl. Microbiol. Biotechnol.* 46, 570–579.
7 Stickler, D., Morris, N., Moreno, M. C., Sabbuba, N. (**1998**). Studies on the formation of crystalline bacterial biofilms on urethral catheters. *Eur. J. Clin. Microbiol. Infect. Dis.* 17, 649–652.

2.4 Preparation Techniques

M. Hoppert and A. Holzenburg[*)]

2.4.1 The Transmission Electron Microscope

The general construction principle of an electron microscope has not changed during the past decades, though numerous extensions for examination of frozen specimens and element analysis have been developed. Like light microscopes, electron microscopes also contain lens groups for illuminating the specimen (condenser lens systems) and for specimen magnification. Light and electron optics share basic principles, though some essential details differ, which in practice causes more problems with electrons than with visible light as illuminating medium. The wavelength of electrons is five orders of magnitude shorter than the wavelength of visible light, however, and so instruments operating with electrons rather than light have considerably higher resolution, although in fact, because of technical restrictions, only about three orders of magnitude better than light microscopes.

Because of its intensive interaction with matter, formation of an electron beam is only possible under high vacuum conditions. The electron source is composed of a wire filament and a Wehnelt shield. Whereas these elementary building blocks of a transmission electron microscope have been present in nearly all instruments since the 1950s, the application potential of the microscopes (Fig. 2.36) has been broadened

*) Director, Microscopy and Imaging Center
Professor of Biology and
Biochemistry & Biophysics
Texas A & M University

by numerous additional features. The electron source in standard instruments is a tungsten or lanthanum hexaboride (LaB_6) filament. LaB_6 filaments have longer lives and produce a higher image brightness. The filament is heated by a small electric current, which leads to thermionic emission of electrons at the tip of the filament. The filament tip is placed close to a circular aperture in the Wehnelt shield. For operation, the filament heat is brought up to a point at which an equilibrium exists between the filament and the shield potential, which is then between 100 to 500 V negative to the filament (due to an automatic biasing circuit). The electrons are accelerated between the Wehnelt shield and an anode plate by an "accelerating" potential, between 60 and 100 kV for biological specimens. All the emitted electrons pass through a plane of minimum cross-section where electron density per unit area is highest. This gun

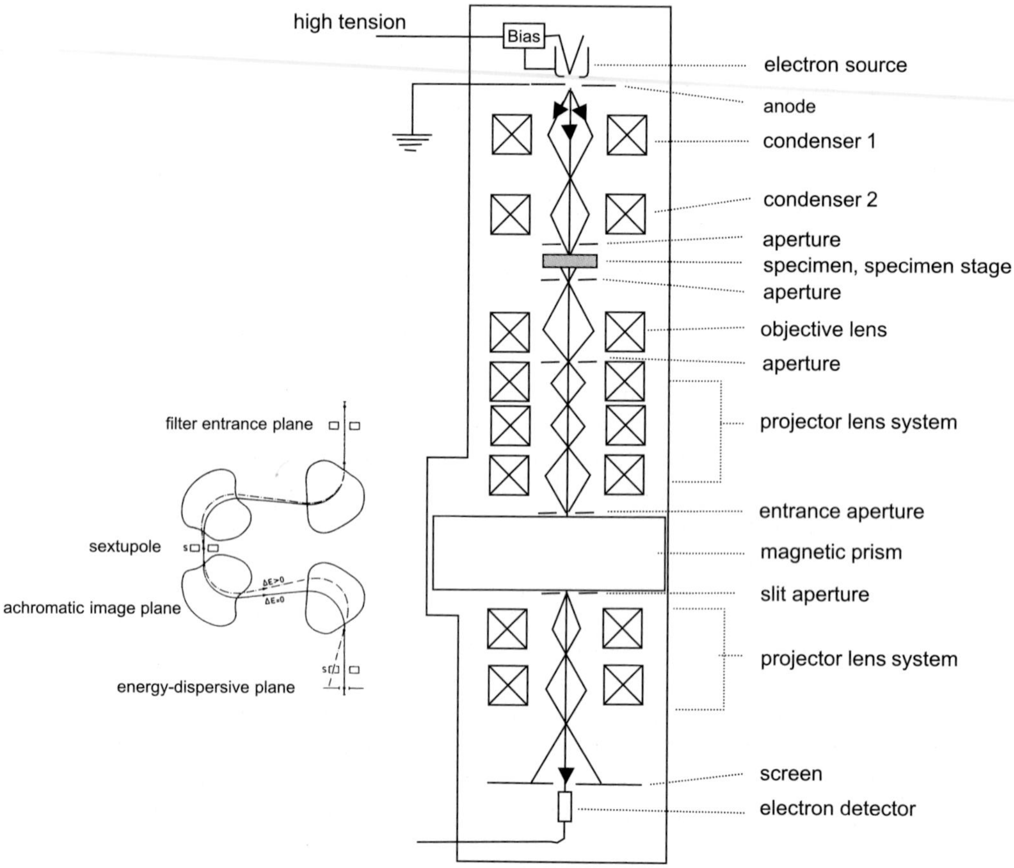

Fig. 2.36 Schematic view of an electron microscope (LEO 912 Omega, LEO Electron Microscopy, Ltd.) containing the building blocks of conventional instruments and an additional energy selecting unit. The unit consists of four magnetic prisms; a sextupole magnet is a necessary element for beam correction. The electron beam is bent in an Ω-shaped trajectory. The electrons follow different trajectories according to their energy losses E as indicated by the closed and the dashed lines

crossover is used as the actual electron source. The electron beam passes through a small hole in the anode plate and enters the electron-optical part of the column. Field emission guns (FEGs) are required whenever a small diameter of the electron beam or high coherence are necessary, such as for highest resolution in scanning transmission electron microscopy and electron crystallography. The electron source is a needle-point single crystal of tungsten. The filament is either cold or (usually) heated (Schottky type). A very strong electric field is needed to extract ("tunnel out") electrons from the tip.

Magnetic lenses consist of concentric copper coils, generating a magnetic field when a current passes through the copper wire. This field acts as a convex lens, bringing off-axis rays back into focus. Changing the strength of the current alters focal length. Analogously to light waves, electrons of different wavelength (velocity) leaving a point in the specimen will not be brought to a focus at the same point in the image. This "chromatic aberration", correctable in light microscopes with appropriate combination of lens groups, cannot be compensated in a comparable way in electromagnetic lenses. The resolution of light microscopes is therefore much closer to the theoretical limit represented by the wavelength of light than the resolution in electron microscopes, relative to the wavelength of accelerated electrons.

Two condenser lenses are regularly used as illuminating system. The first condenser lens creates a demagnified image of the gun crossover. By adjusting the current of the first condenser lens, preadjustment of the minimum limit of the spot size in the rest of the condenser system is possible. The condenser lens current is often fixed, because this adjustment is of limited practical use. The second condenser lens affects the convergence (i.e., the angle describing the cone of illumination) and the diameter of the illuminated area on the specimen. It therefore helps to control the intensity of illumination.

Positioned immediately below the specimen, the objective lens forms an inverted initial image. As in the light microscope, this lens is the most important single component for proper imaging, since any defects in this lens will be further magnified by the rest of the optical system. Focusing of the specimen image is accomplished by varying the focal length of this lens. To support focus adjustment in electron microscopes, there are additional controls, indicating the precision of the adjusted focus, present in several instrument types. The following lenses represent the projector system, consisting of either two or three additional lenses. The first lens of this system can be focused on the initial image formed by the objective lens or the rear focal plane of the objective lens. This determines whether the viewing screen of the microscope shows an electron diffraction pattern of the specimen or the real image. The additional projector lenses are used to magnify the intermediate images further. The final image is projected on a fluorescent screen or can be directed to a plate camera or a video camera.

Because of the need to operate under high vacuum, potential for manipulation of the specimen during inspection under the electron microscope is limited. Numerous types of specimen stages have been developed in order to overcome these obstacles:

Goniometer stages are used to tilt a specimen by selectable angles. A goniometer stage is indispensable for obtaining the structural information required for three-dimensional reconstruction of protein structures in single-particle analysis and electron crystallography, as well as for reconstruction of topography of objects in thick sections. Under normal conditions when the specimen is tilted, the image moves out of focus. A eucentric stage keeps the specimen area of interest near the tilt axis and minimizes changes in focus.

Multiple specimen stages are useful for observing several specimens in succession or a set of serial sections.

Cryo stages are cooled by thermal contact with liquid nitrogen or by a constant flow of cold nitrogen gas. These stages are widely used for examination of frozen-hydrated specimens. For some applications that require the highest resolutions, stages cooled by liquid helium are also used. Temperatures down to 10 K can be achieved.

In **environmental chambers** or hydration stages, the specimen is enclosed in a cell that is sealed with small apertures above and below the specimen to restrict gas flow into the microscope vacuum, but allowing the electron beam to pass. The specimen can be kept in a hydrated state under low vacuum conditions by maintaining a relatively high humidity. Although these stages are widely used in scanning electron microscopes, they are only rarely applied in TEM.

2.4.1.1 Image contrast formation in the transmission electron microscope

In spite of its numerous analytical features (see below), the electron microscope is primarily an instrument to display magnified images of a specimen. Thus, all preparation techniques, as well as construction and operation procedures, for the electron microscope aim primarily at the formation of an image that provides enough contrast to visualize all resolvable structures.

In bright-field light microscopy, one common way to obtain image contrast is the application of a differential stain that absorbs light, resulting in visible differences in the specimen. Under a TEM, for biological specimens of normal thickness (around 100 nm), the proportion of the beam electrons absorbed is minimal. Instead, image contrast is brought about by electrons that are i.) deflected (scattered) in such a way that they become masked out, and ii.) that pass the specimen without interaction.

The amount of scattered electrons on a particular area of specimen is directly proportional to its mass density (mass per unit area = density × thickness). For a 50 nm thick carbon film illuminated with 50 kV electrons, for example, 34 % of the beam is undeflected, while 11 % is elastically scattered and 55 % is inelastically scattered. An electron of the incident beam is elastically scattered when it is attracted by the positively charged nucleus of an atom. It is deflected at a large angle from its original trajectory, but loses no energy. Electrons with large scatter angles are masked out of the image by an objective aperture diaphragm. When an electron strikes a shell electron of an atom in the specimen, inelastic scatter may occur. The incident electron becomes decelerated, energy being transferred to the shell electron. The inelastically scattered electron is only slightly deflected from its original trajectory, and so is very likely to pass through the objective aperture diaphragm and does not contribute to contrast formation. High energy losses of electrons due to several inelastic scattering processes lead to visible chromatic aberration, impairing contrast and resolution, the probability of scattering increasing with increasing mass thickness. Since biological specimens consist mainly of carbon, nitrogen, oxygen, and hydrogen atoms, weak contrast is a limiting problem in the imaging of biological specimens. Inherent contrast may be increased by the addition of chemicals that differentially increase mass thickness, preferably substances such as heavy metal ions that specifically bind to specimen structures or by heavy metal shadowing of surface reliefs.

Other ways to improve the "scattering contrast" (or "amplitude contrast", due to the loss of amplitude – i.e., electrons – from the beam) are to work with lower acceleration voltages and with smaller objective apertures. Lower accelerating voltages intensify the scattering. The higher chromatic aberration at lower voltages due to inelastic scattering may, however, lead to unacceptable loss of resolution. As the size of the objective aperture is reduced, more scattered electrons are masked out and contrast is improved. The phases of electron waves are also affected during image formation, and contribute to generation of contrast. A marked increase in contrast can be achieved by slight defocusing, especially by underfocusing an image, whereas the contrast at the true focus is minimal. This increase in contrast can be explained in terms of interference of electron waves. When an electron beam passes through a thin scattering foil and another beam from the same source passes through a nearby hole, the two beams will be in a condition to interfere due to a phase delay in the beam passing through the matter. In a fully transparent (i.e., no variation in refractive index) object, there are no phase differences and, hence no phase contrast in the image. Defocusing, in which path lengths for scattered beams are changed more than those for the unscattered beam, can be used to enhance phase contrast. So-called Fresnel fringes serve to enhance such points or edges by delineating them with a bright line in the underfocused position (objective lens too weak) or with a dark line in the overfocused position (objective lens too strong). The fringes virtually disappear at the point of "exact" focus, and image contrast due to phase effects is at a minimum. It is common practice to defocus (underfocus) slightly to improve image contrast with virtually all types of specimens, as long as they are not to be subjected to image analysis (Fig. 2.37).

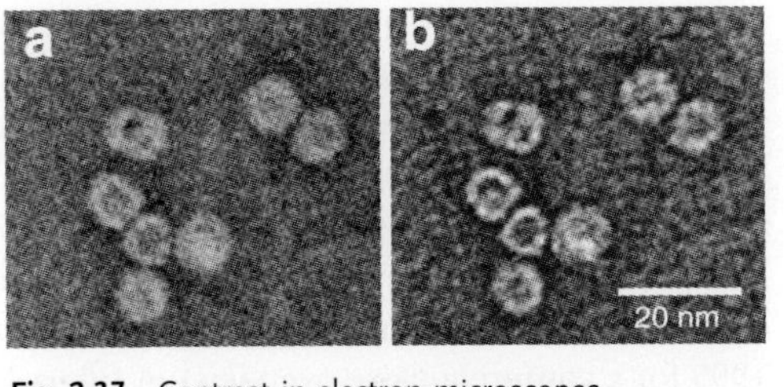

Fig. 2.37 Contrast in electron microscopes.
(a) Negatively stained ferritin molecules. In the focused specimen, contrast is mainly brought about by the uneven distribution of the heavy metal stain surrounding the particles and the iron core of ferritin (scatter contrast).
(b) The underfocused particles exhibit considerably increased contrast, but the apparent granularity makes small structures difficult to interpret

For thin unstained objects, contrast generation due to phase difference is often indispensable to provide an image with discernable details.

At slight overfocus the image is generally confusing. Initially, at slight defocus, resolution is not significantly reduced, but further defocusing reduces resolution considerably. Suitable focusing is a skill learned by experience. Excessive defocusing may lead to spurious details in the image, which may be erroneously interpreted as real structural details in the specimen.

2.4.1.2 Analytical features in transmission electron microscopes

Electron energy loss spectroscopy (EELS) and electron spectroscopic imaging (ESI) are techniques for analysis of the energy distribution, relative to the primary beam energy, in the transmitted electrons. These are primarily electrons that have been inelastically scattered by the specimen. They contribute to reduction in resolution in standard instruments, due to chromatic aberration of the electromagnetic lenses as mentioned above. Their energy loss is element-specific and, if the specimen is thin enough for electrons to experience only one inelastic collision, it may be resolved by "electron velocity" (or "electron energy loss") spectrometers. The required thickness is about 30–50 nm for biological material viewed with electrons at an accelerating voltage of 80 kV. Since electrons are deflected by magnetic fields at angles that depends on their energy, these spectrometers produce spectra of electrons with discrete energy losses. These spectra may be recorded either step-by-step, by adjustment either of the accelerating potential of the incident beam or of the magnetic field of the spectrometer, or by recording the whole spectrum at once with a sensitive CCD camera. While the step-by-step procedure is more accurate, the parallel recording of spectra is more rapid and so reduces the exposition time of the specimen. The energy loss spectrum contains information relating to the element composition and also on chemical bonding, molecular structure, and dielectric constant of the specimen. The electrons may also be selected so as to record an element-specific image (ESI) or to filter the chromatic aberrant images in order to obtain a final image with improved contrast and resolution. The former is especially important for images of unstained frozen-hydrated specimens, the latter for thick specimens. Several construction principles (i.e., post-column or in-column spectrometers) are used.

Post-column instruments are attached beneath the observation screen. The energy-filtered image or spectrum is recorded with a CCD camera. In-column spectrometers are incorporated in the projector lens system (Fig. 2.36). In this setup, several spectrometers are arranged in such a way that the electron beam is deflected back into the optical axis of the instrument (Castaing-Henry and Ω-filters; Bauer, 1988; see Reimer, 1995 for a comprehensive overview). A filtered image may either be observed directly on the fluorescent screen, or may be recorded with digital cameras or electron micrographic emulsion. This is especially advantageous for imaging of unstained specimens: electrons with chromatic aberrations that blur the final image may be filtered out, and only electrons with no energy loss contribute to the final image, enhancing contrast. In-column energy-filtering TEMs are often used for direct imaging of unstained (e.g., frozen-hydrated) specimens. Both instrumental setups allow depiction of the distribution of elements as a real image, a gray value being assigned to a relative element concentration (i.e., element-specific distribution in an image (image-EELS) may be performed). The resolution of energy loss spectra can be improved by the use of field emission guns instead of standard electron sources.

Scanning transmission electron microscopes (STEMs), although a distinct class of electron microscopes, may be technically "merged" with transmission electron microscopes (of particular practical relevance for specimens in life sciences). The electron source and the condenser lenses remain in operation, while the objective and projector lens systems are switched off. An additional condenser lens is necessary to provide a focused electron beam (probe spot), together with probe scan coils to deflect the electron beam for scanning. In advanced STEMs, an electron velocity spectrometer is an integral part of the imaging system, and so the instrument is also used for element analysis. The focused electron probe spot is raster scanned across a thin specimen, as with a TEM specimen. Resolving power is determined by the diameter of the probe spot. Electrons scattered by interaction with the specimen at wide angles are collected by an annular detector system, whilst unscattered and low-angle (inelastically) scattered electrons are treated in the spectrometer and subsequent detector systems. The great advantage of the STEM is its ability to analyze the image into separate signal channels, all of which are free of chromatic aberration (due to the absence of lenses) and which can be processed separately. The annular detector signal can be used as a measure of mass and is applicable to, for instance, measurement of mass distribution in unstained biological macromolecules. The filtered bright-field image, together with elemental analysis, allows high-resolution mapping of elements beyond the resolution limit of TEMs.

Another technique for analysis of the elemental composition of specimens in TEM is X-ray microanalysis. X-rays are emitted as a result of loss of energy from the incident electron to the outermost electron shells of the specimen atoms as it is slowed down during its passage through the specimen. This emission manifests as quanta of soft X-ray irradiation. Since the slowing process is continuous rather that stepwise, a continuous or "white" spectrum of soft X-rays is emitted. The wavelengths of the emitted X-rays are from 10 – 100 nm. The electrons of the incident beam also interact with the inner shells of specimen atoms. This causes an orbital electron to jump from a lower energy level shell to a higher energy shell. When it drops back, a "hard" or short-wavelength X-ray is emitted. These transitions and the respective X-ray irradiation are characterized by distinct energy quanta. Because the transition energies differ for individual atomic species, the emitted wavelengths are characteristic of the atoms making up the specimen. An electron microprobe X-ray microanalyser is used for recording for qualitative and quantitative assessment of element composition. The X-ray spectrum from a comparatively small volume of the specimen may be analyzed for all the detectable elements present. In the scan mode, the detector is set to an X-ray spectral line or an energy corresponding to a selected element, and the probe beam is scanned in a raster across the desired area of the specimen. In this way, an image of the element distribution is obtained.

2.4.2
Scanning Electron Microscopy

Scanning electron microscopes are primarily used for imaging of surface features. Since most electrons do not penetrate deeply in the specimen, information from below the surface cannot be collected, and so penetration of intact specimens and volume rendering techniques such as tomograpy cannot be performed. However, scanning electron microscopic images do show – at high depths of field – the surface reliefs of objects, and so are invaluable tools for surface analysis. Although scanning electron microscopes are more important for materials science than for life sciences – especially due to their element analysis features – SEM has great potential for application at, for instance, the cutting edge between organisms and surfaces or when relatively rigid biological material (wood, bone, etc.) is under investigation. In order to obtain good images of most non-conductive specimens under the SEM the sample must first be covered with a thin metal coating. This increases conductivity, and also reduces thermal damage and increases secondary and back-scattered electron emission.

As in transmission electron microscopes, the electron beam is produced by an electron gun, often based upon a lanthanum hexaboride filament. Higher resolution can be achieved with FEGs (field emission guns). The beam is condensed by the first condenser lens and then focused by the second condenser lens into a thin, coherent beam. A set of coils then scans or "sweeps" the beam in a grid-like fashion. The scan coils are essentially electromagnetic coils, arranged in two pairs around the beam. One pair of coils is responsible for controlling movement of the beam in the X direction, while the other pair moves it in the Y direction, triggered by a scan generator. The scan generator is connected to other components, the magnification module and the cathode ray tube, displaying the final image.

The final lens (objective lens) focuses the scanning beam onto a small spot on the specimen surface. The beam electrons penetrate into the specimen and travel for some distance before encountering an atom and colliding. In this way, the electron beam produces a region of (primary) excitation. Because of its shape this region is also known as the "tear-drop" zone. A variety of signals are produced from this zone, and it is its size and shape that ultimately determine the maximum resolution. Various signals (secondary electrons, back-scatter electrons, Auger electrons, characteristic X-rays, and cathodoluminescence) are now produced and may be collected with suitable detectors.

Secondary electrons are low-energy electrons ejected from the sample as a result of inelastic collisions with beam electrons. They are produced near the surface of the sample, from a small area around the beam. The resolution of the image is therefore very high, because the incident beam is only a few nanometers in diameter. The topographic features of the final image depend on how many of the secondary electrons actually reach the detector. Specimen areas emitting secondary electrons that do not reach the detector do not contribute to the final image. These areas appear as shadows or darker in contrast.

Back-scattered electrons are primary electrons emitted as a result of elastic collisions with specimen electrons. The emission intensity is a function of the specimens atomic number: the higher the atomic number, the brighter the signal. For this reason, a sample surface composed of two or more different elements differing significantly in their atomic numbers will produce an image that shows differential contrast of the elements despite a uniform topology. Elements of higher atomic number will produce more back-scattered electrons and so will appear brighter than neighboring elements.

The detectors for back-scattered and secondary electrons are similar. Both are scintillator and photomultiplier tube designs. Because secondary electrons possess low energies, the secondary electron detector is surrounded by a positively charged anode, which tends to attract many of the secondary electrons towards the scintillator. The back-scatter detector is not equipped with such an anode. Only those electrons that travel in a straight path from the specimen to the detector contribute to formation of the back-scattered image. In order that enough electrons to produce an image are collected, many SEMs use multiple back-scattered detectors positioned directly or nearly above the specimen. SEM images are combinations of both back-scattered and secondary electron signals in varying proportions. Since the x/y positions of the signal is determined by the scanning coils, a two-dimensional map of these signals represents the final SEM image.

When the beam interacts with the sample, electromagnetic radiation is released over a variety of wavelengths. This energy is the result of interaction of beam electrons with the inner electron shells of an atom. Radiation characteristic of the atoms electron shells energy levels is therefore released. The X-rays generated are collected and depicted as a spectrum. Each peak on the spectrum represents a transition with a characteristic energy – every element has its own "fingerprint". It is possible to calculate absolute concentrations of individual elements by comparing spectra with others collected from samples of known composition. Not all X-rays are characteristic of the elements present in the specimen; many belong to the background continuum (Bremsstrahlung) generated as the beam electrons slow down when they hit the sample. Maps of relative concentration for multiple elements can be generated by using the SEM. The electron beam is scanned across the sample, stopping at regular intervals to count the number of X-rays within a predefined energy window arriving at the detector. The number of counts at each stop (pixel) can then be displayed as a map. High-resolution maps can be produced by maximizing the number of points at which X-rays are counted, and by scanning the beam over the sample many times.

Phosphorus-containing materials show high luminescence, emitting visible light photons to build up a luminescent image. By collecting these photons with a light pipe and photomultiplier similar to those utilized by secondary electron detectors, it is possible to detect and count these photons. The best possible image resolution by this approach is estimated to be about 50 nm.

Environmental scanning electron microscopes overcome a main obstacle of electron microscopes: operation with a specimen under high vacuum conditions. In ESEMs the specimen chamber can be operated in a gaseous environment. The chamber itself is separated from the electron optical system by a system of apertures. When the primary electron beam hits the specimen, secondary electrons are emitted. These secondary electrons are attracted to the positively charged detector electrode. During their passage through the gaseous environment, collisions occur between electrons and gas particles, resulting in emission of more electrons and ionization of the gas molecules. This increase in the number of electrons effectively amplifies the original secondary electron signal. The positively charged gas ions are attracted to the negatively biased specimen and compensate charging effects. As the number of secondary electrons varies, the amplification effect of the gas varies. If a large number of electrons are emitted from a position on the specimen during a scan, there is a strong signal. If only a small amount of electrons are emitted, the signal is less intense. The difference in signal intensity from different locations on the specimen allows an image to be formed. The gas itself can be altered and may be, for example, water vapor, air, argon, or nitrogen (Li et al., 1995).

In this manner, the sample no longer needs to be coated in order to make it electrically conductive, and there is no longer the need for ultra-high vacuum. This means that samples can be imaged in a gas or water vapor at full SEM resolution with no dehydration. The sample may be observed in its natural environment with no sample preparation required, even for insulating samples, and X-ray analysis can also be performed on non-conductive samples.

2.4.3 Specimen Support

2.4.3.1 Grids

The universal "microscope slide" for transmission electron microscopes is the specimen grid. The standard diameter of grids used in TEM is 3.05 mm. Usually, the grid has to be covered with an electron-translucent specimen support film prior to use (Tables 2.1 and 2.2). The grid type should be selected with respect to the specimen and its preparation; some examples are listed in Table 2.1. The stability of the specimen and the size of the field of view are two parameters that have to be taken into consideration when making the choice. For most preparations, copper grids provide sufficient stability. Nickel, gold, or gold-coated copper grids are recommended for procedures involving the application of copper-oxidizing chemicals.

For standard preparations, grids with small square or hexagonal meshes are most frequently used. Grids with single holes or slots give maximum field of view, but provide little specimen support. With the so-called 'finder' grids, individually labeled meshes facilitate the re-location of interesting areas of the specimen.

On the one hand, specimen support films should provide optimal stabilization of the specimen during preparation and under the electron microscope, while on the other hand, minimal interaction with the incident electron beam is required in order to obtain images with low background noise. Therefore, thin (20 nm or below), structurally homogenous support films of maximum strength should be used. One also needs to consider whether the surface properties of the support film (e.g., its hydrophilicity) facilitate efficient adhesion of the sample.

Tab. 2.1 Grids and specimen support films for different specimen preparations.

Specimen	Support film	Preparation/ sample characteristics	Grid specifications
Protein molecules, sub-cellular structures, single cells	Carbon film (very thin carbon films need to be supported by a perforated plastic film)	Negative staining, unidirectional metal shadowing	Small square or hexagonal mesh (700–300 mesh/in) copper
	Perforated carbon, which has been rendered hydrophilic	Frozen-hydrated specimens	Small square or hexagonal mesh (700–300 mesh/in) copper
Single cells tissue, and cellular aggregates	Formvar	Ultra-thin sections of resin-embedded samples (70–80 nm thick)	
		– small specimens (e.g. single bacterial cells)	Square/hexagonal mesh (400 mesh/in) copper
		– large specimens (e.g. large aggregates or tissue samples)	100–300 mesh/in copper
		– ribbons of serial sections	Single slot, multiple slots copper
		– subsequent treatment (e.g. immunocytochemical localization	nickel
	No support film	Very thin sections 40–50 nm thick) for elemental analysis	Small square/hexagonal mesh (700–400 mesh/in) copper
	Formvar, carbon-coated Formvar	Cryosections	Square/hexagonal mesh (400 mesh/in) copper
Large specimens with small areas of interest	No support film	Specimens for scanning electron microscopy (SEM)	Finder grids of different form and dimension

2.4.3.2 **Plastic support films**

Plastic films are decomposed by the electron beam and tend to drift during the first seconds of irradiation. However, they usually provide sufficient stabilization for ultra-thin sections of resin-embedded samples (see below). Carbon coating (see below) stabilizes the structure of plastic films. Generally, plastic films are not suitable for the visualization of specimens at high resolution. Though Formvar and carbon-coated Formvar are the most widely used support films for transmission electron microscopy, numerous other plastic materials are suitable, especially when the physical properties of the support film (and therefore the interaction between the support and the sample) need to be varied. These films are based on, for example, collodion

Tab. 2.2 Preparation of Formvar films by casting on glass

1.) Clean a microscope slide with detergent and water. When the slide is dried, a thin residual layer of detergent remains on the surface of the slide to facilitate Step 6.
2.) Place the slide inside a cylindrical glass funnel (diameter: 3.2 cm; length: 8 cm; fitted with a tap and drain tube), filled with a 0.5 % (w/v) solution of Formvar in water-free chloroform or ethylene dichloride. The slide should not be completely submersed but by a maximum of 3/4 of its length.
3.) Cover the funnel with a lid.
4.) After ~30 s, open the tap and let the solution drain off.
5.) After a further 30 s, remove the slide with forceps and allow it to dry.
6.) Scratch the edges of the glass slide with a razor blade and breathe on the film prior to immersion.
7.) Float the film off onto a clean water surface. This is achieved by slowly lowering the short edge of the slide into a dish filled with double-distilled water (almost up to its rim) at an angle of 30–45°.
8.) Place grids on top of the film, starting at each of the four corners in order to facilitate visualization of the area to be covered.
9.) A suitably sized piece of Parafilm® **backing** paper (not the Parafilm itself!) is now lowered onto the grids and gently pressed home. By using forceps, the whole "sandwich" is then removed from the water surface, turned upside-down, placed on a piece of filter paper, and left to dry overnight.

Comment: The thickness of the film is essentially dependant on two parameters: the concentration of the Formvar solution and the time allowed for the solution to drain off.

(solvent: isoamyl acetate), cellulose acetate (solvent: acetone, ethyl acetate), methyl methacrylate (solvent: chloroform), or polystyrene (solvent: benzene). Table 2.2 describes the experimental procedures for production of support films in detail.

2.4.3.3 **Carbon and carbon-coated plastic support films**

Carbon is the most widely used film for the direct adsorption of small specimens such as protein molecules. Very thin carbon films may be supported by a perforated film (see Table 2.2). Carbon films are produced by resistance evaporation and subsequent sublimation of carbon onto plastic support films on grids or freshly cleaved mica as an intermediate carrier. In the latter case, carbon-only support films are produced. In order to achieve uniform distribution of small carbon particles and improved smoothness, the surface can be prepared by *indirect* carbon coating, during which the evaporated carbon is reflected from a glass slide at an angle of 45° to the target surface (mica). Sublimation of non-reflected carbon (direct coating) is prevented by a metal shield between the evaporation source and the mica (Fig. 2.38 a). The carbon film is subsequently floated off onto a clean water surface and grids are placed on top of the film. Mounting is completed as in the method detailed for Formvar films.

2.4.3.4 **Preparation of hydrophilic films by glow-discharge**

To increase the hydrophilicity of the support film surface, grids may be glow-discharged immediately prior to use (the hydrophilicity of the grid surface is then stable for about half a day) under reduced atmospheric pressure (approximately 0.1 Torr) in air (resulting surface charge is negative) or in the presence of alkylamines (resulting surface charge is positive). Glow-discharging improves sample adhesion and spread-

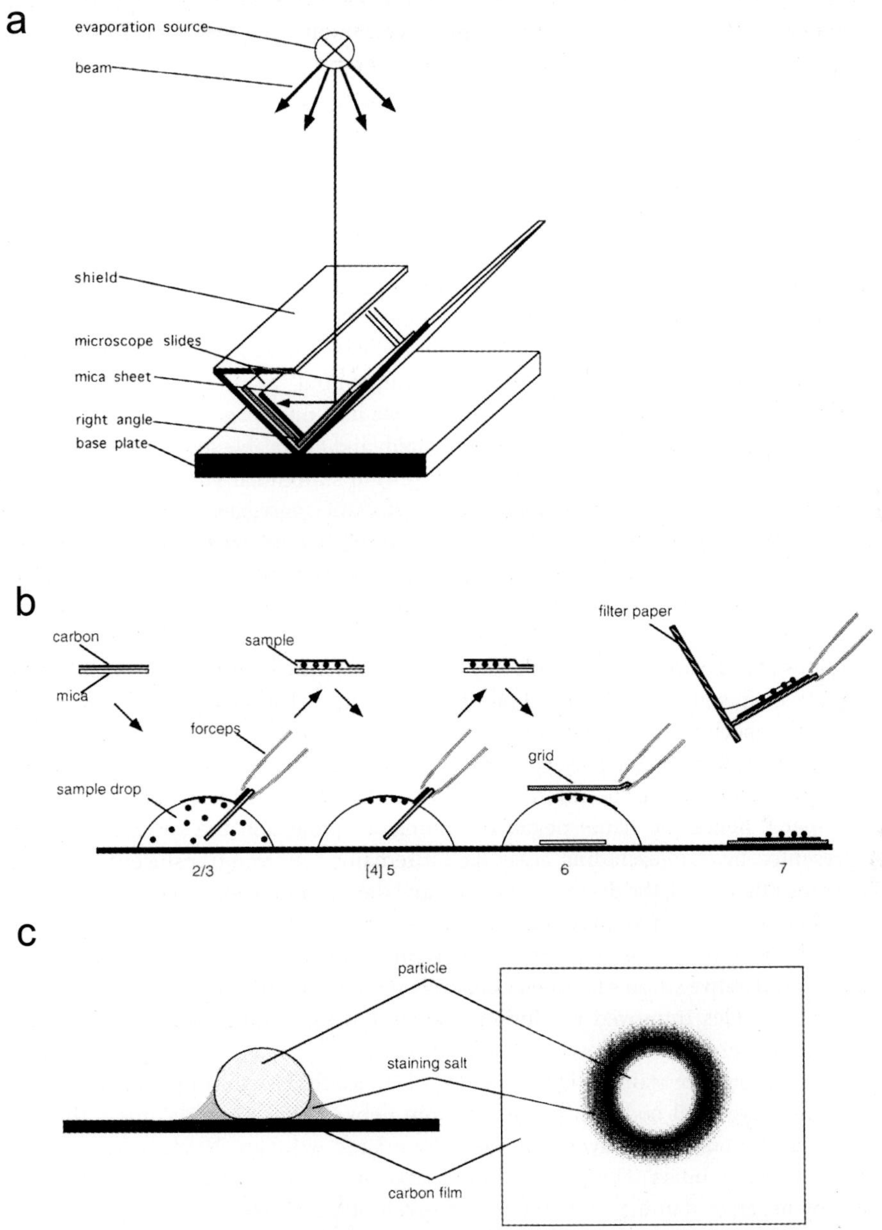

Fig. 2.38 Carbon films and negative staining.
(a) Indirect carbon coating. The simple apparatus, preformed from sheet steel, is fixed on a metal base plate. A shield protects the mica plate from direct exposition to the carbon beam. The beam is deflected as indicated by the large arrow (redrawn from Robinson et al., 1987).
(b) Schematic drawing of the negative staining procedure referring to steps 1–5 in Table 2.4.
(c) Sample particle attached to a carbon film and surrounded by dried stain. Left: side-on view; right: face-on view of the particle mimicking the corresponding projection as observed in the TEM

ing of the staining solutions (Dubochet, 1971). The use of hydrophilic support films is essential for the preparation of frozen-hydrated samples and extremely useful when preparing macromolecular structures and membranes.

2.4.4 Specimen Preparation Techniques

Negative staining. Since biological specimens generally consist of elements of low atomic mass, which interact with electrons only very weakly, contrast is conventionally imparted to the final image by the addition of heavy metals. For good contrast, the specimen support film needs to be as "electron-neutral" as possible in order not to introduce significant levels of background noise into the image.

The technique is convenient, extremely rapid, and applicable to objects ranging in size from protein macromolecules (several nm) to subcellular structures and whole organisms such as bacteria (several μm). Surfactant aggregates and liposomes are also suitable objects. The technique allows resolution of structures between 0.7 and 1.5 nm to be obtained with quite simple preparation procedures: the shape, size, and surface structures of protein molecules may be analyzed, as well as subunit stoichiometries and symmetry in oligomeric complexes. Staining is achieved by use of solutions of heavy metal salts (Table 2.3, Fig. 2.38 b, c), which form a glassy cast around the objects upon drying. The metal salt solution occupies hydrated regions around and within the specimen and penetrates into cavities. Since the stain replaces the water in these cavities, the specimen escapes exposure to surface tension forces, thereby minimizing distortion of the specimen ultrastructure. Ideally, the stain should not infiltrate specimens; hence the name negative staining. In projection the specimen is then represented by stain-excluding areas (i.e., it remains "electron-translucent").

On the other hand, the drying of the stain and the specimen, together with remnants of buffer solutions, may produce artifacts. A tendency to form microcrystals in an otherwise homogeneous deposit may obscure ultrastructural details of the specimen. Comparative studies between negatively stained and frozen-hydrated specimens revealed, besides improved resolution, a small but easily observable change in the structure of protein domains in some specimens.

Whereas negative staining is the method of choice for visualization of bacterial appendages, large cell bodies often suffer from dehydration. They become artificially flattened, the outermost layer often exhibiting large wrinkles. Nevertheless, a rigid S-layer remains intact. The preparation of large and even more sensitive eukaryotic cells by negative staining is therefore of limited practical use.

Because of the ease with which negative staining may be performed, it is a powerful preparative tool not only for structural analysis, but also for rapid screening and quality monitoring (e.g., checking the homogeneity of cell and protein preparations, or contamination of cell cultures by viruses; see below).

Several variations of the negative staining technique (originally by Robert Horne, see Brenner and Horne, 1959) have been developed (Valentine et al., 1968; Haschemeyer

Tab. 2.3 Selected heavy metal solutions suitable for negative staining

Stain (reference)	Solutions and selected properties	Suitable for the visualization of:
Uranyl acetate (Van Bruggen et al., 1960, 1962)	Aqueous solution (0.5–4 % w/v), pH (~ 4.2) adjustable to ⩾ 2 with HCl or ≤ 4.8 with 0.1 M KOH (titrate carefully, above pH 4.8 insoluble $UO_2(OH)_2$ is formed), by addition of EDTA it is possible to adjust the pH up to 7.4 with a concomitant increase in grain size, store in the dark	Macromolecules, viruses, cellular appendages. Note: damage due to acidic pH (e.g., loss of membrane-bound F1-ATPase) may be possible
Uranyl formate (Lebermann, 1965)	Aqueous solution (~ 0.75 % w/v), similar to uranyl acetate but smaller grain size, pH 3.5 without titration; pH can be titrated to 4.5–5.2 with ammonia, very unstable in solution (always prepare fresh), highest density uranyl salt	
Uranyl oxalate (Mellema et al., 1967)	0.5 % (w/v) uranyl acetate in 12 mM oxalic acid, pH 5–7, pH adjustable with ammonia, stable at pH 6 for up to 2 days and at pH 7 for a few hours (at 4 °C), (solutions are light-sensitive store in the dark)	
Uranyl sulfate (Estis et al., 1981)	Aqueous solution (2 % w/v), pH 3.6, pH adjustable to 4.5 with dilute ammonia, titration to pH⩾ 5 causes precipitation, stable for weeks when stored in the dark at room temperature, characterized by high radiation resistance (!), second highest density uranyl salt	
Phosphotungstic acid or corresponding Na and K salts (Brenner and Horne, 1959)	Aqueous solution (0.5–3 % w/v), useful pH range 4–9, adjustable with 1 M KOH or NaOH, for the lower pH range use free acid, tendency to positively stain an object (especially at lower pH)	Viruses, subcellular structures, membrane proteins, bacterial cells*)
Methylamine tungstate (Faberge and Oliver, 1974)	Aqueous solution (typically 2 % w/v), useful pH range 3–10, can be used with buffers	Macromolecules, viruses, acellular appendages subcellular structures, whole bacterial cells
Ammonium molybdate (Muscatello and Horne, 1968)	Aqueous solution (2–10 % w/v), pH range from 5.2–7.5, titrate with NaOH/KOH or ammonia	Viruses, bacterial cells; particularly suitable for cell membranes, membrane-bound systems, protoplasts and other osmotically labile samples, since the tonicity can be adjusted over a large range*)
Sodium silicotungstate	Aqueous solution (1–4 % w/v) adjustable with 1 M NaOH**)	Viruses

Tab. 2.3 continued

Stain (reference)	Solutions and selected properties	Suitable for the visualization of:
Mixtures of carbohydrates and negative stains (Harris and Horne, 1994)	Aqueous solution of 4% (w/v) uranyl acetate or 5% (w/v) ammonium molybdate and 1% (w/v) trehalose, ammonium molybdate-trehalose at pH 6.9 with NaOH, improved spreading behavior (for details see reference)	Large macromolecules; there may also be possible advantages with bacterial cells, cellular appendages, etc.
Salts of light elements (Massover and Marsh, 1997)	Aqueous solutions of 2% (w/v) sodium tetraborate (pH 9.6), potassium aluminium sulfate (pH 3.4) or ammonium borate (pH 8.6)	Alternative to heavy metal stains (low electron beam doses required)

*) When staining macromolecules this stain may result in a low contrast.

**) Only available as silicotungstic acid. For best results, when preparing the corresponding salt, the following procedure is recommended. Solutions of silicotungstic acid and sodium hydroxide are combined in equimolar amounts. In practice, the NaOH solution is added dropwise and the pH monitored. The final pH should be 7. Now the solution is rapidly injected into ethanol at –20 °C and left overnight. The white precipitates containing solution is filtered, the white deposit appearing on the filter subsequently washed with cold (–20 °C) ethanol, dried into a cake (overnight), and finally ground to a fine powder. This powder can be stored and solutions are made up fresh as required at 1–4% (w/v) in double-distilled water.

and Myers, 1972, and references therein). The procedures in Tables 2.4 and 2.5 describe two of the more successful modifications.

Positioning of a specimen between two carbon films results in smaller variations in stain thickness around the specimen, since the specimen is slightly squeezed between the carbon layers. The sandwich technique is used if a more uniform stain distribution throughout the entire structure is desired. Variation in stain depth may provide additional information on the specimen structure. A "sandwich" preparation is achieved by using a rectangular piece of carbon-coated mica (measuring, say, 3×6 mm) and turning the grid when picking it up from the staining solution (step 6) in such a way that the film folds back onto itself, thereby trapping the staining solution and the particles adsorbed between two layers of carbon film. While this effect may be desired under certain circumstances and when imaging macromolecules, it often tends to produce rather poor images, especially with large objects such as whole bacteria. With highly symmetric viral capsids, two-sided staining can give rise to Moiré effects, and subsequently lead to a false impression of the arrangement of capsomers. For these reasons, unintentional folding back of the carbon film should be avoided.

In cases in which the above protocol does not give satisfactory results, it is recommended that protocol 2 (Table 2.5) is tried. It is particularly suitable for membraneous samples and detergent-solubilized membrane proteins.

Tab. 2.4 Negative staining protocol 1

1.) Pipette 50–100 µl droplets of sample, [fixing], washing, and staining solutions onto a clean Parafilm® surface or place solutions in small wells (e.g., microcentrifuge tube lids). The volume can be varied according to the amount of material available. However, it is important that, whatever volume one uses, a pronounced convex meniscus is visible. As far as the concentration is concerned, trial and error is the best way to achieve optimal results. As a guideline, isolated protein molecules with a molecular mass of 500 kDa should be used at concentrations between 70–100 µg/ml (the lower the molecular mass, the lower the concentration required), viruses at about 10^{10}–10^{12} particles/ml, and with bacteria an aliquot from the early log phase is adequate.
2.) With sharp, pointed scissors cut an approximately 3×3 mm square from the carbon-coated mica. Holding the mica at one end with forceps, introduce it into the sample suspension at an angle of 30–45°. The carbon film will start to float off, but is held in place by the forceps. During this step, particles are being adsorbed onto the carbon film. The adsorption time (usually in the range of 10–60 s) and/or the sample concentration determine the final particle density on the film.
3.) Remove the mica slowly, allowing the carbon to fall back into its original position. Drain off **excess** liquid only with filter paper. Do not blot dry!
4.) By the same procedure (step 2), introduce the mica into a fixing solution (0.1–0.5 % v/v buffered glutardialdehyde). Drain off **excess** liquid.
N.B. Fixation may be omitted with most objects. The uranyl salt used in step 6 acts as a fixative.
5.) By the same procedure, introduce the mica into a drop of distilled water or a suitable buffer in order to obtain a cleaner background. Drain off **excess** liquid.
6.) Completely float the carbon off the mica and onto the staining solution (e.g., 4 % (w/v) uranyl acetate, see Table 2.3) for 30 s. With forceps, pick up a grid and touch it to the surface of the carbon film. When the film attaches, lift up the grid and turn it around before blotting on filter paper. Complete removal of the staining solution results in a shallowly stained specimen. When some part of the staining solution remains on the carbon because of incomplete blotting, a deeply stained specimen is obtained.
7.) Allow the specimen to air-dry.

For the visualization of virus particles, sodium and potassium phosphotungstate have traditionally most routinely been used. However, one should bear in mind that these stains may disrupt membranes. When preparing negatively stained specimens of enveloped viruses, it is therefore advisable to try methylamine tungstate or a solution of sodium silicotungstate at pH 7. The success of the staining procedure, particularly with respect to enveloped viruses, is influenced not only by the stain em-

Tab. 2.5 Negative staining protocol 2

1.) Onto the clean surface of a strip of Parafilm, pipette 15–25 µl sample droplets (i) followed by two droplets of double-distilled water (ii, iii) and one droplet of the staining solution (iv).
2.) Place carbon-coated collodion grids (freshly glow-discharged for 30 s under a reduced atmospheric pressure of 0.1 Torr – optional), carbon side towards the sample, onto (i) for 20–50 s and quickly blot to remove **excess** liquid. This is the adsorption step.
3.) Place grid onto (ii) and (iii) for about 1 s each (each followed by a blotting step – but without blotting dry). This is the washing step.
4.) Stain specimen by placing it on (iv) for 30 s.
5.) Blot dry with the grid perpendicular to the blotting paper or use modifications according to protocol 1 above.
6.) Allow specimen to air-dry.

ployed but also by the surface properties of the support film. Uranyl acetate, for instance, is not normally the preferred stain for enveloped viruses, but results are improved by using it in connection with a negatively glow-discharged support film.

Electron microscopy was used early on in the elucidation of the nature of viruses and for rapid viral diagnosis (e.g., smallpox versus chicken pox; see also Almeida, 1980). Nowadays, despite all molecular biological and genetic tools, electron microscopy is still useful, as long as the concentration of virus in samples is high enough for detection by electron microscopy. This is the case in fermentation processes, for example, of bacterial cells, where bacteriophages may be detected within some minutes.

2.4.5 Sample Preparation by Rapid Freezing

It is well known that, although negative staining salts and chemical fixatives act as fixing and stabilizing agents, macromolecules and whole cells suffer from distortion caused by dehydration. In order to preserve samples in their fully hydrated state, they must be frozen at very high cooling rates and, from then onwards, be kept at temperatures below $-140\,^{\circ}C$ until further processing occurs. Rapid freezing methods have overcome one of the crucial problems in electron microscopy: dehydration under high vacuum conditions. Nearly all variations of the technique allow coagulation, decomposition, shrinkage, flattening, and other artifacts caused by dehydration to be avoided. The preparation of vitrified specimens is therefore the method of choice for high-resolution methods such as electron crystallography.

Slow cooling rates (below 1 K/s) result in the formation of large ice crystals outside cells, which progressively dehydrate the cells during the crystal growth. This results in increasing concentrations of solutes inside cells and, of course, in severe structural changes (shrinkage). Nevertheless, the high solute concentration prevents formation of large intracellular ice crystals and leads to high cell survival rates. Cooling rates of between 1 and 1000 K/s are too fast for cellular dehydration by formation of extracellular crystals, and large ice crystals are also formed inside cells, leading, for example, to membrane disruption, resulting in reduced survival rates. Cooling rates faster than 1000 K/s give rise to smaller ice crystals. Above 10,000 K/s, cells are structurally well preserved and show high survival rates. The object is embedded in microcrystalline ice or, under ideal conditions, amorphous ice (i.e., the sample is surrounded by a water network similar to liquid water, immobilized in a "frozen-hydrated" state also termed "vitrified water"), which does not damage its structure, the prerequisite for electron microscopy of these samples. Since heat-transfer rates in biological specimens are low, the cooling rates are influenced primarily by sample size, which has to be as small as possible. High cooling rates are therefore relatively easy to achieve with suspensions or thin layers of biological macromolecules, viruses, or single microbial cells, but difficult in complex tissue. Because of the poor heat conductance of water, cryofixation without pretreatment of the sample with cryoprotectant and without use of high-pressure freezing (HPF) devices is only achievable for specimens up to 20 μm. Cryoprotectants overcome the problem of ice crystal formation in large specimens or when HPF devices are not available (Skaer, 1982). When penetrating agents (glycerol, dimethyl sulfoxide, ethylene glycol) are used as cryoprotectants, artifacts such as irreversible plasmolysis, swelling, phase separation, and membrane rearrangements may occur (Sleytr and Robards, 1981). Electrolytes and other solutes are redistributed. Non-penetrating agents (polyvinylpyrrolidone, hydroxyethylstarch, dextran) may induce shrinkage artifacts through dehydration (Allen and Weatherbee, 1979). Like cryoprotectants, the surfaces of the specimens themselves also have a cryoprotecting effect, through their influence on the structure of the surrounding water over distances of 2 – 3 nm. This produces a local cryoprotection effect, preventing water from crystallizing in channels or cavities smaller than about 5 nm in diameter. Specimens fixed by cryo-immersion can be freeze-sectioned, freeze-dried, freeze-fractured, etc., or they may be directly visualized under an electron microscope with continuous cooling below the devitrification temperature (130 K). For comprehensive treatises on this subject, the reader is referred to Dubochet et al. (1982, 1988), Robards and Sleytr (1985), Sitte (1984), and Stuart (1991).

2.4.5.1 **Pretreatment of subcellular components**

Rapid freezing of protein molecules, viruses, and subcellular compartments leads to excellent structural preservation at the molecular level. Rapid cooling rates can only be achieved when thin films of particle suspensions (< 200 nm) and appropriate cryogens allowing rapid cooling rates are used. Only liquefied gases that do not display "film boiling" (Leidenfrost phenomenon, causing reduced cooling rates because of the production of a gas layer around the specimen) can be employed as cryogens.

There are several alternatives for producing very thin films of particle suspensions. Grids with hydrophilized perforated films or bare hydrophilized grids (G 400–G 600 mesh) are used (Adrian et al., 1984; see Table 2.6).

Tab. 2.6 Performance of thin specimen films

Specimen film on a thin support film: Place a small droplet (3 – 10 µl) of the particle suspension on the grid surface and suck off excess liquid. Only a thin film of liquid should remain on the grid surface. The grid is immediately processed by plunge-freezing (see below).
Bare grid method: Apply a small droplet of suspension to a bare hydrophilic grid. Suck off excess liquid, so that the grid squares are filled by thin aqueous films. Further thinning of the films is achieved by evaporation of the liquid for several seconds. The procedure may be checked with a stereomicroscope (installed on the quick-freezing apparatus): when about 50 % of the squares have lost their films, the grid is immediately plunged into the cryogen (see below). Vitrified films are cooled below the devitrification temperature (130 K) with a cryotransfer device and directly visualized under electron microscopes equipped with cryostages. Cryostages for most modern electron microscopes are available, but their operation needs some experience. Visualization at ambient temperature is possible after freeze-drying and metal-shadowing (see below).

2.4.5.2 **Pretreatment of cells**

Thanks to their small size, single cells – especially prokaryotes – need no elaborate preparation before freezing to avoid artifacts. If ultra-rapid freezing methods are applied, washed cell suspensions may be subjected to cryofixation without any further treatment. Chemical fixation and cryoprotection may be omitted, so chemically induced artifacts can be avoided. This factor is important for immunocytochemical localization of some aldehyde-sensitive antigens. Larger samples, especially tissue, need pretreatment by chemical fixation and cryoprotection; freezing would otherwise induce artifacts. If cryofixation without cryoprotection and/or chemical fixation does not give satisfying results or is not applicable, proceed as outlined in Table 2.7.

If cell structure or interactions between cells are influenced by centrifugation steps prior to fixation and/or freezing (e.g., loss of cellular appendages), an alternative method for cell concentration can be applied (Hohenberg et al., 1994; Rieger et al., 1997). Very high cell densities may be achieved by concentration of the cells in cellulose capillary tubes. The tubes are filled with cell suspension and incubated in growth medium. The capillaries allow the growth medium to diffuse freely; the organisms are retained and may be cultivated to higher cell densities than in regular cultures. The capillary tubes are then cut into approximately 3 mm segments and subjected to rapid freezing. Since the capillary tubes are too large to permit rapid sample cooling rates by normal means, they have to be subjected to high-pressure freezing, requiring specialized devices, or chemical fixation prior to plunge freezing (see Rieger et al., 1997).

Embedding in gelatin stabilizes single cells, and also biofilms and tissue samples – especially after the fixation step (6 in Table 2.8). However, the gelatin may be visible under the electron microscope as a dense fibrillar network. It may thus be omitted for small tissue blocks (i.e., steps 3–5 should not be applied).

2.4.5.3 **Freezing in liquid nitrogen**

This method is used mainly for specimens to be subjected to cryosectioning. Only samples that have been completely infiltrated with cryoprotectant (sucrose/polyvinylpyrrolidone; see Table 2.8) may be frozen directly in liquid nitrogen. Since cooling rates are low, unprotected samples would be destroyed by ice crystal formation. The method does not require any specialized equipment.

After the sample has been attached to the appropriate specimen holder (depends on the cryo-ultramicrotome used, see below), the sample is placed in a cryo cap (or screw-capped microcentrifuge tubes) and covered with liquid nitrogen. The sample is stored in liquid nitrogen until use.

2.4.5.4 **Plunge freezing**

Plunge freezing devices for small specimens, with entry velocities of around 1 m/s, are most widely distributed and may also be used for rapid freezing of cryoprotected specimens. In a simple plunge-freezing apparatus, a cryogen (Freon 12, Freon 22, ethane, propane) is placed in an aluminium cylinder surrounded by liquid nitrogen. Care has to be taken that the cryogen remains liquid (propane: m.p.: –189 °C, b.p. –42 °C; ethane: m.p. –172 °C, b.p. –89 °C). Stirring of the cryogen during plunge fixation facilitates heat-transfer. The essential design of a plunge-freezing device in liquid propane or ethane is shown in Fig. 2.39 a, b, and the procedure is outlined in Table 2.9. The metal cylinder is attached to a steel rack, cooled by liquid nitrogen. The cylinder itself is not immersed in nitrogen, to allow electrical counter-heating for exact temperature adjustment. The outer insulating container traps an atmosphere of nitrogen gas (produced from liquid nitrogen

Tab. 2.7 Cryoprotection of cells

1A)	For cells with high water content (algae, filamentous fungi, small tissue samples), chemical fixation prior to glycerol infiltration is necessary. Incubate cells with buffered glutardialdehyde solution (1 % w/v–4 % w/v) for 90 min at 0 °C. Usually, a 1 % w/v solution provides sufficient fixation. If the cells are to be subjected to immunolocalization procedures, incubate with a mixture of 0.2 % w/v paraformaldehyde and 0.3 % glutardialdehyde for 90 min at 0 °C (see below).
1B)	For prokaryotes and yeasts, attempt to culture the organisms in the presence of 20 – 30 % glycerol in the growth medium. Alternatively, resuspend cells in 20 – 30 % buffered glycerol solution (adjust pH and ionic strength to the respective values of the growth medium) and incubate for up to 1 h. Cells should not be plasmolyzed after the procedure.
2)	Centrifuge the cells (not the tissue samples). When no cryoprotectant has been used, proceed with a densely packed cell pellet.

A cryoprotection procedure allowing the use of low cooling rates (see below, "freezing in liquid nitrogen") and requiring chemical pre-fixation of cells is mainly used for samples subjected to cryosectioning and immunolocalization (Tokuyasu, 1986; Table 2.6).

Tab. 2.8 Cryoprotection of cells and tissue according to Tokuyasu (1986)

1.)	Incubate a washed cell pellet or a small tissue sample in a mixture of 0.2 % w/v paraformaldehyde and 0.3 % glutardialdehyde in an appropriate buffer (50 mM potassium phosphate is recommended if there are no special requirements) for 90 min at 0 °C (see below).*)
2.)	Centrifuge the cells and resuspend in a small volume of a liquefied 10 % (w/v) gelatin solution (in buffer). Infiltrate a tissue sample for 30 min with this solution.
3.)	Pour the solution onto a flat, chilled surface. For tissue samples, allow the infiltrated blocks to solidify by chilling.
4.)	Cut the solidified gelatin into small prism-shaped blocks (this shape facilitates the trimming procedure prior to sectioning; see below).
5.)	Incubate the small blocks in buffer for 1 h.
6.)	Incubate in a mixture of 0.2 % w/v paraformaldehyde and 0.3 % glutardialdehyde for 8 h or overnight and then incubate in a solution of 1.6 M sucrose and 25 % (w/v) polyvinylpyrrolidone for 8 h or overnight.
7.)	Attach the sample to an appropriate specimen holder (depends on the equipment used, see below) and then either freeze in liquid nitrogen or plunge freeze (see below).

*) Higher fixative concentrations (1 – 4 % (v/v) glutardialdehyde) are necessary for large tissue samples. When immunolocalization will follow; do not exceed 1 % (v/v) glutardialdehyde concentration. Eventually, omission of glutardialdehyde is necessary. Then raise paraformaldehyde concentration to up to 8 % w/v; refer to Tables 2.9 and 2.33

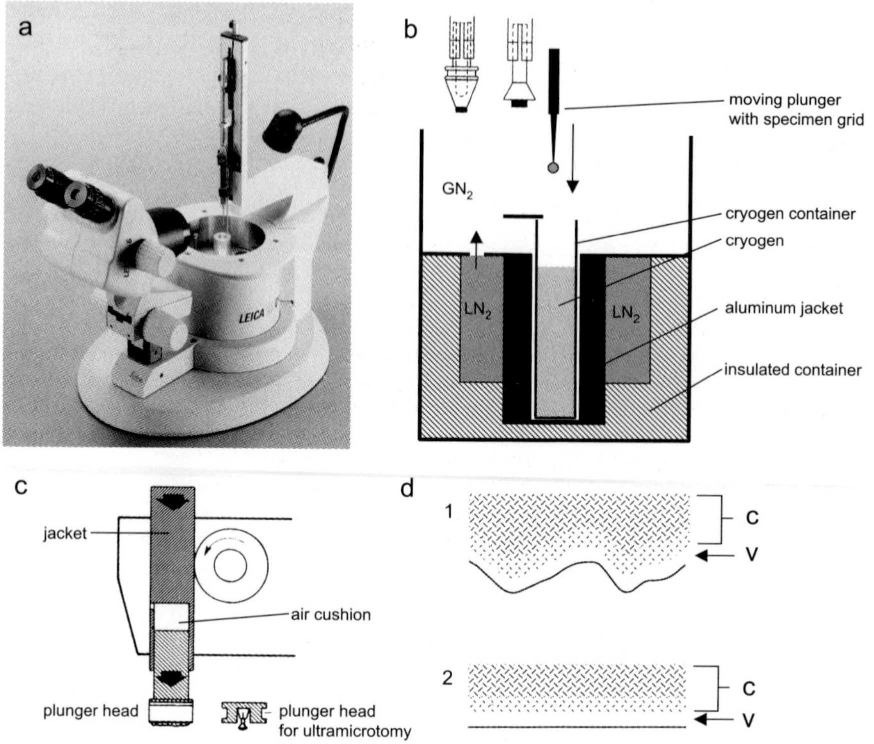

Fig. 2.39 Rapid freezing.
(a) Apparatus for plunge freezing of thin specimens, originally according to Sitte, 1984, manufactured by Leica (Leica Inc.).
(b) Simplified scheme of the cryochamber in a plunge freezing apparatus. Forceps with a specimen grid are used as plunger head (usually for preparation of small particles in cryoelectron microscopy). Two additional plunger heads are depicted for cells in suspension and tissue (Insets: Leica Inc.).
(c) Head for cold block freezing (slamming). The plunger head is actively propelled against a cold block (installed in a cryochamber instead of a cryogen container as depicted in b). The specimen velocity and pressure after impact may be adjusted to avoid bouncing of the specimen. For subsequent ultramicotomy, a specimen holder that fits directly into an ultramicrotome may be used (Leica Inc.).
(d) Distribution of ice crystals (c; small dashes) in samples prepared by plunge freezing (A) and cold block freezing (B). After slamming, a smooth vitrified layer (v) is present on the sample surface. This simplifies further processing by, for example, ultramicrotomy

by a small heated wire submerged in nitrogen) that surrounds the cylinder and all tools to be pre-cooled in the chamber and prevents entry of moisture, so ice contamination is avoided. Temperature control by sensor circuits increases the reproducibility of plunge fixation. The gas is ducted into the nitrogen-chilled cylinder and is immediately liquefied. Ethane is kept at 103 K, propane at 83 K. In order to avoid contamination by moisture, the cylinder has to be closed with a lid. Operation with propane requires extensive safety measures. A mechanical spring-driven device for injecting specimens into liquid cryo-

Tab. 2.9 Plunge freezing procedure

1.) Before freezing, prepare the plunge-freezing device and the necessary instruments. Liquefy the cryogen (see above). Pre-chill all tools that will come into direct contact with the frozen sample (forceps, cryotransfer tools, etc.) in liquid nitrogen. Ensure that that the chilled parts of the instruments are free of ice crystals. This is achieved by keeping the instruments in the layer of nitrogen gas settled above the liquid nitrogen. Transfer times of the instruments in air have to be reduced to a minimum. Contact of the frozen specimen with air has to be avoided throughout the complete procedure.
2.) Apply a small drop of the concentrated cell suspension or gelatin-embedded block to a bare specimen grid, a small (3 mm^2) piece of filter paper, a small plunger, or a wire loop (it depends on the subsequent preparation steps) and mount it on the plunger of the plunge fixation device (see Fig. 2.39 b).
3.) Inject the sample into liquid propane.
4.) Withdraw the frozen sample from the propane vessel. If a drop of liquid propane covers the sample, remove it with a small piece of chilled filter paper kept in the cryochamber. Residual propane may cause problems during the subsequent preparation procedures (especially during preparation of frozen-hydrated samples). All operations with the sample have to be done gently, since the frozen sample and the support are very brittle.
5.) Transfer the sample according to the processing steps given below. Samples to be processed by freeze-substitution or cryosectioning may be kept in liquid nitrogen for several days.

gens is depicted in Fig. 2.39 a. Various injection mechanisms exist, driven by electromagnets, compressed air, or (most simply) gravity.

Liquid cryogen residues on samples may interfere with subsequent preparation steps (e.g., cryo-electron microscopy of frozen-hydrated samples, cryosectioning). These problems are avoided by the use of sub-cooled nitrogen as cryogen. Though liquid nitrogen at its equilibrium boiling point is not a good coolant, “film boiling” does not occur with sub-cooled nitrogen. A device for the preparation of sub-cooled nitrogen has been described by Umrath (1974). A container of liquid nitrogen, built as a vacuum chamber, surrounds a copper container also filled with liquid nitrogen, but open to the atmosphere. On evacuation, the nitrogen in the surrounding vacuum chamber cools the liquid nitrogen in the copper container, which can therefore be continuously maintained at a temperature of around 59 K and used for plunge fixation.

Besides sub-cooled nitrogen, nitrogen slush (a mixture of solid and liquid nitrogen) is also used as a cryogen. Its preparation is very simple, but the reproducibility of cooling rates in this cryogen is low because of the heterogeneity of the solid-liquid mixture. The nitrogen container is placed in a vacuum chamber and evacuated under a vacuum provided by a one-stage rotary pump. Nitrogen solidifies at a temperature of 63 K and a pressure of 13.5 kPa. After readmission of air, the mixture of liquid and solid nitrogen remains stable for some minutes.

2.4.5.5 **Freezing on cold surfaces (cold block freezing, “slamming”)**

Heat transfer between a cooling agent and the sample may be increased by use of a solid (instead of a liquid as secondary “cryogen”). The heat-transfer rates of very cold metals (especially copper) are very high, and so are suitable for specimen freezing. The optimum temperature for rapid cooling on a cold copper block is 19 K, and so liquid helium must be used instead of liquid nitrogen for the fastest cooling rates. The technique is primarily used for the cryofixation of tissue. However, the formation of a planar sample surface during preparation, as occurs during the technique, is advantageous for any sample type. A plane layer of 10 – 20 μm parallel to the surface is vitrified; deeper regions are damaged by large ice crystals and compression shock (Fig. 2.39 d).

In principle, slamming is performed by a modified plunge fixation procedure. A plunger head for slamming is depicted in Fig. 2.29 c (Van Harrefeld and Crowell, 1964; Van Harrefeld et al., 1974). A highly polished, scratch-free metal block (metal mirror) is chilled with liquid nitrogen. The mirror is kept ice-free by an atmosphere of nitrogen gas. The plunger can be driven by gravity, a spring, or an electromagnet. The specimen has to be cushioned from the full force of impact and must be prevented from bouncing on impact. The specimen support may be a small piece of filter paper or freshly cleaved mica attached to the foam with double-sided adhesive tape. Care has to be taken that the metal mirror is not damaged during operation, because scratched surfaces reduce heat transfer rates.

2.4.5.6 High-pressure freezing

Because of the poor heat conductance of water, specimens, even frozen at the fastest rates, are only satisfactory frozen in a layer between 5 and 20 μm. At a pressure of 2100 bar the melting point of water drops down to –22 °C. Since the water is 1,500 times more viscous than at atmospheric pressure, the rate of ice crystal growth is drastically reduced. Hence, specimens of up to 0.5 mm in thickness are adequately frozen at low freezing rates without requiring addition of cryoprotectants (Studer et al., 1989). The apparatus, provided by several suppliers (Bal-tec, Leica), is of course, far more complex than devices that work at ambient pressure. The specimen does not need to be pretreated by fixation, but may be directly subjected to freezing after sampling. The specimen is placed in the high-pressure chamber of the device, automatically frozen under high pressure, and subsequently stored under liquid nitrogen for further processing, or subjected to freeze substitution (see below).

2.4.5.7 Freeze-fracturing and freeze-etching

These techniques are used for preparing replicas from frozen surfaces of biological specimens (Robards and Sleytr, 1985) in order to gain information about the three-dimensional structure of the sample as judged from surface reliefs along fracture planes. Fracturing occurs preferentially between the two leaflets of a lipid bilayer and structural inhomogeneities such as inclusion bodies or spores. The applications are numerous and have been reviewed extensively (see, for instance, Holt and Beveridge, 1982; Hawes and Martin, 1995; Torrisi and Manzini, 1996; Takizawa and Robinson, 2000; Meyer and Richter, 2001). Freeze-fracturing and freeze-etching have been perhaps the most widely applied cryotechniques for elucidation of the structures of biological membranes and remain reliable tools for characterization of bilayer membranes – not only of cellular origin, but also for liposomes in emulsions and other lipid-phase systems. Again, the sample is processed by plunge freezing, a sample droplet being loaded onto a specimen carrier. The frozen sample is then transferred into a freeze-fracture apparatus (procedure in Fig. 2.40 and Table 2.10). Very thin layered

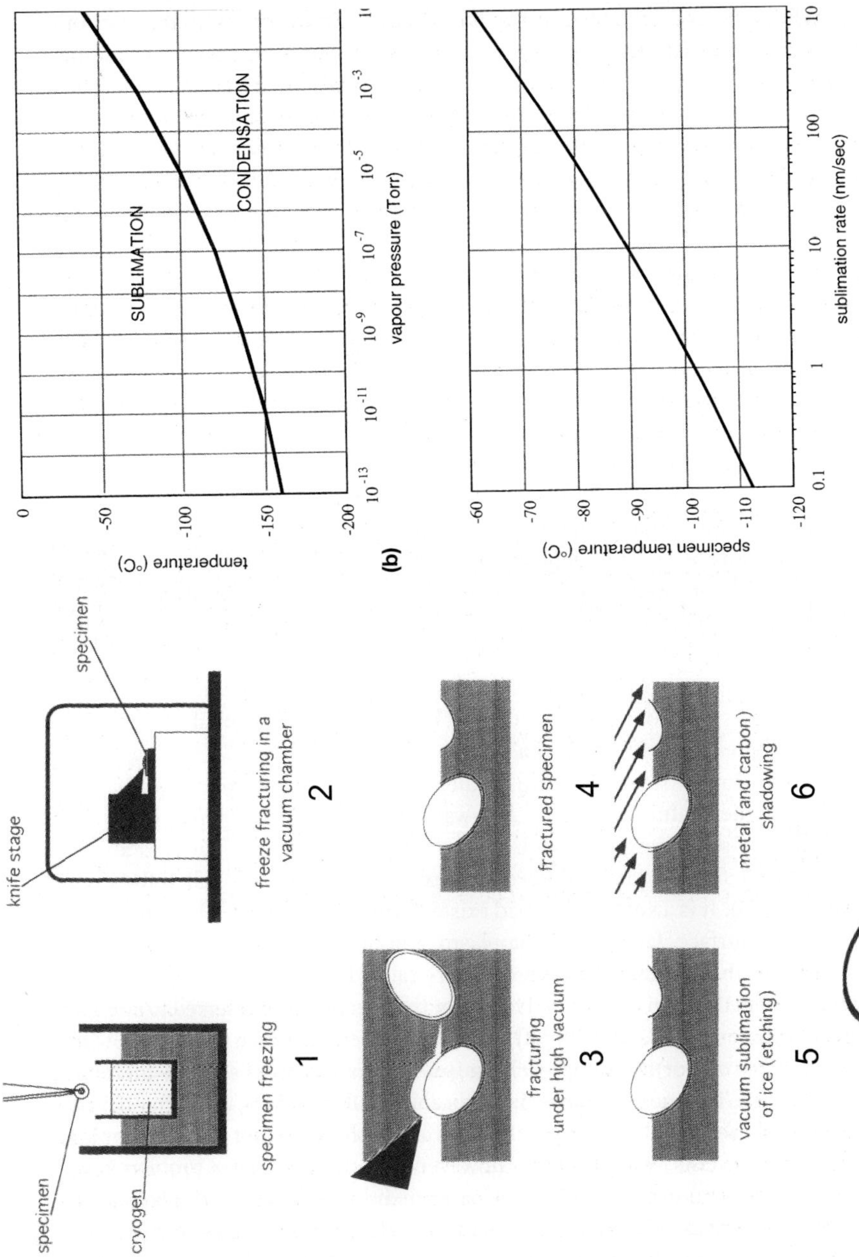

Fig. 2.40 Freeze-fracturing I.
(a) Basic steps of the freeze-fracturing procedure (according to Shotton and Severs, 1995). See Table 2.8.
(b) Diagram showing the relationship between sublimation/etching rate and temperature, assuming an appropriate pressure.
(c) Diagram of the saturation pressure of water

samples, such as monolayers, can easily be fixed by cryoimmersion. With thicker samples, such as pelleted bacterial suspensions, it may be necessary to apply chemical prefixation (such as 2 % v/v glutardialdehyde in an appropriate buffer for 3 h) and cryoprotection (by, for instance, gradual addition of glycerol (final concentration 30 % v/v) to the sample). It has to be noted that glycerol-penetrated specimens cannot be etched (see below) after fractionating, and one may therefore wish to consider the use of volatile cryoprotectants (see below). A better way forward is to use as small a volume of the sample solution as possible or to sandwich the sample between two specimen supports (as used for the double replicas; see step 3.b in Table 2.10).

At low temperature and under high vacuum, the specimen is sectioned with a cooled blade (or by the action of a double replica holder) in such a way that fractured surfaces are exposed. With the double replica holder, two complementary fracture faces are generated. Cell surfaces below the plane of fracturing can subsequently be exposed by controlled sublimation of the surrounding ice. During this sublimation process, commonly referred to as etching, about 100 nm of ice is lost every minute (at 2.10^{-6} Torr and –100 °C). A replica is formed by deposition of Pt-C or W-Ta, at an angle, onto the frozen surface. This is then stabilized by deposition of a supporting layer of carbon from above. After deposition of the carbon, the chamber is vented, the replicated specimen removed, and the remains of the specimen are dissolved from the back of the replica with suitable reagents (e.g., chromic acid, sulfuric acid). After cleaning, the replicas are transferred to grids and can be visualized under the electron microscope without further treatment. W-Ta replicas need to be stored in a desiccator/ vacuum chamber, since contrast is lost upon oxidation. The essential steps of the procedure are represented in Fig. 2.40 a.

To avoid confusion in freeze-fracture studies, it is particularly important that the correct replica nomenclature is used, as follows (see also Fig. 2.41 e): Upon fracturing of a membrane by separating the two leaflets, four rather than two surfaces are obtained. The ("old") surface, facing towards the extracellular space, is called the extracellular surface (ES). It is a surface that had existed prior to fracturing. The same holds true for the "old" surface, facing the protoplasm, and this is called the P surface (PS). The new surfaces, those created by fracturing, are called fracture faces, and – depending on which leaflet they can be ascribed to, either the E leaflet (also referred to as E half of the lipid bilayer) or the P leaflet (P half) – they are termed the E face (EF) or the P face (PF). The sequence of (sur)faces approaching from the extracellular space is therefore ES, EF, PF, and PS. This nomenclature is applicable to all (sur)faces throughout a cell that can be traced back to either an E or a P origin and is therefore not disturbed by any membrane fusion events. With bacteria, however, one can encounter problems, because the cells are surrounded by cell walls, outer membranes, surface (S-) layers etc., entities which can themselves undergo fracturing. Therefore, in cases in which the usage of the P/E nomenclature by itself is not sufficient, one should attempt to supply additional descriptions that appropriately reflect the individual circumstances (see, e.g., Varga and Staehelin, 1983)

Typical causes of artifacts (see also Fig. 2.41 a–d) are described below, together with strategies for avoiding them. This section should therefore always be regarded as essential reading when planning a freeze-fracture experiment:

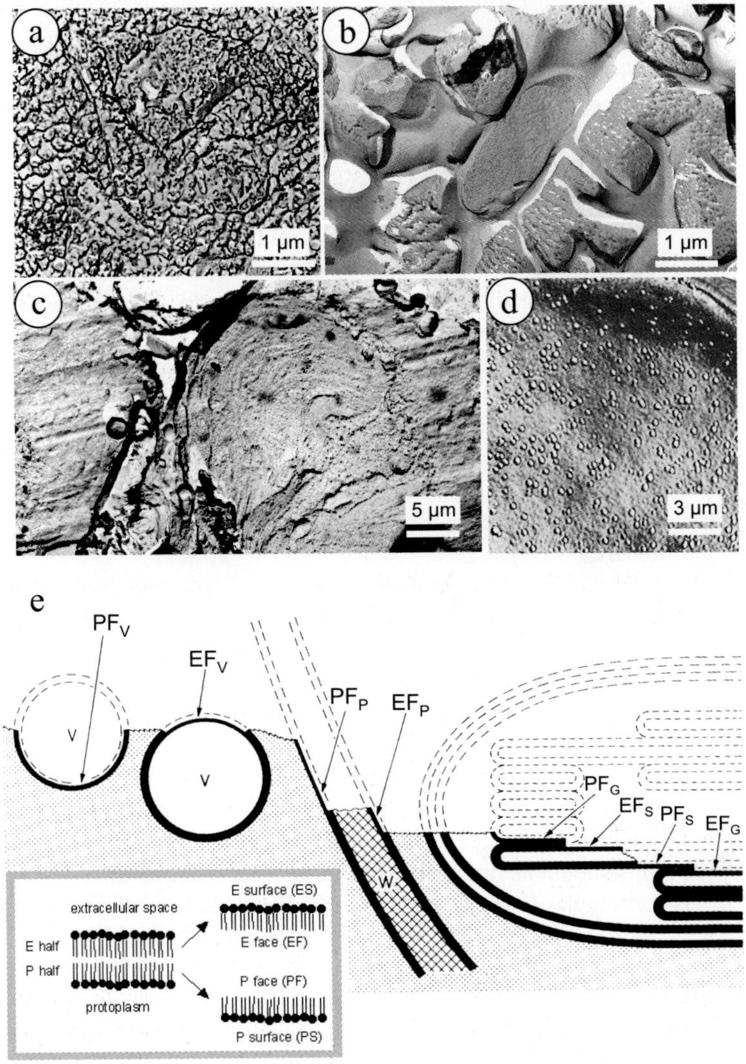

Fig. 2.41 Freeze-fracturing II.
(a–d) Common artifacts. a) Too low freezing speed. The specimen is destroyed by crystallization of intracellular water (Bal-tec Inc., Liechtenstein). b) Cells drop off the replica due to too deep etching (Frank Mayer, Univ. Göttingen). c) Knife marks ("wheel tracks") on the fracture plane due to use of a blunt knife and low cutting speed (Frank Mayer, Univ. Göttingen). d) Effect of condensation of water vapor. The small droplets are sometimes difficult to distinguish from true specimen structures (Bal-tec Inc., Liechtenstein).
(e) Nomenclature for freeze-fractured membranes. The fracture splits the membrane into two halves, an E half (adjacent to the extracellular space) and a P half (adjacent to the protoplasm). When applying the nomenclature to intracellular membranes, "P" (to denote protoplasm) comprises cytoplasm, nucleoplasm, mitochondrial matrix, and, as depicted here, the stroma of chloroplasts. E is used to designate the space within single membrane-bound organelles, such as vesicles, and also the lumens of the ER or the thyakoids as depicted here. "F" denotes the fracture face (i.e., the exposed membrane interior). An exposed natural surface is usually achieved after etching and is denoted "S" (Robinson et al.,1987, with modifications)

1. If cells are frozen at too low rates (<100 K/s), intracellular water may crystallize and destroy any fine structure (Fig. 2.41 a). However, the overall shape of the cell may not be influenced. The use of volatile (no interference with the etching process) cryoprotectants is the best remedy in the absence of a means of improving the freezing rate. Examples of volatile cryoprotectants are methanol and ethanol.
2. Etching too deeply can cause the ice level to decrease to such an extent that cells become completely exposed and lie loosely on the surface (Fig. 2.41 b). These cells may be removed during subsequent preparation steps, leaving holes in the replica. Etching (sublimation) rates are provided in Fig. 2.40 b.
3. When fracturing with a knife, the knife edge should not touch the fracture surface, so that knife marks ("wheel tracks", Fig. 2.41 c) on the surface are avoided. The risk of knife marks may be reduced by avoiding sweeping actions, use of sharp knives with non-jagged edges, and by increased cutting speeds. A small knife advance can be used to avoid steps on the fracture surface.
4. Plastic deformation can occur with materials that are still deformable at low temperature. Examples of such materials are polyphosphate and PHB inclusions. PHB inclusions can be pulled out of the fracture surface into conical protrusions (Sleytr and Robards, 1977).
5. Any residual water vapor that condenses on the specimen before replication gives rise to the formation of artifacts that look like intramembrane particles (small "blobs", Fig. 2.41 d). Water vapor can be avoided:
   - by selecting the temperature carefully (at a given pressure, see Fig. 2.40 b),
   - by protecting the specimen sufficiently with a cold trap, like the cooled knife and the cold shroud (the temperature of which must be below that of the specimen !) or, in cases in which sufficient protection cannot be established, by immediately coating the freshly fractured surface (any etching effect will be negligible in this case),
   - by ensuring that a good, clean vacuum exists around the specimen (diffusion pumps should be fitted with cold traps; rotary pumps should be run on gas-ballast at regular intervals and for sufficient periods of time),
   - by thoroughly degassing the evaporation sources (guns) prior to each experiment.
6. Inadequate (little contrast and no detail) or excessive shadowing (fine details are masked) can be avoided by utilizing a deposit thickness monitor (for standard values, see Table 2.10). If commercial film thickness measuring devices are not available, one can improvise, with pieces of filter paper or gold foil, for example. The former will increasingly darken (make sure part of the paper is folded, to create a reference area without shadowing material) and the latter will assume thickness-dependent interference colors.
7. Superficial cleaning of the replicas or the adoption of unsuitable cleaning procedures and chemicals (e.g., insufficient purity) will lead to smudges across an otherwise perfectly good specimen. A specimen cleaning procedure is provided in the step-by-step protocol. If contaminated water is used during the final cleaning steps, large areas of the replica may become covered with precipitates. As a rule: always use double-distilled water.

Tab. 2.10 Freeze-fracturing procedure (to be used with reference to the instruction manual for the freeze-fracture apparatus)

The sample solution should be as concentrated as possible (barely pipettable).

1. Clean the gold or copper specimen supports in 50 % (v/v) sulfuric acid, followed by double-distilled water (sonication is recommended), and finally acetone, in which they should be left until use. Copper supports may be pretreated with finishing grade (400 – 600) sanding paper. This cleaning procedure improves the friction between sample and support.
2. Turn on cooling water for all equipment that requires cooling (e.g., pumps, guns). Follow the manual to initiate pumping down and cooling down (specimen table, knife). A temperature of around –150 °C should be reached.
3. a) For a single replica: Remove the support from the acetone, allow it to dry, and apply 1 µl of the sample solution so that a pronounced convex meniscus protrudes from the surface of the support. Quick freeze, mount into the pre-cooled specimen holder, and transfer into the recipient. Pump down.
   b) For double replicas: Remove two supports suitable for double replicas from the acetone and allow them to dry. Apply 1 µl of the sample solution onto the first carrier and place the second one upside-down on top of the first one, pressing firmly together. Any excess liquid should be blotted off with filter paper. Quick-freeze this sandwich assembly and mount into a pre-cooled double replica holder. Transfer into the recipient and pump down.
4. Fracturing, etching, and shadowing: Increase the temperature to a value suitable for freeze-etching (at a pressure of approximately 10^{-7} Torr, for example, a suitable temperature is –98 °C – see also Fig. 2.40 c); ensure that the knife is cooled down to <–190 °C. At a temperature of –98 °C, wait a further 15 min in order to avoid temperature gradients between the specimen table and specimen supports. Then fracture the specimen with the cold knife or by flipping open the double-replica holder.
   Immediately position the knife as a cold trap directly above the specimen. The specimen is now etched (i.e., ice sublimes). The time is monitored exactly so that the desired effect is achieved: deep-etching (>30 min), regular freeze-etching (3 – 5 min), or predominantly freeze-fracturing (approximately 30 s). In the latter case, the etching time is kept to a minimum.
   N.B. The times provided are only guidelines for the above temperature-pressure combination.
   Remove the cold trap, and shadow the specimens immediately, first with a contrast-imparting metal (e.g., 1 nm thick Pt/C at 45°), and then with carbon (10 – 20 nm thick C at 90°). Follow the instruction manual of your particular freeze-fracture apparatus. Fracturing, etching, and shadowing ought to be seen as a group of activities carried out in quick succession, unless deep-etching is desired.
   Remove the specimen from the recipient.
5. Clean the specimen as follows: Float the replicas off on 70 % sulfuric acid and leave for 12 h. If there are problems with this first step, float off on double-distilled water and then on solutions containing increasing concentrations of sulfuric acid for a few hours prior to the final step (70 % sulfuric acid, 12 h). With a platinum loop, transfer the replica either into a 6 % (w/v) sodium hypochlorite solution or into 20 % chromic acid and leave for 1 – 2 h. Clean the replicas on double-distilled water (3 × 20 min). Transfer onto a grid and, with pointed filter paper, carefully blot off any liquid from the replica. Grids can be used coated or uncoated, but one should ensure that the surface is hydrophilic (e.g., by glow-discharging; see above). With coated grids, there is generally less chance of making the replicas burst during the blotting step.

Quick check-list for how to recognize an artifact-free freeze-fracture replica:

Look out for:
a very fine grain of the shadowing material,
an overall crisp contrast of the replica,
a good correlation between the complementary fracture faces,
smooth contours around cells and organelles,
a fine granular texture of the frozen medium around the sample (non-etched preparations only),
an even or mosaic distribution of particles when dealing with fracture faces of membranes.

2.4.6 Sample Preparation for Thin Sectioning

Ultra-thin sections of objects provide insights into their internal structures without the need to assess this by analysis of images along fracture planes and surfaces as revealed by the techniques described above. Tissue, cells, subcellular compartments, and vesicles can be visualized in thin section studies, although the technique is unsuitable for resolution of the structures of even the largest macromolecular assemblies. Ribosomes are visible in ultra-thin sections as dark dots. The resolution limit is mainly determined by the thickness of the section. Numerous procedures for the fixation, dehydration, embedding, and staining of biological specimens to achieve preparations with minimum artifacts have been developed (see, e.g., Hayat, 1981). An excellent overview concerning the preparation of tissue for biomedical applications (including a systematic survey of artifacts) has been presented by Maunsbach and Afzelius (1999).

2.4.6.1 Chemical fixation

Although chemical fixation procedures have been replaced by cryofixation techniques in several fields of electron microscopic preparation, they remain the most widely distributed techniques for specimen fixation, due to its relative simplicity and good reproducibility (once a procedure has been found to work well with a specific type of specimen). After chemical fixation, the specimens are usually dehydrated in organic solvent and embedded in resin, but various low-temperature embedding techniques may also follow. Some fixatives inactivate enzymes or mask antigenic determinants, but aldehydes in particular allow subsequent application of specific labeling techniques. Widely used chemical fixatives are described in Table 2.11.

Cells from numerous bacterial species (such as *Escherichia coli*, *Pseudomonas*, and *Bacillus* strains) may be chemically fixed in unbuffered aqueous solutions without any obvious loss of fine structure compared to cells fixed in buffered solutions. Nevertheless, for most other types of specimens, more careful handling is recommended. Generally, application of non-isotonic and unbuffered fixative solutions leads to a less stained appearance of the cytoplasmic ground substance and swelling of organelles. Numerous fixation procedures for all varieties of specimens have been reported. Improper fixation leads to heavy artifacts (e.g., disruption of membraneous organelles) and wrong interpretation of results.

Fixatives such as osmium tetroxide cause a drop in the pH in the sample. Appropriate buffering of the fixation steps in order to avoid artificial change in the specimen caused by acidification is therefore recommended.

Sensitive organisms should be fixed in a physiological medium to prevent swelling of the cells or other damage. The optimal osmolarity for these specimens has to be determined by trial and error, but the composition of an appropriate growth medium is a very helpful basis. If transfer of the organism from its original growth medium to any other buffered solution leads to damage, the organism may also be pre-fixed with aldehydes in the original growth medium. It has to be considered in this case that aldehydes also react with components of the medium. Protein components (yeast extract, fetal calf serum, etc.) in particular bind to all surfaces and form a disturbing layer surround the specimen. After pre-fixation, the stabilized organisms should be washed in buffer before the subsequent preparation steps. It has to be kept in mind that the osmotic pressure of the solution changes when fixatives are added: 0.1 M phosphate buffer has an approximate osmolarity of 210 mOsm, while a 4 % (v/v) glutardialdehyde solution raises osmolarity to 710 mOsm. For fibroblasts, for instance, an appropriate fixation medium consists of 90 mM sucrose, 10 mM $MgCl_2$, 10 mM $CaCl_2$, 3 % (v/v) glutardialdehyde, 5 % (w/v) paraformaldehyde in a buffered medium (0.1 M sodium cacodylate buffer, pH 7.0). After fixation of the specimen in osmium tetroxide (OsO_4), the organisms are no longer osmotically active (even after fixation with aldehydes, obvious osmotic damage of the sample is rare). Osmotic stabilization is therefore no longer necessary after this step. Standard fixation of single cells is carried out in an appropriate buffer including up to 5 % (v/v) glutardialdehyde for 1 – 3 h at room temperature. After washing, the cells are resuspended in a small aliquot of agar. When cut into small blocks of about 1 mm^3 the sample is easier to handle than cell suspensions. Moreover, the agar protects the cells against mechanical disruption. Cells grown in culture may be fixated in situ and removed from the plate after fixation or may be left on their substratum for all other subsequent steps. For this purpose, polyethylene terephthalate (PET) membranes are available as inserts for cell culture plates (Becton Dickinson Biosciences). The membranes may be ultra-thin sectioned together with the surrounding resin.

Tab. 2.11 Fixatives for biological specimens

Fixative	Main effect	Solution (buffered unless indicated otherwise) special features
Glutardialdehyde (glutaraldehyde) (Sabatini, 1963)	Cross-linking of proteins	0.5 % (v/v)–5 % (v/v) 3 % (v/v) for most procedures inclusion of formaldehyde (up to 4 % w/v, freshly prepared from paraformaldehyde) is useful, especially when rapid fixation of specimens is necessary, since it improves the penetration; a 0.2 % (v/v) glutardialdehyde/0.3 % (w/v) formaldehyde solution is suitable for fixation of bacterial cells in immunocytochemistry
Acrolein (Luft, 1959)	Cross-linking of proteins	To be used as alternative to the other aldehyde fixatives only when rapid fixation is of paramount importance
Osmium tetroxide (Millonig, 1961)	Polymerization of hydrocarbon chains (lipids)	1 % (w/v) solution (inclusion of 0.5 – 1 % w/v potassium ferricyanide may provide enhanced staining of membranes)
Permanganate (Luft, 1956)	Fixation of biological membranes	1 % (w/v) aqueous solution; only membranes are preserved, other cell components are damaged or removed

Large specimens with various tissues or organs, such as seeds or insects, contain layers that are difficult to penetrate by fixatives. In particular, water-repelling plant surfaces may be a barrier for fixatives and embedding agents. Infiltration of plant material (e.g., leaves) should be performed under vacuum (rotary or water jet vacuum pump) to remove the air from intercellular spaces. An infiltrated leave is transparent in a liquid, while non-infiltrated leaves have a silver-grayish opaque appearance. It may be necessary to vacuum-infiltrate the specimen not only for fixation, but also during the subsequent dehydration and resin embedding procedure. Water-repellent bark and rigid seedlings, however, are still not penetrated. These specimens must be cut into appropriate pieces in such a way that all tissue layers are exposed and infiltrated. The bodies of small animals such as insects may be perfused by injection of fixatives. Organs subjected to preparation may be removed from the animal prior to fixation or fixed in a dissected animal (in situ). Mild vacuum infiltration of the whole specimen or an organ may also be helpful, but it should be noted that animal tissue is much more sensitive to mechanical stress than plant tissue. It has to be kept in mind that tissue structures may change rapidly after the death of the animal. Fixation of organs in large animals requires fixation methods geared to rapid perfusion of the organs as soon as possible after the death of the animal. The entire procedure consists of a quick sequence of narcosis, removal of possible contents of hollow organs that are subjected to preparation (such as the gut) by flushing with a physiological buffer, and, finally, perfusion with the fixative. Perfusion fixation of vertebrates requires specialized knowledge of anatomical dissection. Otherwise, the success of proper fixation remains questionable.

2.4.6.2 **Dehydration, infiltration, and embedding**

Removal of free water from a specimen before embedding in a suitable resin (see Table 2.12) is achieved by incubation of the specimen in a graded series of organic solvents (methanol, ethanol, or acetone). The dehydration time should be as short as possible in order to keep extraction of tissue components at a minimum, but should allow complete exchange of water by the solvent without induction of swelling artifacts. Prolonged incubation times often lead to extraction of the cytoplasmic background, but this strongly depends on the preceding fixation procedure. Specimens fixed with aldehydes and osmium tetroxide are resistant to modifications in the dehydration procedure and sometimes tolerate even direct transfer into the absolute solvent without formation of significant artifacts.

A typical dehydration procedure involves incubation of fixed samples in a graded acetone series for 3–15 h (see Table 2.13 for details). After dehydration, the sample is infiltrated with an embedding resin (see Table 2.12). A vast number of variations of the procedure given below have been described (though systematic investigations into the effect of variations in the embedding procedure are rare; see, for example, Maunsbach and Afzelius, 1999). For most unicellular organisms, wide variations in fixation time, concentration, and temperature during fixation, dehydration time, and temperature have little or no effect on the preservation of ultrastructure (although enzyme activity and antigenicity may be affected). Conditions should, of course, be kept constant for samples to be subjected to direct comparison of, say, cell volume or dimensions of subcellular compartments, but may be adapted to suit other requirements (such as sampling time, storage, shipping, etc.). The time required for infiltration of single cells with resin is dependent on resin viscosity (and therefore, indirectly, temperature) but is in general 3–18 h. Rigid specimens may require up to 72 h, but ineffective infiltration of hard tissue samples cannot be overcome with prolonged infiltration times. For embedding of mineralized specimens, several alternatives should be considered. Mineral deposits often have to be preserved to allow qualitative and quantitative analysis. Small mineral deposits may be surrounded by the embedding resin and persist in ultra-thin section, though they are also a source of potential sectioning artifacts. Edge fragments of calcified material (e.g., bone) may be embedded in an appropriate resin such as methyl methacrylate/butyl methacrylate in such a way that the section is still surrounded by resin, which provides additional stabilization.

Tab. 2.12 Resins for standard embedding procedures

Resin	Constituents*)	Polymerization conditions
Epon**) (Kushida, 1967)	16 g Epon 812 8 g dodecenyl succinic anhydride 8.7 g methyl nadic anhydride 0.4 g benzyl dimethylamine	24–72 h at 60 °C
Araldite (Glauert et al., 1956)	29 g Araldite CY212 24 g dodecenyl succinic anhydride 0.5 g 2,4,6-tris(dimethylaminomethyl)phenol	5 h at 45 °C and 12 h at 60 °C
Spurr (1969)	10 g vinylcyclohexene dioxide (ERL 4206) 6 g diglycidyl ether of polypropylene glycol 26 g nonenyl succinic anhydride 0.4 g dimethylaminoethanol (a mixture of lowest viscosity is composed of hexenylsuccinic anhydride and Araldite RD 2 instead of nonenyl succinic anhydride and the diglycidyl ether)	8–12 h at 70 °C
Nanoplast (Frösch and Westphal, 1989)	10 g hexamethylolmelamine-methyl ether (MME 7002, 70 % v/v in water) 0.2 g *p*-toluenesulfonic acid (acid catalyst B 52)	Desiccated for 48 h at 40 °C 48 h at 60 °C
Methyl methacrylate/ butyl methacrylate (MBM), pre-polymerized	10 g *n*-butyl methacrylate 2.5 g methyl methacrylate 0.15 g benzoyl peroxide (as mixture with an equal weight of dibutyl phthalate)	12 h at 65 °C

*) quantities of the components may be varied in order to obtain harder or softer polymerized blocks, some mixtures need to be pretreated in several ways; follow purchaser's instructions

**) the term remains established irrespective of the expiry of the trademark "Epon" in 1978; the resin is available under various names (e.g., SPIpon 812, EMbed 812)

Tab. 2.13 Outline procedure for the embedding of samples in Spurr resin

1.) Wash cells by filtration (recommended for filamentous cells such as molds or certain cyanobacteria) or centrifugation (15 min at 2000×g for most unicellular organisms). Appropriate buffers for the washing procedure are 100 mM phosphate buffered saline [PBS] (pH 7.0–7.5), cacodylate buffer (pH 7.0–7.5), or 100 mM HEPES (pH 7.0–7.5) containing 2 mM $MgCl_2$. 100 ml of bacterial culture (OD_{540} about 1) provides enough cells for one preparation.
For some fragile organisms that would not withstand the washing procedure, perform prefixation (step 2) in the culture medium prior to the washing procedure.
For tissue cultures, gently decant the culture medium, wash in situ by applying an appropriate buffer or immediately proceed with prefixation (Step 2).

2.) (Prefixation). Prefixation is very useful when the sample has to be stored or transported. The fine structure of prefixed bacterial cells is preserved for several months. If prefixation is not necessary, proceed with step 4. Add glutardialdehyde (usually available as a 25 % [v/v] solution) to a washed cell suspension or directly to the liquid culture to a final concentration of 0.5 % (v/v) and incubate for 90 min at room temperature. Remove the glutardialdehyde solution by filtration, decanting (for tissue cultures), or centrifugation of the cell suspension.

3.) Store the sample refrigerated in buffer (see Step 1). Do not freeze!

4.) The second chemical fixation step is performed in 3 % glutardialdehyde. If transportation or storage of the sample is not necessary, the prefixation may be omitted. Add glutardialdehyde solution to the sample to a final concentration of 3 % (w/v) and incubate for 2–3 h at room temperature. Wash the pellet in buffer three times (see Step 1). Resuspend the pellet in a small volume of buffer (just enough to permit mixing with molten agar according to Step 5).
Gently remove cells from tissue culture from their substratum, centrifuge at 500×g

5.) Dissolve agar in buffer at 1.5–2 % and keep at 45 °C. Add 1–1.5 volumes of agar to one volume of pellet and mix rapidly. Incubate the agar suspension at room temperature or in an ice bath until the agar is solid. Remove the agar block from the centrifuge tube and cut it into small (approximately 1 mm^3) pieces. Transfer the cubes to small glass vials. The small cubes are easy to handle during the subsequent fixation and embedding procedures.

6.) Wash the cubes once with an appropriate buffer to remove small pieces of agar and free-floating cells. Since the cubes are too large to slip into the pipette tip, no centrifugation or filtration is necessary for this procedure or for the subsequent incubation steps.

7.) Incubate in 1 % (w/v) osmium tetroxide solution for 1 h (the incubation time may be extended to up to 4 h to enhance staining of the sample). Wash the cubes (now a dark brown color) at least three times with distilled water.

8.) Dehydrate the sample by incubating the cubes in 10 % (v/v), 30 %, 50 %, 70 %, 90 % and 100 % acetone solutions (10–15 min per step, 0 °C [ice bath] incubation temperature, acetone solution prepared in distilled water). Improved staining of the specimens is achieved if the 70 % acetone incubation is replaced with one in 2 % (w/v) uranyl acetate in 70 % acetone (3–12 h). Addition of $CuSO_4$ or $CaCO_3$ as desiccant to the 100 % solution some hours prior to use ensures it is water-free. Avoid contamination of the sample with desiccant. Repeat the 100 % acetone incubation step twice.

9.) The subsequent steps are performed at room temperature. Add 1 volume of Spurr resin to 2 volumes of 100 % acetone containing the samples. Mix and incubate for 30 min. The infiltrated agar cubes sink to the bottom of the vial. Add 2 volumes of resin to the sample and incubate for 1.5 h. Replace the resin-acetone mixture with pure resin and incubate for 8–16 h. Replace the resin with fresh resin and incubate for 1–3 h.

10.) Transfer single cubes into small gelatin or plastic capsules that are full of fresh resin. The cubes should sink to the bottom of the capsules after some minutes. De-gas the capsules in a vacuum chamber (the vacuum provided by a one-stage rotary pump is sufficient) for 5–10 min until the development of gas bubbles stops.

11.) Polymerize the capsules at 70 °C for 8 h.

Rapid freezing in conjunction with freeze substitution minimizes mineral loss during the embedding procedure.

Highly calcified specimens usually have to be decalcified before embedding. The decalcification procedure is routinely applied to highly calcified bone, but also to, for example, biofilms in tight association with a mineral substratum. Decalcification kits, especially for bone, are commercially available. If the specimen decomposes during decalcification, then fixing, dehydration, and embedding of the specimen according to a standard procedure prior to decalcification is recommended. The specimen will then be mechanically fixed. After embedding, the calcified face of the specimen has to be exposed by sawing or grinding the specimen. Decalcification may take up to several days in a medium containing EDTA (10 %, w/v, pH 8.0) or EDTA and 2 % acetic acid. After washing in buffer, the specimens are re-embedded in resin. Exposure of the specimen to EDTA may of course induce artifacts, but the overall cellular structure is preserved after fixation in glutardialdehyde.

The resins Epon, Araldite, and Spurr are soluble in acetone but not in ethanol. If ethanol is employed as a dehydrating agent, a transitional solvent, propylene oxide, must be used. Infiltration of the embedding medium occurs by gradual replacement of the solvent by the liquid resin. A typical embedding procedure for unicellular organisms in Spurr resin (including an appropriate fixation and dehydration procedure) is presented in Table 2.13.

2.4.6.3 Nanoplast resin embedding

Embedding in the highly water-soluble Nanoplast melamine resin does not require dehydration of the sample by organic solvents (Frösch and Westphal, 1989; Table 2.14), which may be advantageous for overcoming problems with extractions in organic solvents and is also beneficial for cells that suffer ultrastructural change as a result of dehydration. Though conventional fixation with glutardialdehyde may be performed, treatment with osmium must be omitted. Pretreatment of the sample

Tab. 2.14 Outline embedding procedure for Nanoplast resin

1.)	Prepare the Nanoplast resin according to the supplier's instructions shortly before use (10 g MME 7002 and 0.2 g B 52 are recommended for samples that are appropriate for the preparation of conventional ultra-thin sections).
2.)	Prepare the sample according to Steps 1–6 in Table 2.13.
3.)	Place a small sample or a small agar cube in an embedding mould, filled to a depth of approximately 5 mm (not more, because otherwise the desiccation procedure has to be extended) with Nanoplast resin.
4.)	Place the capsule in a desiccator containing silica gel and incubate for 2 days at 40 °C. Continue incubating the specimen at 60 °C (without further desiccation). If the resulting resin block is too soft, incubate the capsule at 80 °C for 5–10 h. During the embedding and polymerization process, the resin volume shrinks markedly.
5.)	If the resulting resin block is difficult to handle because of extensive shrinkage, embed the capsule in a conventional resin. After polymerization, the surrounding resin provides a mechanical support for the Nanoplast resin.

before resin embedding by this method is short and simple, but the desiccation and polymerization procedures are time-consuming.

2.4.6.4 Embedding media for immunocytochemistry

The embedding media described above are not recommended for immunocytochemical localization of cellular constituents. Suitable resins for immunocytochemical studies are given in Table 2.15.

Low temperatures (–35 °C to –70 °C) during the dehydration and infiltration steps result in excellent preservation of structure, antigenicity, and enzyme activity. Lowicryl resins are therefore the most widely used embedding media when these features are desired. A detailed procedure for embedding with Lowicryl K4M resin is given in Table 2.16. The sensitivity of the antigen to the fixative used determines the concentration and composition. Insensitive antigens usually tolerate 1 % glutardialdehyde, which generally provides proper fixation. Sensitive antigens are fixed in glutardialdehyde/formaldehyde mixtures, either with low glutardialdehyde concentrations (0.1–0.2 %, w/v) or with formaldehyde alone. Formaldehyde is a less effective fixative than glutardialdehyde and is partially extracted from the specimen during prolonged dehydration. Nevertheless, formaldehyde concentration may be low (0.3%) when used for

Tab. 2.15 Resins suitable for immunocytochemistry

Resin	Special conditions for embedding and polymerization	Special properties
Glycol-methacrylate (Leduc and Bernhard, 1967)	Water (instead of organic media) as solvent; polymerization under UV light possible	Sections relatively unstable to the electron beam
London resins	Dehydration in methanol or acetone/ethanol; complete dehydration of the sample not necessary; polymerization under UV light possible	Very low viscosity (infiltration of rigid specimens)
LR White (Newman and Hobot, 1987)	Use at room temperature	
LR Gold (Shires, 1990)	Use at low temperature (–4 to –25 °C)	
Lowicryl resins (Carlemalm et al., 1982)	Dehydration in methanol or ethanol, low-temperature embedding, polymerization under UV light possible	Low viscosity at low temperatures, excellent preservation of structure and antigenicity of the sample
K4M	Embedding temperature as low as –35 °C, polar resin (residual amounts of water in the sample are tolerable)	
HM 20	Embedding temperature as low as –70 °C, non-polar (no residual water is tolerated)	

fixation of, for example, bacteria, which are small and easily penetrable, but must be raised considerably (up to 8 %, w/v) when used as a single fixative for tissue. Fixation with osmium tetroxide is usually deleterious to antigenicity and should be omitted.

Some samples (especially dilute suspensions) are difficult to detect in polymerized resin, because the intense osmium stain is absent. Embedding of a visible marker (best included in the agar medium; see Step 5 of Table 2.13) to enable easy localization may be helpful, but the marker should not interfere with the sample and must be insoluble in the resin. Fine carbon particles from Indian ink co-entrapped in the agar medium are helpful (centrifuge the ink before use and resuspend in the buffer used during the

Tab. 2.16 Outline embedding procedure for samples in Lowicryl K4M

1.) Prepare a fresh 5–10 % (w/v) formaldehyde solution from solid paraformaldehyde by suspending paraformaldehyde in distilled water and heating to 80 °C. Slowly add drops of 1 M sodium hydroxide until the solution becomes clear.

2.) Wash cells by filtration (recommended for filamentous cells such as molds or certain cyanobacteria) or centrifugation (15 min at 2000×g for most unicellular organisms). Use 100 mM phosphate buffered saline (pH 7.0–7.5) for the washing steps. Cacodylate buffers are not recommended, because they may reduce the antigenicity of the specimen. 100 ml of bacterial culture (OD_{540} about 1) provides enough cells for one preparation.
For some fragile organisms that would not withstand the washing procedure, the chemical fixation (see Step 3) should be performed in the culture medium prior to the washing procedure.
For tissue cultures, gently decant the culture medium wash in situ by applying an appropriate buffer.

3.) Add glutardialdehyde (optional) and formaldehyde solutions to a washed cell suspension or directly to the liquid culture (Step 2 is omitted) to appropriate final concentrations*) and incubate for 90 min in an ice bath with gentle shaking. Centrifuge three times for 15 min at 2000×g in PBS containing 10 mM glycine.

4.) Embedding in molten agar as described in Step 5 in Table 2.13.

5.) Dehydrate the specimen by incubating the agar cubes in 15 % and 30 % methanol (15 min per step) at 0 °C, in 50 % methanol (15 min) at –20 °C, in 70 %, 90 %, and 100 % methanol (30–60 min per step) at –35 °C.**)

6.) Add 1 volume of prechilled Lowicryl resin per volume methanol and incubate for 1 h at –35 °C. Add 0.5 volumes of resin per volume of mixture and incubate for 1 h at –35 °C. Replace the resin-methanol mixture with fresh resin and incubate overnight at –35 °C. The infiltrated agar cubes sink to the bottom of the vial. Replace the resin with fresh resin and incubate for 2 h at –35 °C.

7.) Transfer the agar pieces into gelatin capsules. The cubes will sink to the bottom of the capsule within 30 min. Polymerize for 40 h at –35 °C and 3 days (might be unnecessary) with UV light (370 nm, 15 W, distance: 30–40 cm. To provide diffuse illumination, a right-angle reflector is suspended below the UV lamps).

*) Check the sensitivity of the antigen to the aldehydes by Western blot or dot blot. Most antigens withstand 0.2 % (v/v) glutardialdehyde and 0.3 % (v/v) formaldehyde, which is an appropriate fixative for unicellular organisms and other single cells in suspension. Glutardialdehyde may have to be omitted and the formaldehyde concentrations raised.

**) The incubation times are sufficient for unicellular organisms, but may be longer for larger specimens. Be aware that components may be extracted during these procedures if glutardialdehyde fixation has had to be omitted.

embedding procedure). The stain should result in no more than a light gray appearance of the sample. Higher concentrations would prevent sample polymerization by ultraviolet light! It has to be borne in mind that the carbon particles produce a background of irregular dark spots in the ultra-thin section.

The use of Lowicryl resin requires cooling devices capable of producing an ambient temperature of –35 °C or lower for a long period of time for optimal results. At this temperature, denaturation of proteins in organic solvents is considerably reduced.

2.4.6.5 **Freeze-substitution**

Cryofixation of the samples can be coupled with (low-temperature) resin embedding procedures by freeze-substitution. This technique is used when conventional dehydration procedures are to be avoided, such as when water-soluble substances (small ions) should not be removed, but the frozen water has to be replaced with an anhydrous solvent, typically followed by resin infiltration. With the aid of freeze-substitution, the advantages of sample vitrification are combined with procedures for resin-embedded specimens that are easier to handle and store than frozen specimens. Besides resin embedding, samples may be processed by subsequent freeze-drying or critical point-drying methods. See Hippe-Sanwald (1993) for a review and Paul and Beveridge (1993) for examples of application.

Samples may be prepared by the method of Tokuyasu (Table 2.8), by plunge-freezing (Table 2.9), or by high-pressure freezing, and then transferred to a freeze-substitution system in which they are gradually warmed up to a temperature that will permit resin embedding (–40 °C for subsequent low-temperature resin embedding, or ambient temperature for conventional resins). During the freeze-substitution process, the ice in the sample is replaced by an organic solvent. For subsequent X-ray microanalysis of the specimen, water has to be removed completely by the use of water-free absolute methanol (dried over molecular sieves and with the addition of molecular sieves during freeze-substitution). Several authors recommend chemical fixation (glutardialdehyde, osmium tetroxide) during methanol infiltration in order to stabilize the cellular constituents after the frozen water has been removed (see Robards and Sleyter, 1985, for survey). However, this process may destroy the superior structural preservation achieved by rapid freezing (Sjöstrand, 1990).

A simple setup for cryofixation and freeze substitution devices is described by Sitte (Fig. 2.42; see Robards and Sleytr, 1985; Roos and Morgan, 1990). A commercially available device is based upon the instruments developed by Sitte and Edelmann (Sitte 1984; Leica AFS, Leica Inc.).

The principal setup of the device is as follows: An aluminum container is placed over liquid nitrogen (not submerged) in a Dewar. Cooling is provided by the nitrogen gas surrounding the container. Temperature regulation is achieved by adjusting a thermosensor/heater-circuit. The bottom of the metal container is insulated to prevent unnecessary nitrogen loss. Frozen specimens are placed on a metal grid inside the container, which is filled with a substitution medium continuously agitated by a stirrer. A standard procedure for freeze-substitution is detailed in Table 2.17. Prior to this procedure, cells are chemically fixed as described for Lowicryl embedding in Table 2.16

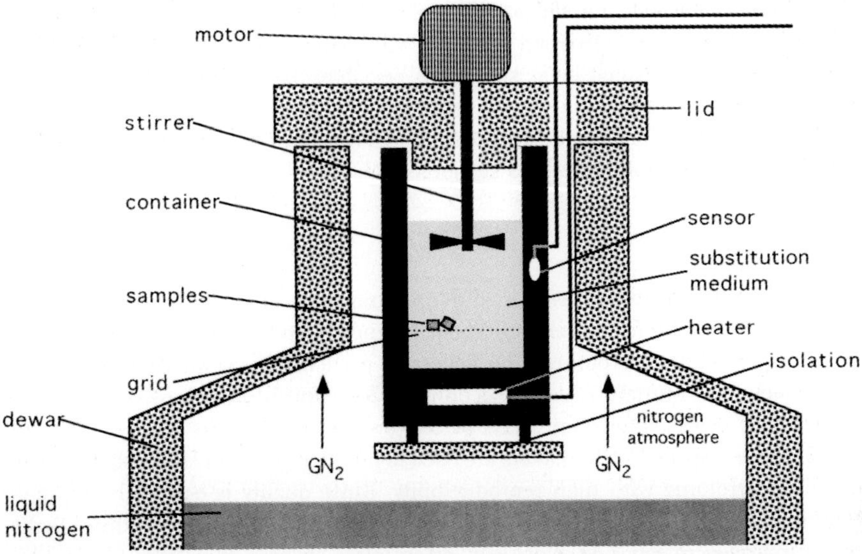

Fig. 2.42 Simplified representation of a freeze-substitution apparatus. The setup is essentially according to Sitte (1984). The commercially available instruments allow control of the substitution process by presetting a temperature gradient and also allow final polymerization with an ultraviolet light source

Tab. 2.17 Freeze-substitution of samples

1.) Fill the container of the freeze-substitution device with absolute methanol and adjust the temperature to –80 °C. Allow the system to equilibrate (continuous loss of nitrogen in simple devices may cause an undesired temperature gradient).
2.) Transfer the sample from liquid nitrogen (see Table 2.9; plunge freezing) to the pre-chilled methanol.
3.) Allow substitution to occur for 18 h at –80 °C, followed by gradual warming to –40°C over a 60 h period.
4.) Transfer the sample in the chilled metal container to a refrigerator maintained at –40 °C and then proceed from Step 5 in Table 2.16 (embedding in Lowicryl resin). Although the sample itself may be completely invisible in the embedding resin, the support (grid or filter paper) allows it to be localized.

and subjected to plunge freezing as described in Table 2.9. The cells are then embedded in Lowicryl resin.

2.4.7 Ultramicrotomy

2.4.7.1 Trimming

Sections of specimens embedded in resin, as described in the previous chapters, should be between 60 nm and 90 nm thick to allow proper imaging under the electron microscope. Samples are therefore subjected to trimming and ultra-thin sectioning. Trimming of the (resin-embedded) samples is done so as to obtain small, flat-topped pyramids with a face of about 0.2–2 mm^2. Trimming of the firmly held resin block may be performed with the aid of a sharp razor blade or, more accurately, with a rotating milling cutter (specimen trimmer with a diamond cutter). The flat top of the pyramid (mesa) should be as smooth as possible. Its shape is of the utmost importance: ribbons of sections (e.g., for serial sections) are produced most easily when the trimmed surface is a trapezium. A square is also usable, but unsuitable for production of ribbons more than 4–5 sections long.

2.4.7.2 Knives

Sectioning of the sample is performed in an ultramicrotome with the aid of glass or diamond knives. The former are sufficient for most standard sectioning procedures (including cryo-ultramicrotomy), the latter are particularly useful when sectioning hard objects, for numerous serial sections or for very thin sections. Glass knives are produced by cleaving square glass pieces (broken from a glass strip) into two triangular knives. Special knife-makers are used to produce glass knives for standard and cryo-ultramicrotomy with high reproducibility. Knife quality is essential for proper sectioning and needs to be carefully checked. Important features of knife quality are summarized in Fig. 2.43. Normally, 40 cm glass strips (6 mm thick) are used to produce 16 squares with right-angled corners. By breaking the squares slightly off the diagonal axis, two usable knives are obtained: the square is scored in such a way that the scores stop at some distance from the corner. When pressure is applied under the score, the fracture extends towards the corners of the square, following the line of the score. At each end of the score, the fracture deviates from the line of the score to curve away from the corner towards the edge of the square ("free" break). This results in the real included angle of the knife being somewhat greater than the angle of the score. The real angle of the knife increases as the score is moved further from the diagonal: that is, when the counter-piece (of the **opposite** knife!) is larger. A counter-piece with a width of 0.4 mm with an adjusted knife angle of 45° results in a real angle of 55° (+/–2°) at the cutting edge. Increasing the width of the counter-piece by adjusting the knife maker results in larger angles, reducing it results in smaller angles, closer to 45°. The smallest knife angles lead to very sharp, but brittle cutting edges. For sectioning of ultra-thin sections from resin-embedded samples, knives with counter-pieces of 0.3 mm–1 mm give the best results. Knives with very small counter-pieces (0.1 mm and smaller) are recommended for cryosectioning. The overall appearance of the knife edge and the length of the useful cutting edge vary as indicated in Fig. 2.43 d. Flaws or frills are easy to detect by inspection of the cutting edge in dark-field illumination under a stereomicroscope. Diamond knives have lifespans of several years, even when used daily. The knives are delivered with a recommended cutting speed and clearance angle, suitable for resin-embedded specimens of various hardnesses and section thicknesses, as well as for cryosectioning. Cleaning is necessary for best section quality and is performed with a styrofoam (polystyrol) rod, or a sharpened wooden toothpick, wetted with double-distilled water. Cleaning with ethanol or other organic solvents is possible, but care has to be taken that the cementing material of the diamond (or the cleaning tool itself) does not become dissolved. Usually it should be resistant to ethanol.

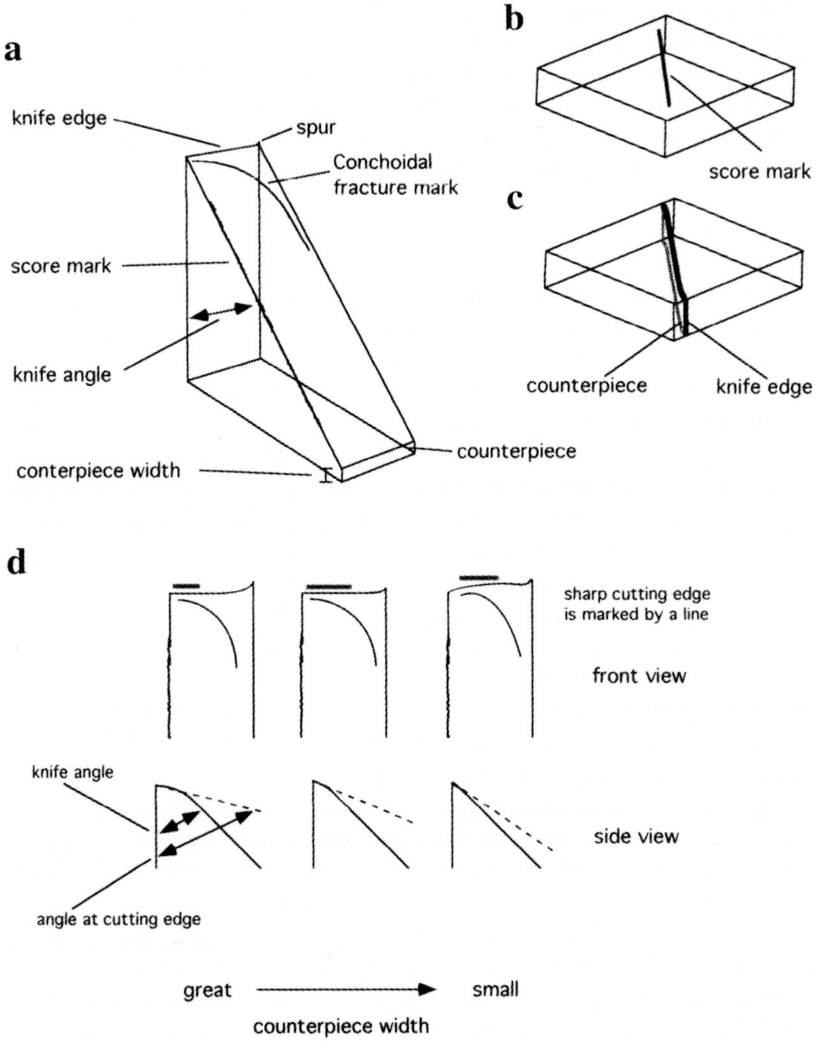

Fig. 2.43 Appearance and quality of glass knives. A. General appearance of a glass knife (without trough). B, scored, C, broken glass square. D, Glass knives of different qualities (Redrawn after Meek, 1976)

For sectioning of resin-embedded samples, a small fluid-filled (normally with particle-free water) trough is attached around the knife edge. Troughs may be made from a variety of materials, such as plastic insulating tape or metal foil, or they may be obtained commercially. Troughs are sealed with warm dental wax.

2.4.7.3 Ultra-thin sectioning

Operating procedures for commercially available ultramicrotomes differ and are described in detail in the respective user's manuals. The procedures included here (Tables 2.18, 2.19) refer to adjustment procedures common to all known standard instruments, but do not supplant the manuals.

The thickness of sections floating on the water surface during microtomy can be estimated by adjusting the microtome viewing system illumination so that the sections appear colored on the water surface (see Table 2.18). The sections are picked up with a (Formvar-coated) grid (see above).

For some elemental analysis procedures (electron spectroscopic imaging, ESI), sections of minimal thickness (below 50 nm) are required (see Table 2.19). Very thin sections are best cut with knives with very sharp cutting edges. Diamond knives provide the best results.

In order to overcome the minimal thermal expansion of the instrument, switch on the instrument illumination 1 h prior to use; knife holder and specimen holder should already be mounted. When sections of very low thickness are to be cut, the microtome

Tab. 2.18 General procedure for ultra-thin sectioning

1.) Mount the trimmed block.
2.) Mount the knife and select a clearance angle between 1–5°.
3.) **Perform ultra-thin sectioning:** All adjustments need to be made by observation through the microtome binocular: set the cutting edge parallel with the mesa of the resin block. This is most easily done if the mesa has a smooth, reflective surface.
   - Switch on microtome lights.
   - Bring the mesa face level with the cutting edge by adjusting the microtome arm height.
   - Advance the knife with the coarse control, observing the reflection of its edge in the block face. The reflection aids adjustment of the knife parallel to the mesa surface and estimation of the distance between the mesa surface and the cutting edge.
   - Move the microtome arm up and down and observe the clearance between knife and mesa carefully: if it increases or decreases, the mesa surface is not exactly vertical and the resin block needs to be adjusted.
   - Advance the knife with the coarse and fine controls until the clearance between knife and mesa is barely visible.
   - Fill the trough with water until an almost flat meniscus causes optimum wetting of the cutting edge.
   - Switch on the microtome arm automatic advance and cut some semithin sections in order to obtain a flat surface.
   - Switch over to ultra-thin sectioning.

   Observe interference color (see Table 2.20) and overall appearance of the sections: only sections of uniform color, without scratches, wrinkles, or holes, are suitable for further processing.
4.) Move sections on the water surface with the aid of a mounted eyelash. Do not touch the sections directly during this process, but create gentle currents in the liquid to orient the sections suitable for further preparation so that they can be picked up.
5.) Pick up the arranged sections with a Formvar-coated specimen grid: slowly lower a grid onto the water surface, allow the grids to adsorb the sections for some seconds, and then withdraw the grid.
6.) Drain excess liquid off with a piece of filter paper and store the grid (sections upwards until post-staining is applied). If immunolocalization procedures are to follow, place the grid on a drop of phosphate buffer (50 mM, pH 7.0) until use (overnight refrigeration possible).

Tab. 2.19 Preparation of sections below 50 nm in thickness

1.) Trim the resin-embedded sample in such a way that the flat top of the pyramid (mesa) has a surface area of 0.25 mm^2 or less (it is easier to obtain sections below 50 nm with very small mesas).
2.) If glass knives are used, they should be of the highest quality. Cutting semi-thin sections during knife adjustment reduces the quality of the knife edge. Ultra-thin sectioning requires readjustment of the knife edge to a new position. Cut some 80 nm thick ultra-thin sections ("silver" sections) and check section quality. If the sections are absolutely homogeneous (interference color, no wrinkles, knife marks, etc.), proceed to Step 3. If not, use a new glass knife.
3.) Reduce section thickness to 50 nm. If gray sections are cut, reduce feed in 1 nm steps. When sections of doubled thickness every other cut are performed, increase thickness step-by-step until sections are performed every cut. The sections obtained by this method are of the minimal possible thickness achievable with the setup used.
4.) Sections in a ribbon may be stabilized and, since they are barely visible, "marked" by silver sections: after a section of the desired thickness has been cut, immediately raise the section thickness to 80 nm and cut several silver sections. Then reduce section thickness again.
5.) For subsequent elemental analysis (see below), pick up the sections with bare grids (i.e., grids without any support film, 600–1000 mesh).

Tab. 2.20 Determination of section thickness on the basis of interference color

Interference color	Section thickness	Useful for:
Grey	< 60 nm	Analysis of elemental distribution (e.g., by electron spectroscopic imaging, ESI); use best quality glass knives or diamond knives).
Silver	60–90 nm	"Standard" thickness for most morphological and immunocytochemical studies.
Yellow–gold	90–150 nm	Easier to cut with medium quality knives, sometimes usable for immunocytochemical studies as well.
Purple–blue	150–300 nm	For thick section studies (at 100 kV or with the aid of ESI; see below).
Opaque silver	> 500 nm	For high-voltage electron microscopy and light microscopy (sections to be stained with Toluidine Blue).

itself should not be touched by the operator. All standard procedures to minimize transfer of mechanical vibrations have to be applied.

2.4.7.4 **Post-staining of sections**

Contrast is achieved (or enhanced, when heavy metal staining is applied during the infiltration procedure; see above) in the sectioned samples by floating the grids, with the attached sections facing downwards, on a small drop of the staining solution. Uranyl acetate heavily stains nucleic acids and proteins, while its staining of membranes is only poor. Lead citrate stains membranes, proteins, nucleic acids, and proteins, because lead cations bind to phosphate, carboxyl, and sulfhydryl groups. Osmium stains are enhanced by application of lead citrate staining. Staining with Ruthenium Red solution (in combination with uranyl acetate) leads to improved contrast of wall layers (including slime and capsule), cytoplasmic membrane, nucleoid and some cell inclusions (Vogt et al., 1995). Preparation of staining solutions is described in Table 2.21.

Tab. 2.21 Preparation of staining solutions for ultra-thin sections

Uranyl acetate

Suspend uranyl acetate in distilled water (up to 4 %, w/v) and allow to dissolve over several hours, remove undissolved material by sedimentation in the bottle and centrifugation just prior to use. Store in a light-tight bottle at room temperature. See also Table 2.3

Lead citrate

A) According to Reynolds (1963):

1.) Dissolve 1.33 g of lead nitrate ($Pb(NO_3)_2$) and 1.76 g of sodium citrate ($Na_3(C_6H_5O) \times 2\, H_2O$) in 80 ml of distilled water and shake vigorously for 1 min.
2.) Incubate for 30 min with intermittent shaking.
3.) Add 8.0 ml of 1 M NaOH and 12 ml of distilled water.

The clear solution is ready to use, faintly turbid solutions have to be centrifuged (10 min, 15000×g) before use; store refrigerated.

B) According to Venable and Coggeshall (1965):

Dissolve 10 % (w/v) lead citrate in 0.1 M NaOH (solution in CO_2-free [boiled!] water); the turbid solution should be centrifuged (10 min, 15000×g) before use.

Ruthenium Red (Vogt et al., 1995, and references therein):

Dissolve Ruthenium Red in distilled water or buffer solution. Combinations of Ruthenium Red with borate buffer (pH 9.2), cacodylate buffer (pH 7.0), and glycine buffer (pH 3.0 and pH 9.0) have been described.

Tab. 2.22 Staining with uranyl acetate and lead citrate

1.) Prepare a fresh lead citrate solution (Reynolds stain is more stable against carbonate precipitation) as described in Table 2.21 and centrifuge a uranyl acetate solution (does not need to be freshly prepared).
2.) Place 50 µl drops of uranyl acetate on a clean Parafilm surface, place drops of lead citrate on a Parafilm strip in a Petri dish containing several moistened KOH or NaOH pellets (in order to maintain a CO_2-free atmosphere).
3.) Place the grids on the uranyl acetate drops with the sections facing downwards. Incubate for 3 min (e.g., for Lowicryl-embedded sections) to 15 min. If no further staining is necessary, soak away excess liquid with a filter paper and store the grids (sections upwards). For subsequent Ruthenium Red staining, place the grids on drops of Ruthenium Red solution for up to 10 min and then blot dry. For subsequent lead citrate staining, proceed to Step 4.
4.) Place the grids on lead citrate drops and incubate for 2–10 min.
5.) Wash the grids free of excess stain by repeated dipping in a series of three 20 ml beakers of boiled (i.e., CO_2-free), double-distilled water. Alternatively, wash in a beaker containing 20 mM NaOH in double-distilled water and three beakers of double-distilled water.

Staining with uranyl acetate and either Ruthenium Red or lead citrate may be combined in order to achieve maximum contrast. During lead citrate staining, crystalline precipitation of lead carbonate may occur (often along membranes or cell walls), and so atmospheric CO_2 has to be excluded during the staining procedure (see Table 2.21). Staining with Ruthenium Red does not lead to precipitates.

2.4.7.5 **Cryo-ultramicrotomy**

Thin sections of specimens prepared by rapid freezing methods (especially high-pressure freezing) give a more accurate picture of the overall structure of the object under study, due to superior ultrastructural preservation because of the omission of conventional dehydration and embedding methods. Though preparation of samples by freeze-substitution, followed by resin embedding and conventional sectioning (see above), improves structural preservation, well performed cryosections provide insights into the structure of specimens that have been maintained fully hydrated and uncontaminated since the initial rapid freezing procedure. When antigens cannot be detected in chemically fixed and/or resin-embedded specimens by immunolocalization, application to cryosections is a very promising approach. This, of course only applies for specimens that have been accurately prepared and sectioned. The methods are less tolerant to inaccuracies during the preparation process, especially when the specimen is unfixed.

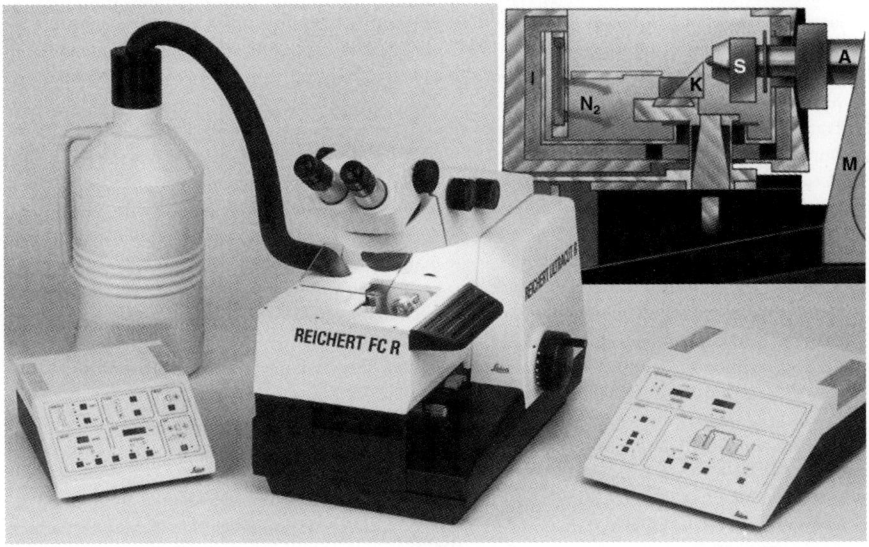

Fig. 2.44 Ultramicrotome for routine applications (Leica Inc.). The microtome is equipped with a nitrogen-cooled cryochamber (see inset). The control panels allow microtome adjustment and temperature control for knife, specimen holder. and nitrogen atmosphere. The inset depicts a section through a cryochamber in side view. I: insulation, N_2: cool nitrogen gas, K: knife, S: specimen holder, A: specimen arm of the microtome, M: body of the microtome (Leica Inc.)

For the production of ultra-thin sections from frozen samples (see above), an ultramicrotome equipped with a cryochamber has to be used. This instrument maintains low temperatures (–130 to –50 °C) throughout the sectioning procedure. Ultramicrotomes with cryochambers are available from various manufacturers (Leica, Dupont-Sorvall; see Fig. 2.44).

Glass knives for cryosectioning should be made in knife-makers according to the "balanced break" method. A glass strip (see above) is broken into two equal halves. With an equal mass of glass on each side of the score, the break is balanced and the freshly fractured surfaces are planar. By continuing to divide each piece produced into two equal halves, up to 16 squares can be made from a glass strip of 40 cm in length. All the squares produced have straight sides and precise right-angled corners. The glass knives should have a real knife angle closest to 45° and be of the best quality for maximum sharpness (Griffiths et al., 1983). Knives should be prepared shortly before use. The stability of the cutting edge may be improved by a light tungsten coating (Roberts, 1975). The knives are normally used without a trough and are kept dry. The specimen is transferred, in liquid nitrogen, to the pre-cooled specimen holder. Sectioning is performed, in principle, according to the conventional procedure with subsequent cooling. The ultra-thin sections are picked up by adsorption of

Tab. 2.23 Cryosectioning

1.) Pre-cool the ultramicrotome and mount the knife. Knife and specimen holder temperature should be in the order of –80 °C to –120 °C for cryoprotected and/or chemically fixed specimens (see above) or lower for specimens frozen in the native state. For samples intended for X-ray microanalysis, temperatures should not be raised above –140 °C. Pre-cool all tools in liquid nitrogen or directly in the cryochamber of the ultramicotome. An atmosphere of cold nitrogen gas in the cryochamber prevents the entrance of moisture.
2.) Transfer the frozen specimen (see, for example, Table 2.9) to the pre-cooled cryo-ultramicrotome. During transfer, the specimen is protected by the liquid nitrogen that remains in the cryo-cap. Most simply, "pour out" the specimen holder with the LN2 in the cryochamber.
3.) Mount the specimen in the specimen holder, using the pre-cooled tools.
4.) Trim the samples with an edge of the mounted knife or with trimming tools mounted on the knife holder. Form a flat-topped pyramid, but be aware that, if no cryoprotection has been applied, only a thin surface layer will be well vitrified, and so careless trimming will remove the vitrified parts of the specimen (freezing the specimen by the slamming technique would overcome this problem, since the technique produces a flat top; see chapter 2.4.5.5).
5.) Adjust specimen and knife as described for conventional sectioning (see above). The sections are placed, with the aid of an eyelash, mounted on the end of, for example, a Pasteur pipette, on the knife surface. Electrostatic interactions between the tool and the sections will lead to loss of some sections, but proper orientation is essential for the subsequent step. Use an antistatic pistol to reduce loss of sections.
6.) Use a drop of 2.4 M sucrose solution in water in a wire loop of about 3 mm in diameter to pick up the section placed on the knife surface. Transfer the drop (with the section attached to its surface) to room temperature. The procedure has to be finished before the sucrose drop is frozen.
7.) Pick up the sections with a carbon-coated Formvar grid. If immunolocalization is to follow (see Table 2.33), a carbon-coated Formvar nickel grid has to be used.
8.) Wash away the sucrose solution by placing the grid on the surface of distilled water. Repeat the washing procedure three times.

Tab. 2.24 Staining of cryosections

(A) Stabilized with methylcellulose:*)

1.) Prepare a 2 % (w/v) methylcellulose solution in water, centrifuge for 90 min at 250,000 × g. (The solution may be stored refrigerated for several weeks.) For preparation of the staining solution, add the desired amount of a uranyl acetate stock solution (4 % [w/v]). A final concentration of 0.3 % (w/v) uranyl acetate is suitable for the first trial. Optimize the uranyl acetate concentration by checking the contrast of the specimen in the electron microscope.
2.) Place grids onto two drops of the staining solution for several seconds each and then onto a third drop of staining solution for three minutes.
3.) Pick up the grids with forceps or a wire loop and soak away excess staining solution. Allow the grids to dry for 20 min.

(B) Non-stabilized:

1.) Transfer the grids after the washing procedure (Table 2.23, Step 8) for 1 min on a heavy metal staining solution (uranyl acetate or phosphotungstic acid are recommended, concentrations between 0.3 and 1 % [w/v] provide sufficient, but not excess, contrast)
3.) Pick up the grids with forceps and blot dry.

For subsequent elemental analysis, wetting of the sections is not recommended. Once sectioned, the specimens are further processed by mounting, by pressing the sections, in a specially designed device, onto a grid surface (Tvedt et al., 1984), and then subjected to subsequent freeze-drying (Geymayer et al., 1977).

*) Methylcellulose is usually recommended for further section stabilization. At least for the preservation of sections with small specimens (bacteria, yeasts), the simplified staining procedure (B) may be applied.

a drop of sucrose in a small metal loop onto the surface. During this transfer, the section is warmed up to room temperature and transferred to a (carbon-coated) Formvar grid.

A detailed outline for a cryosectioning procedure is given in Table 2.23. The preparation method is suitable for subsequent morphological studies of the sections and immunolocalization procedures.

For subsequent immunolocalization, place the grids on ice-cooled gelatin (2 %, w/v, aqueous solution) and proceed as described in Table 2.33. Alternatively, the sections may be stained as described in Table 2.24.

2.4.8
Localization and Structure of Macromolecules

Single epitopes of enzymes in pure preparations can be localized by epitope mapping with the aid of specific antibodies. The localization of proteins in organisms is performed by application of labeling procedures based on specific binding of colloidal gold or other electron-dense markers. A survey of the techniques used is given in Table 2.25.

Tab. 2.25 Techniques for the localization of biological macromolecules by electron microscopy

Problem	Specimen treatment	Marker system (technique; references)
Localization of specific proteins (antibody-gold) or carbohydrate complexes (lectin-gold) inside cells	Ultra-thin sections of resin-embedded samples (Lowicryl, LR gold) subjected to immunolocalization	Antibody-colloidal gold, lectin-colloidal gold (Post-embedding technique; Roth et al., 1980; Horisberger, 1985)
Localization of specific proteins on the cell surface or subcellular particles (modified technique useful for SEM-studies)	Incubation of cells or subcellular fractions with the marker system prior to conventional embedding in, for example, Spurr resin	Antibody-colloidal gold, antibody-ferritin, lectin-colloidal gold (pre-embedding technique; Rohde et al., 1988)
Localization of specific proteins on the surface of, for example, membrane vesicles	Subcellular fraction adsorbed onto a carbon-coated Formvar-grid subjected to immunolocalization	Antibody-colloidal gold, antibody-ferritin, lectin-colloidal gold (whole-mount-technique; Acker, 1988)
Detection of viral surface antigens	Aggregation of virus particles by incubation with specific antibodies (specific to surface epitopes, for example) followed by negative staining	Virus particle and a specific antibody – no additional markers (Miller and Howell, 1997)
Localization of single epitopes (subunits) and their relative position to each other on the surface of the object	Homogeneous protein preparation (especially useful for large proteins or viruses) subjected to localization of epitopes	Antibody, Fab-fragment, (epitope-labeling; Hermann et al., 1991)
Localization of biotin-containing enzymes, localization of the biotin-site	Subcellular fraction, homogeneous enzyme preparation subjected to localization	Avidin-gold (Mayer and Rohde, 1988), avidin (Däkena et al., 1988, and references therein)
Detection of enzyme activities	Incubation of cells or subcellular fractions with the marker system prior to conventional embedding in, for example, Spurr resin	Specific enzyme substrate coupled to precipitation of electron-dense marker (Hayat, 1973ff, Wohlrab, and, Gossrau, 1992)
Detection of macromolecular components acting as enzyme substrates (such as starch, nucleic acids)	Labeling of enzyme substrates with gold particles coated with the relevant enzymes in ultra-thin sections of samples	Enzyme-colloidal gold (Bendayan, 1985)
Localization of metabolites in the cell	Incubation of metabolically active cells with the precursor, embedding in conventional resin or by application of cryotechniques, treatment with a specific photoemulsion	Metabolic precursors (such as sugars or aminoacids) labeled with radioactive markers (autoradiography; Gregg and Reznik-Schüller, 1984)
Localization of specific DNA or RNA (preferably mRNA) sequences in cells	Hybridization DNA/RNA of fixed samples with the marker system followed by embedding and sectioning. Detection of the marker system by autoradiography or with streptavidin-colloidal gold	RNA or DNA oligonucleotide labeled with different markers (in situ hybridization; Egger et al., 1994)

2.4.8.1 Preparation of marker systems for localization procedures

The most widely used localization method is labeling with the aid of a specific (monoclonal or polyclonal) antibody and colloidal gold (5 – 20 nm in size) as an electron-dense marker system. The most common immunocytochemical localization procedures use colloidal gold particles coupled to proteins that bind to the specific ("primary") antibody. These proteins are either Protein A or a "secondary" antibody directed against the Fc-fragment of the specific antibody. Preparation of the marker systems is performed according to established procedures (Rohde et al., 1988, and references therein; Roth et al., 1978); the markers are also available from various commercial sources.

Monodisperse colloidal gold preparations (both coupled and uncoupled) may also be prepared easily in sizes ranging from 2 to 150 nm (Frens, 1973), and even smaller particle sizes for special applications are possible. The size of the gold particles depends mainly on the strength and quantity of the reducing agent used. Phosphorus, ascorbic acid, sodium citrate, and tannic acid are widely used reducing agents. To date, a wide range of proteins have been successfully coupled to colloidal gold. Geoghegan and Ackerman (1977) found that adsorption of proteins to colloidal gold depends on the pH of the solution and the isoelectric point of the protein to be coupled. Adsorption is also influenced by the interfacial tension and the solubility of the molecule (see Tables 2.26, 2.27 and 2.28 for preparation of colloidal gold and the coupling procedure).

Since colloidal gold-coupled probes are the most widely used marker systems (antibodies, biotin, lectins, etc.) for localization in electron microscopy, they are available from numerous suppliers. Preparations from various suppliers still show wide variations in quality. Both uncoupled and coupled monodisperse colloidal gold of 5 – 15 nm in size has a dark red to wine red color. Even slight violet or blue violet coloration indicates clustering of the gold particles, and clustered particles are useless for electron microscopic applications (although they may still function in light-microscopic histochemical or biochemical tests). The gold probes should therefore be obtained from manufacturers with proven experience (British Biocell, Dako, Nanoprobes for ultra-small particles).

However, self-prepared colloid allows the experimenter to design new specific probes in a one-day procedure with a minimum of laboratory equipment. The quality of the preparations (particle size distribution, clustering) has to be carefully checked by electron microscopy before they are coupled to proteins or other reagents, and before use. In double-labeling experiments, the use of two differently sized gold particle preparations coupled to secondary reagents permits differentiation between the locations of two defined binding sites.

During the immunolocalization procedure, the specimen is incubated with the first antibody (preferably IgG), and this is then detected by means of the marker system, through binding of Protein A or the secondary antibody. Most protein types may be detected by this method if the density of epitopes is high enough. When epitopes of interest are exposed down to a depth of 20 nm from the specimen surface (this is a maximum value achievable by etching of the section surface), the background is zero, and all exposed antigens are marked, at least 100 antigens of a cytoplasmic protein

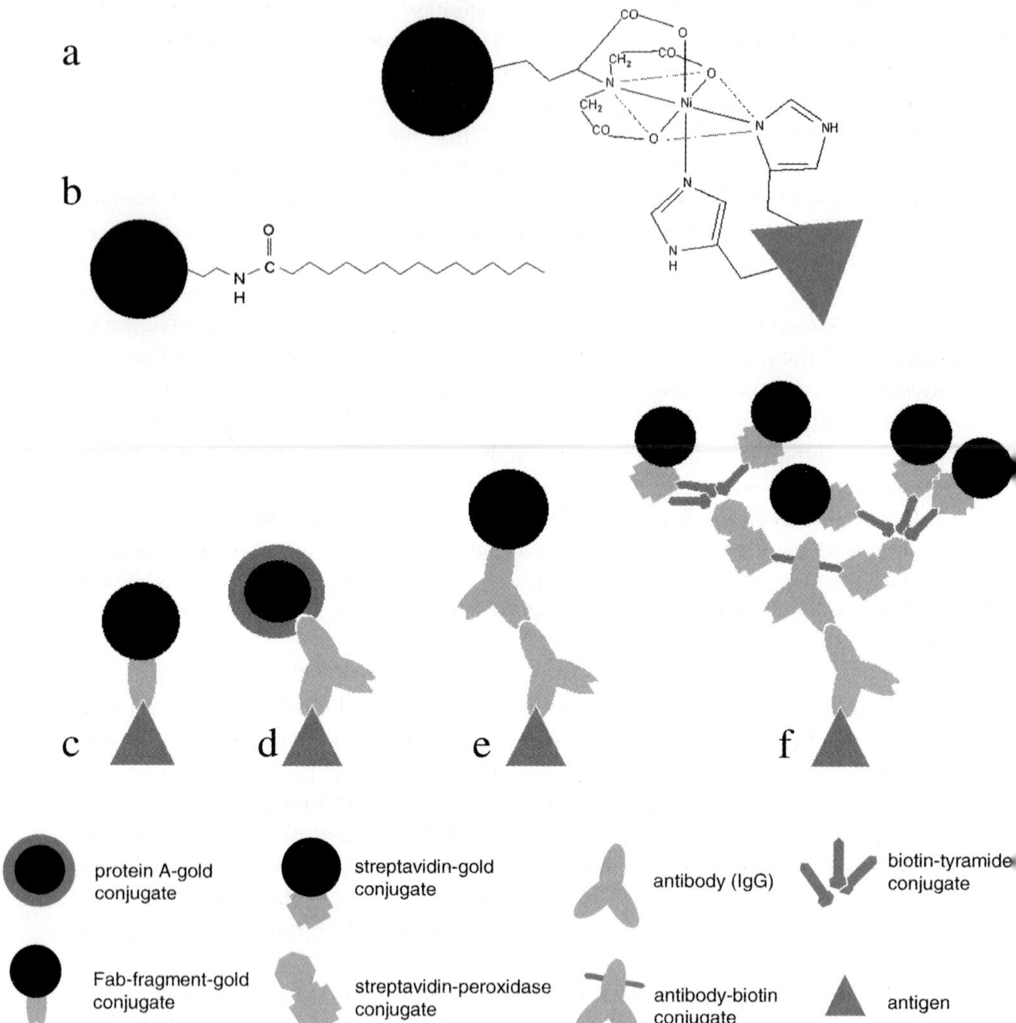

Fig. 2.45 Various marker systems based upon colloidal gold as an electron-dense, detectable marker.
(a) Gold particles coupled to nickel-nitrilotriacetic acid (Ni-NTA) specifically binding to a His-tagged protein as manufactured by Nanoprobes Inc.
(b) Palmitoyl conjugate for reconstitution in liposomes, as manufactured by Nanoprobes Inc.
(c) The specific component of the marker sandwich (e.g., an enzyme for detection of substrates, a lectin, or a Fab fragment) may be directly coupled to gold colloid. There is only a short distance (3–5 nm) between the detectable marker and the marked molecule.
(d, e) The most widely used marker sandwiches for routine applications are protein A gold bound to the primary antibody (d) or a gold-coupled secondary antibody bound to the primary antibody (e).
(e) Signal amplification by catalyzed reporter deposition (CARD; see text for further explanations)

should be present in an "average" bacterial cell to provide around five markers on a section. Higher local densities of antigens at membranes, in small vesicles, etc. may also produce specific signals of lower antigen numbers. In practice, only a small fraction of the exposed antigen will be detected, due to modification of the individual antigen during preparation, masking of the antigen, and – at high antigen densities – steric hindrance between the markers. The resolution of the technique is no greater than 15 nm, as the colloidal gold particle is held at a distance of 10 – 15 nm from the epitope, due to the space occupied by primary (e.g., antibody) and secondary (e.g., Protein A-shell) reagents (see Fig. 2.45 c–f).

Several more recent developments can increase the resolution and sensitivity of the method. Dodecaborane clusters may be coupled to antibodies or Fab-fragments through polylysine dendrimers. The boronated antibodies have a lower detection limit in immunolocalization experiments than conventional gold conjugates. Detection of boron in immunolabeled sections (less than 50 nm thick) is performed by element-specific imaging in energy-filtering TEMs. The experimental is described in outline by Bendayan et al., 1989, Qualmann et al., 1996, and Kessels et al., 1996. Very small gold particles ("undecagold" or "nanogold") of around 0.7 – 18 nm in diameter can be used for similar purposes (Hainfeld et al., 1989, 1995, 1996, 1999); several types are available for high-resolution mapping of targets, examples being depicted in Fig. 2.45 a and b. The small gold clusters may also be enlarged with silver or gold salt solutions for immunocytochemisty.

GFP-tagged proteins may be located in TEM preparations with (commercially available) anti-GFP-antibodies, but also cytochemically on formation of photooxidized, electron-dense diaminobenzidine (DAB) precipitates at sites where GFP is present (Monosov et al., 1996). Catalyzed reporter deposition (CARD) is a technique for amplification of a signal, originally designed for the Western blotting technique, but modified for light and electron microscopy. A specific primary antibody that has bound to the antigen is coupled to a secondary, biotin-tagged antibody. A steptavidin peroxidase conjugate is then bound to the biotin tag. Peroxidase catalyzes the deposition of tyramides. The tyramide itself is bound to a detectable marker (e.g., biotin again), and may be localized by use of streptavidin-coupled colloidal gold. The method may be modified by exchanging the elements of this multiple marker sandwich. Some variations, however, tend to introduce a high nonspecific background. A modification of the technique may also be used for in situ hybridization (Schäfer et al., 1997; Mayer and Bendayan, 1999).

If the sequence of a section of the protein is known, epitope-specific antibodies may be produced. Several software packages (such as the Wisconsin GCG sequence analysis software package; see Devereux et al., 1984) allow the determination of possible epitopes in a given amino acid sequence (derived from the DNA sequence). The epitope (10 – 15 amino acids) is synthesized, and the resulting peptide is then coupled to a carrier molecule and used for immunization.

Coupling of lectins to colloidal gold is performed as described in Table 2.28, but pH adjustment has been reported to be essential for proper complex formation (Horisberger, 1985). In practice, the yield of usable gold conjugates depends on the lectin species used. The lectins bandeiraea (from *Bandeiraea simplicifolia*) and concanavalin

A (from *Canavalia ensiformis*) are very well suited for colloidal gold stabilization. Others, such as arachis (from *Arachis hypogaea*) and lentil (from *Lens culinaris*), are less useful. Gold particles tend to precipitate after the centrifugation steps during the coupling procedure (see Table 2.28). The most difficult to handle is wheat germ agglutinin (from *Triticum vulgaris*), which must be cross-linked to BSA prior to gold coupling (chapter 2.4.8.3; Horisberger, 1985, and references therein).

The electron microscopic localization of enzyme activity in cells ("catalytic histochemistry" or "electron microscopy cytochemistry") allows direct detection of catalytic activity. This method has mainly been applied to tissue sections (at the light microscopic level) in biomedical research (comprehensive surveys on preparation procedures for electron microscopy are provided by Hayat, 1973ff, Tauschel, 1988, and Wohlrab and Gossrau, 1992). Some of the detection procedures are applicable to electron microscopic preparations, because the catalytic reactions may be coupled to precipitation of electron-dense agents.

Tab. 2.26 Preparation of colloidal gold (approximately 10 nm particle size)

1.) Siliconize the inner surfaces of all vials and flasks to be used for the preparation and storage of colloidal gold by rinsing with a solution of Sigmacote (Sigma Inc.).
2.) Dry the glassware and microcentrifuge tubes for 2 h at 110–130 °C.
3.) Mix 4 ml of 1 % (w/v) sodium citrate solution and 16 ml of double-distilled water in a 50 ml Erlenmeyer flask.
4.) Mix 1 ml of 1 % (w/v) tetrachloroauric acid solution and 79 ml of double-distilled water in a 300 ml Erlenmeyer flask.
5.) Incubate both solutions in a water-bath (60 °C) for 15–20 min.
6.) Combine both solutions with immediate rotation of the flask or magnetic stirring. The reduction of tetrachloroauric acid is indicated by a violet color. The process is complete after 60 min.
7.) Boil the solution for 5 min; this results in a deep red color.
8.) Store the solution refrigerated in a siliconized glass vessel (stable for several weeks; check by electron microscopy before use).

Tab. 2.27 Preparation of colloidal gold (about 5 nm particle size)

Use siliconized glassware (see Table 2.26, Steps 1 and 2).

1.) Mix 4 ml of 1 % (w/v) sodium citrate solution, 1 ml of 1 % tannic acid solution, 1 ml of 25 mM potassium carbonate, and 14 ml of water.
2.) Mix 1 ml of 1 % (w/v) tetrachloroauric acid solution and 79 ml of double-distilled water in a 300 ml Erlenmeyer flask.
3.) Combine both solutions, with immediate gentle shaking of the flask. The solution should turn red immediately.
4.) Boil the solution for 5 min; this results in a slight deepening of the red color.
5.) Add hydrogen peroxide solution to a final concentration of 0.3–0.5 % [v/v] to remove residual tannic acid.
6.) Incubate overnight at room temperature; then incubate for 30 min at 90 °C to remove hydrogen peroxide.
7.) Store the solution refrigerated in a siliconized glass vessel (stable for several weeks; check by electron microscopy before use).

Tab. 2.28 Determination of optimal protein concentration and coupling procedure to obtain colloidal gold conjugates

1.) Adjust the pH of the colloidal gold solution (as a general rule, 0.5 pH units above the isoelectric point of the protein – pH 5.5–6 for Protein A, pH 9.0 for IgG – is recommended). Use 0.1 N HCl or 0.2 M K_2CO_3.
Dialyze the protein solution against an appropriate buffer of low ionic strength (simple dilution in distilled water is sufficient for Protein A, dialysis against 2 mM borax, pH 9.0 is recommended for IgG).
2.) Mix aliquots (250 µl) of colloidal gold suspension with 50 µl aliquots of various protein dilutions (1:8, 1:16, 1:32,...1:1024). Incubate for 10 min at room temperature.
3.) Add 30 µl of a 10 % (w/v) sodium chloride solution to each protein dilution step. Incubate for 10 min at room temperature; a color change from red to blue (aggregation of colloidal gold particles) can be observed in vials in which colloidal gold is insufficiently stabilized by the protein dilution. Sufficiently stabilized samples retain the wine-red color.
4.) For preparation of the bulk amount of protein-gold complexes, add protein solution in a 10 % excess of the minimal stabilizing concentration determined from the stability test (Steps 2–3). Incubate for 10 min at room temperature.
5.) Check the stability of the solution by adding 10 % (w/v) sodium chloride solution to a small sample as described in Step 3. The sample should retain its wine-red color (if so, continue with Step 6). If it does not, the protein concentration is too low; check the results of the optimal protein concentration determination (Steps 2–3).
6.) Add 5 ml of 1 % (w/v) polyethylene glycol solution (PEG 20,000; recommended for Protein A and other proteins) or add 7.5 ml of 8 % (w/v) BSA (recommended for IgG; adjust to pH 9.0) to 50 ml of the protein-gold solution. Incubate for 5 min at room temperature.
7.) Centrifuge for 30 min at 30,000×g (10–15 nm particle size) or for 45 min at 85,000×g (5 nm particle size) and 4 °C. Centrifugation results in the formation of a dark red sediment consisting of fully stabilized protein-gold complexes. Particle aggregates or incompletely stabilized gold particles precipitate onto the centrifuge tube wall, forming a dark spot.
8.) Do not decant the supernatant, since this disturbs the sediment! Suck the sediment off carefully with a pipette and then dilute it in 10 ml of potassium phosphate buffer, 50 mM (pH 6.9), containing 0.9 % (w/v) NaCl and 0.02 % (w/v) PEG 20,000 (recommended for Protein A and other proteins). Alternatively, dilute it in 20 mM Tris-HCl buffer, pH 8.5, containing 0.5 % (w/v) NaCl and 1 % (w/v) BSA (recommended for IgG).
9.) Recentrifuge the resuspended sediment under the same conditions (Step 7). Suck the sediment off and store at 4 °C (do not freeze!). Add some sodium azide crystals as preservative.

2.4.8.2 Preparation of colloidal gold

Tables 2.26 and 2.27 describe the procedure for preparation of colloidal gold of particle sizes 5 and 10 nm with narrow size range distribution. These preparations are useful for double-labeling methods. The subsequent coupling of protein molecules is described in Table 2.28 and mainly refers to coupling of Protein A and IgG, but may be also be adapted to coupling of other proteins (see also Beesley, 1989).

2.4.8.3 Preparation of wheat-germ agglutinin/bovine serum albumin (BSA)

Cross-linking may be performed successfully with a solution containing 2 mg/ml wheat-germ agglutinin and 2 mg/ml BSA in 10 mM potassium-phosphate buffer at pH 7.0. After addition of glutardialdehyde (0.25 % [v/v] stock solution) to a final concentration of 0.025 %, the mixture is kept at room temperature for 1 h with gentle shaking. Any precipitate is then removed by filtration (0.2 μm pore size sterile filter).

2.4.8.4 Pretreatment of primary antibodies

The main requirement for immunolabeling methods is high antibody specificity. It is recommended that antisera be purified by affinity chromatography with Protein A- (for rabbit antisera) or Protein G- (for mouse antisera) sepharose materials. Use of whole antisera for immunolocalization is possible, but may result in nonspecific background labeling. Purification of antisera by affinity chromatography has the advantage that intact antibodies of the IgG-subspecies remain in the resulting antibody solution. Thus, no perturbing antibody species or Fab-fragments, which may react with the antigen but not with the marker, are present. This purification method is best used in conjunction with subsequent utilization of a Protein A-labeled marker.

Polyclonal antisera may contain antibodies directed against "common antigens" in the cell walls of Gram-negative bacteria and yeasts, because some pathogenic members of these groups may cause infections during immunization. This often results in the labeling of bacterial and, especially, fungal cell walls supposedly directed against quite different antigens. The availability of a mutant organism that does not express the respective target antigen is very helpful in these cases. When an antibody solution is incubated with an extract of the mutant, all antibodies except the one directed against the target will be bound and will not react in the subsequent immunolocalization experiment. Large antigen-antibody precipitates may be even removed by high-speed centrifugation of the solution. The procedure may also work with closely related organisms that do not express the target antigens. Anyhow, an antibody solution pretreated in this way has to be checked carefully, due to its specificity to the correct antigen. Application without additional control experiments (see below) will lead to unreliable results.

2.4.8.5 Localization procedures

The immunocytochemical localization of antigens in ultra-thin sections of cells is performed as described in Table 2.29 (Hoppert and Mayer, 1995; see also the schemes presented in Fig. 2.45). The concomitant localization of two antigens helps to reveal possible interactions and has been successfully applied for, for instance, the elucidation of structure-function relationships in multi-enzyme complexes and metabolic pathways in bacteria (see Freudenberg et al., 1989; and Aldrich et al., 1992). The procedure is described in Table 2.30.

Labeling with lectin-gold complexes is performed as in Table 2.29, but with some modifications: Step 2 may be omitted, and in Step 4 the enzyme-specific IgG-antibody is replaced by the lectin-gold complex. The pH value of the incubation buffer should be slightly above the isoelectric point of the coupled lectin. Steps 5–8 are omitted.

The "pre-embedding" procedure (Acker, 1988; Hoppert and Mayer, 1995) allows enzymes on the surfaces of cells and vesicles to be localized and is mainly used to detect cell surface antigens and extracellular enzymes attached to the cell surface

Tab. 2.29 Immunolocalization procedure in outline for ultra-thin sections

1.) Immediately after the freshly cut sections are picked up, place the grids, with the sections facing downwards, on drops of 50 mM phosphate-buffered saline, pH 7.0 (PBS). This step is carried out to prevent loss of reactivity of – for example – antigens due to drying of the sections.
2.) Place the grids on drops of a 1 % (v/v) hydrogen peroxide for 5 min. This etching procedure results in better exposure of antigens on the surface of the section. Rinse the grids extensively in distilled water.
3.) Place the grids on drops of 0.5 % (w/v) skim milk solution for 5 min. This is done to reduce non-specific reactions of antibodies with the section.*)
4.) Transfer the grids onto drops of PBS containing various concentrations of enzyme-specific IgG-antibodies (incubation time: 2 h). A first experiment may be performed with a 1/10, 1/100, 1/1000 (...) dilution series. Evaporation of water is minimized by covering the samples.
5.) Rinse the grids extensively ("jet wash", i. e., running a stream of buffer from a washing bottle across the grids held with forceps) for 20 s in PBS containing 0.05 % (v/v) Tween 20.
6.) Incubate twice, for 2 min on drops of PBS containing 0.05 % (w/v) Tween 20, and then for 2 min on a drop of PBS without Tween 20.
7.) Repeat Step 3.
8.) Incubate on drops of the appropriate Protein A- (or secondary antibody-) coupled gold dilution for 30 min, prepared in PBS containing 0.5 % (w/v) BSA. A first experiment may be performed with a 1/10, 1/100, 1/1000 dilution series. Dilutions of around 1/30 (10–15 nm gold particles) and 1/80 (5 nm particles) are usually found appropriate.
   Note: The PAG solution has to be centrifuged for 3 min at approximately 20,000×g and at room temperature to remove larger particle clusters immediately prior to use.
9.) Remove the grids from the surface of the drops and rinse extensively ("jet wash") with PBS solution for 20 s.
10.) Incubate twice for 2 min on drops of PBS at room temperature.
11.) Transfer the grids to drops of double-distilled water and incubate for 1 min at room temperature. This washing step removes any residual buffer salts that may cause precipitates during staining. The step should be omitted if the antibody-antigen reaction is weakened by it (as seen in some monoclonal antibody species).
12.) Dry the grids by blotting on filter paper but prevent direct contact between the grid surface and the surface of the filter paper.
13.) Transfer the grids to drops of 4 % (w/v) uranyl acetate solution. Incubate for 3 min at room temperature, then repeat Step 12.

***)** In addition or as an alternative to Step 3, nonspecific antibody-antigen and especially sample-Protein A-gold (PAG) interactions can be quenched by addition of 0.1 % to 2.0 % (w/v) casein hydolysate to the antibody and/or PAG-solution prior to use.

(Rohde et al., 1988; Table 2.31). Smaller isolated cellular constituents may be labeled by application of the whole-mount procedure (Table 2.32).

2.4.8.6 Immunogold labeling of cryosections

Labeling of cryosections may, in principle, be carried out by the corresponding conventional procedure (see also Griffith et al., 1984). Washing steps (in PBS containing 0.02 %, w/v, potassium chloride) performed before and after the incubation with the specific antibody and the colloidal gold solution have to be performed very gently, by successively placing the grid on the surfaces of drops of the buffer solution. Double

Tab. 2.30 Double labeling procedure

1.) Perform the immunolocalization with the first antigen as described in Table 2.29, Steps 1–10. In Step 8, use Protein A-Gold (PAG) of defined size (e.g., 10 +/– 3 nm diameter).
2.) Incubate the grids on drops of Protein A in 50 mM phosphate-buffered saline, pH 7.0 (PBS; 0.25 mg Protein A/ml) for 30 min, in order to saturate any remaining Protein A-binding sites.
3.) Remove the grids from the surface of the drops and rinse extensively ("jet wash") with PBS solution for 20 s.
4.) Incubate twice, for 2 min, on drops of PBS.
5.) Transfer the grids onto drops of PBS containing the second specific IgG-antibody (incubation time: 2 h). Evaporation of water is minimized by covering the samples.
6.) Rinse the grids extensively ("jet wash") for 20 s in PBS containing 0.05 % (v/v) Tween 20.
7.) Incubate twice for 2 min on drops of PBS containing 0.05 % (w/v) Tween 20 and then for 2 min on a drop of PBS solution without Tween 20.
8.) Incubate for 30 min on drops of the appropriate PAG dilution, prepared in PBS/BSA. Use PAG of defined size (e.g., 5 +/– 2 nm in diameter), differing by at least 4 nm from that used in step 1. **Note:** The PAG solutions have to be centrifuged for 3 min at approximately 20,000 × g and room temperature to remove larger particle clusters immediately prior to use.
9.) Continue from Step 9 of Table 2.29.

Tab. 2.31 Pre-embedding procedure

1.) Incubate cells or cellular components with the specific antibody (in excess), for 2 h at room temperature.
2.) Wash the sample twice by centrifugation and resuspension in an appropriate buffer (e.g., 50 mM PBS, pH 7.0–7.5 for cells or components where no osmotic stabilization is necessary; avoid conditions that weaken antibody binding such as low pH or high salt concentration) to remove unbound antibody.
3.) Incubate with the electron-dense marker system (e.g., Protein A-coupled gold or ferritin coupled to a secondary antibody) for 2 h at room temperature. The optimal dilution is generally 10 times higher than that used in post-embedding experiments (see above).
4.) Proceed from Step 1, Table 2.13, with the standard embedding procedure, or perform low-temperature embedding in Lowicryl resin, if additional post-embedding localization is desired (Table 2.16). After glutaraldehyde fixation, antigen-antibody binding is irreversible.

Tab. 2.32 Whole-mount procedure

1.) Allow the cell components to adsorb directly onto an electron microscopic support film: place carbon-coated Formvar grids on the surface of a drop of the sample for 30–45 s.
2.) Blot off the liquid on the grid surface completely or partially, depending on the particle concentration in the sample or the desired particle density. Isolated particles (e.g., membrane vesicles), not attached to each other and equally scattered over the grid surface, are preferable for significant labeling signals.
3.) Before the liquid is completely removed, place the grid onto a drop of the IgG solution.
4.) Proceed as described for ultra-thin sections in Table 2.29, or, for double labeling, Table 2.30, performing washing steps after the antibody and protein-gold incubation very gently by placing the grid successively on the surfaces of two drops of PBS solution containing 0.01 % (v/v) Tween and a third drop of PBS without Tween.

Note: Generally, all specific and nonspecific binding reactions obtained in the whole-mount procedure are stronger than those obtained during immunocytochemical localization with ultra-thin sections. The presence of casein hydrolysate (up to 2 % w/v, final concentration) in the antibody and colloidal gold solution prior to the incubation steps serves as an appropriate quenching agent. Generally, higher dilutions of the antibody and especially of colloidal gold are recommended for the whole-mount procedure. Staining should be performed with sodium phosphotungstate (3 %, w/v, phosphotungstic acid adjusted to pH 7.0 with NaOH solution).

labeling and labeling with colloidal gold coupled to lectins or other markers may also be performed. Blocking of the sections to avoid nonspecific reactions must be enhanced. A complete outline is given in Table 2.33. For the cryosectioning method, see Table 2.23.

2.4.8.7 Control experiments in immunocytochemistry

To exclude false negative or positive results in immunolocalization experiments, several controls are necessary. These controls check the specificity of the antibody-antigen binding as well as the binding of the marker system to the primary antibody.

- Incubate the specimen exclusively with the marker system. Any resultant binding is nonspecific binding of the marker system to the sample. If this occurs, use a different marker system or try to reduce nonspecific binding by the addition of quenching agents (e.g., casein hydrolysate) to the solution.
- Incubate the primary antibody with the purified antigen. The specimen should not subsequently be labeled by this mixture. Labeling of the specimen indicates the presence of antibodies targeted against other antigens or that bind nonspecifically. Remove these antibodies by affinity chromatography with the antigen as ligand (specific antibody binds and has to be eluted in a second step), or use monoclonal antibodies.
- Incubate the specimen with the specific antibody, followed by incubation with the secondary antibody or Protein A alone, followed in turn by incubation with the antibody or Protein A coupled to the electron-dense marker. If the marker system interacts specifically with the primary antibody, no labeling should occur. Avoid nonspecific binding by using a different marker system or try to reduce nonspecific binding by the addition of quenching agents (e.g., casein hydrolysate) to the solution.

Tab. 2.33 Immunolocalization procedure for cryosections

1.) Warm the gelatin up to room temperature, and then leave the floating grids for a further 10 min incubation.
2.) If the samples were subjected to chemical fixation prior to rapid freezing, incubate on three drops of 20 mM glycine in PBS for 10 min each. Glycine will react with excess aldehyde groups that may otherwise interfere with the protein solutions used in the subsequent incubation steps.
3.) Carefully suck of excess liquid (do not blot dry) and transfer the grids onto drops of dilutions of the specific antibody (prepared in PBS containing 1 % w/v BSA).
4.) Incubate four times for 1 min on drops of PBS/BSA.
5.) Incubate on drops of the appropriate colloidal gold marker dilution (e.g., Protein A-gold prepared in PBS/BSA) for 30 min.
6.) Incubate four times for 5 min on drops of PBS.
For most applications, the binding of antibodies and colloidal gold marker are stable enough to withstand the subsequent procedures. If not, incubate for 10 min on drops containing 0.2–1 % v/v glutardialdehyde in PBS (without BSA!). This step will cause covalent bonds to form between antigen, primary antibody, and the colloidal gold-coupled protein.
7.) Incubate three times for 1 min on drops of distilled water and blot dry after the final incubation step.
8.) Stain the sections as described in Table 2.22.

- Incubate the sections with pre-immune serum (or any serum that does not contain the specific antibody), followed by the marker system. This reveals nonspecific binding of antibodies to the specimen. Try to reduce nonspecific binding by addition of quenching agents (e.g., casein hydrolysate) to the solution, or use monoclonal antibodies.
- Use a specimen that does not contain the antigen of interest. It is most appropriate if the same organism is used both for the specific labeling and for the control experiment, but for the control to be grown under conditions under which the cell component to be localized is not expressed. In cases in which cell wall labeling in Gram-negative bacteria or fungi is carried out with polyclonal antisera, it is strongly recommended that the specificity of the labeling is checked with other Gram-negative bacteria or fungi that do not contain the epitopes of interest.

2.4.8.8 Labeling of epitopes on isolated protein molecules

Localization of epitopes on the surfaces of (large) protein molecules may be performed with the aid of specific monoclonal or polyclonal antibodies directed against the epitope(s) in question. The protein sample is mixed with a monoclonal or polyclonal antibody at a concentration appropriate to give binding site equivalence. The sample is then incubated (a first experiment might entail an incubation time of 1 h at room temperature) and diluted by a factor of 10 to 100 with an appropriate buffer solution (such as 20 mM Tris-HCl, pH 7.0–7.5), and the specimens are prepared by negative staining. Alternatively, the antibody-protein complexes may be separated from unbound protein and antibody molecules after incubation by gel permeation chromatography prior to sample preparation. The antibody fraction may be replaced by Fab-fragments or gold-labeled Fab-fragments of antibodies (see Hermann et al., 1991, and references therein for some applications).

2.4.8.9 2-D crystals of membrane proteins

2-D crystals are ordered, two-dimensional protein or protein-lipid arrays that constitute a molecular monolayer or two layers exactly in register (Fig. 2.12 c, 2.16 c). For reconstitution of membrane proteins into 2-D crystals, the idea is to produce detergent-lipid-protein mixed micelles and remove the detergent by e.g., dialysis and thereby inducing the formation of 2-D crystals, i.e., ordered 2-D protein-lipid arrays. Sometimes, 2-D crystals may also be observed after simple isolation procedures. Thus, if it is possible to obtain membranes which are enriched in one particular protein, then an in situ approach should be considered. In the following, a description of the reconstitution experiments and the different parameters involved is given.

It is crucial for a successful reconstitution to find the best possible **lipid component** for a particular protein. While the search for such lipids is mainly based on trial and error, a number of important considerations should be taken into account: the transition temperature of the lipid, the shape of the molecule and the length of its alkyl chain.

A lipid should be chosen which exhibits not too high a gel-to-fluid phase transition temperature as to interfere with the stability of the protein transition temperature is strongly dependent on the length and degree of saturation of the alkyl chain: the shorter the chain and the more unsaturated, the lower the transition temperature (See e. g., Cevc and Marsh, 1987). In this context it should also be noted that certain lipid phases, like the pre-transition phase can be mistaken for 2-D crystals of the membrane protein when viewed in the TEM. While control experiments (in the absence of protein) as well as the undulatory nature and periodicity (often around 2 – 3 nm) of these solely of lipid consisting, transition temperature-depending phases may clarify the situation, it is best to avoid such problems by using lipid mixtures like soybean PC (phosphatidylcholine), bovine liver PC, DMPC: POPC (dimyristoyl phosphatidylcholine, palmitoyl-oleoyl phosphatidylcholine) etc. These 2-D crystal-like lipid phases are not observed with e.g., DMPC, if the protein readily inserts and crystallises.

The shape of a lipid is, mainly defined by the cross-sectional area of the alkyl chains (larger for unsaturated and smaller for saturated alkyl chains), the cross-sectional head group area, head group hydration and charge, hydrogen-bonding, and counterions like calcium. Phosphatidylcholine, for instance, assumes a cylindrical molecular shape and would, therefore, favour the formation of planar bilayers while phosphatidylserine in water would assume the shape of a truncated cone and support the formation of flexible bilayers or vesicles. The addition of calcium ions to phosphatidylserine renders its shape cylindrical due to a decreased head group area. The pH determines the state of dissociation and hence the head group area, e.g., with phosphatidylserine a smaller head group area will be present at a pH lower than 4 than at neutral pH, since the more dissociated a head group the larger its area. Generally, phosphatidylcholines should definitely be on the top of the list to try.

The alkyl chain length can be optimised using the following rationale. In an ordered 2-D lattice, the protein molecules interact specifically with each other. These specific contacts are of hydrophilic as well as hydrophobic nature. While the hydrophobic regions should always be encased by the lipid (or detergent), the hydrophilic regions should remain accessible. Hence, a lipid should be found which, because of its alkyl chain length, encases the hydrophobic centre region but does not cover up any specific hydrophilic contact areas. Some proteins continually withstand any crystallisation attempts due to the fact that they do not possess sufficiently large extramembraneous domains that could support hydrophilic intermolecular contacts. In these cases, a possible solution could be to fuse the membrane protein under study with a soluble protein so that the resulting fusion exhibits a larger polar surface area. Hydrophobic interactions, it seems, can also be mediated through the hydrophobic tails of the lipids, so that for these interactions there is less a demand for the 'correct' lipid, except that the lipids need to be sufficiently long to shield the hydrophobic area of the protein. A sensible range for the variation of the chain length is 12 – 18 carbon atoms.

The assembly of naturally occurring or in situ formed crystals, as mentioned above, takes place in native membranes, i.e., in an endogenous lipid environment. These endogenous lipids may prove advantageous for the formation of crystals and should, therefore, also be included in the list of lipids to be tested in reconstitution experiments.

Detergents interfere with protein-lipid and protein-protein recognition processes, the former of which are required for the formation of a protein-lipid bilayer, and the latter of which is almost certainly a prerequisite for the formation of 2-D crystals. The optimal detergent should be compatible with the formation of 2-D crystals and conserve the activity of the protein. As a general rule: nonionic detergents with large headgroups and long alkyl chains are milder than zwitterionic detergents with smaller headgroups. Maintaining the native structure and activity of a(n) (enzyme) protein is also a function of time. Thus, detergents with low cmc (critical micelle concentration) may require extended periods of dialysis (weeks) to lower the detergent concentration below the cmc. If an activity-maintaining detergent does not prove suitable for 2-D crystallisation, the addition of small amphiphiles might provide a solution. The addition of heptyl- or octyldimethylamineoxide (at 3%, w/v, or 5%, w/v, respectively), or heptane-triol (at 3%, w/v) to the primary detergent improves protein-protein interactions via a decrease in micelle size (Welte and Wacker, 1991; Timmins et al., 1991). If no single detergent seems suitable, one should try binary mixtures of detergents. This is true for the solubilisation of the protein as well as the preparation of the detergent-lipid mixed micelles. The cmc of such mixtures is given by the expression $1/\text{mix} = a/\text{det1} + (1\text{-}a)/\text{det2}$ with a being the molar fraction of detergent 1 in the mixture and mix, det1 and det2 being the cmc of the mixture, detergent 1 and detergent 2, respectively.

The detergent used for solubilisation can differ from the one used for reconstitution. To achieve this, the detergent can either be changed prior to or during reconstitution. The latter strategy, the efficiency of which needs to be established for each individual system, is based on the utilisation of a different detergent for the preparation of the mixed micelles (see Table 2.34, and Zhuang et al., 1999). Zhuang et al. solubilised the lipids in 1:1 (v/v) mixtures of 2% OD (octyl-beta-D-glucopyranoside) and 0.2% DM (decyl-beta-D maltopyranoside) while the protein was in 0.01% DDM (dodecyl-beta-D-maltopyranoside) prior to dialysis. This also meant that upon mixing the protein with the lipid, the DDM concentration dropped below the cmc. The authors report that as long as the DDM concentration was around 0.005% and in the presence of solubilised lipids, the protein remained in solution. This procedure elegantly circumnavigated problems with dialysis-driven reconstitution caused by the low cmc of DDM.

The **lipid-to-protein ratio** (LPR) is another parameter which needs to be optimised, since it controls, during the course of detergent removal, the balance between protein-protein and protein-lipid interactions as well as their absolute magnitudes, factors which crucially influence the formation of 2-D crystals and the final lattice parameters (packing density and symmetry). Optimum LPRs have been reported to be in the region 0.1 to 10.0 (w/w). It seems that membrane proteins with bulky extramembranous domains require a higher LPR than 'lean' membrane proteins. In initial trials, one should include at least three different LPRs (such as 0.3, 1.0, and 3.0) with a slight bias towards lower LPRs, since one would expect to find better-ordered crystals when the lipid content is low. Additionally, the slow removal of phospholipids from reconstituted membranes can induce crystal formation. This strategy involves the use of phospholipase A2 from bee venom. A detailed account on the usage of this enzyme is given by Mannella (1986). It may also be possible to induce crystallisation in reconstituted membranes by repeated freeze/thaw cycles (freeze in liquid nitrogen and defrost on ice), which also tends to increases vesicle size.

The **temperature** should be chosen so that it is above the gel-to-fluid-phase transition temperature of the lipid(s) in use because this facilitates the incorporation of protein into the bilayer. It should, however, be pointed out that in lipid mixtures, in contrast to a single lipid, sharp transition points do not exist any more (Zhuang et al., 1999). A second temperature-governing factor is the stability of the protein, keeping in mind that the protein might even be stable at unexpectedly high temperatures once incorporated into the bilayer. Once a temperature is found at which protein integrates into the bilayer but only poorly ordered arrays are obtained, increasing the temperature (up to 37 °C) and decreasing it again (Zhuang et al., 1999) might help to increase the order or even induce crystal formation. By inducing (partial) expulsion and reintegration of the protein into and out of the bilayer and allowing the mobility of the proteins within the bilayer to be in-/decreased due to changes in the fluidity, isotropically distributed protein-protein interactions are more likely to be established thus affecting crystalline order. One should also consider employing a temperature ramp following (or during) the aforementioned phospholipase treatment, i.e., incubating the membranes overnight at 37 °C and then letting the membranes cool down slowly.

When choosing the **ionic milieu** for a reconstitution experiment, two primary constraints need to be considered. These are the compatibility with subsequent specimen preparation procedures (such as negative staining) and the compatibility with protein activity (the protein must remain stable/active in the buffer solution employed). A third constraint is the surface charge which is believed to be important for those regions of the protein which protrude from the membrane surface. These charges seem to influence crystal formation and crystalline order and can be experimentally adjusted by adding sodium chloride or magnesium chloride (20–200 mM final concentration) to the buffer system and changing the pH. In cases where the protein can be locked in one conformation or configuration, the addition of a substrate/substrate analogue is very likely to facilitate the formation of ordered arrays.

Upon **detergent removal**, spherical mixed micelles grow by fusion processes into discoidal or cylindrical micelles. These processes are driven by a decrease in edge energy, which is associated with the exposure of hydrophobic regions to the aqueous environment (detergent molecules will cover the rims of the micelles where hydrophobic areas are particularly exposed). At a certain point, the size of a discoidal/cylindrical micelle is such that closure to form a vesicle is the most energetically favourable option, due to the increase in curvature energy (energy associated with the stretching/compression of opposing leaflets) on vesicle closure being less than the resultant decrease in edge energy. According to these considerations, the formation of planar bilayers is favoured over the formation of vesicles when the increase in curvature energy is greater than the drop in edge energy. Given lipids which would, as a result of their shape, favour the formation of planar bilayers, the curvature energy required for vesicle formation would be much greater than the edge energy required for the maintenance of planar bilayers.

The size of the detergent-lipid mixed micelles does not depend on the detergent used while the ultimate size of the vesicles does. This dependence is given through detergent-induced postvesiculation size growth (PVSG) processes (Rotenberg and Lichtenberg, 1991). Vesicle growth by fusion is much more likely to occur within a critical range of lipid-to-detergent ratios (LDRs), regardless of whether this situation is arrived at by reduction of the detergent concentration via dialysis, or addition of detergent to a sample containing none (integrating detergent into lipid bilayers also causes an increase in bilayer-fluidity and this also favours fusion). In order to obtain large vesicles, one needs to make sure that the PVSG processes are rapid with respect to the rate of detergent removal. This can be achieved by (i) slow removal of detergent, and (ii) fast PVSG processes. Although a fast detergent removal (e.g., by Bio-Beads SM2, Bio-Rad, Inc.) may in principle yield very good results, a slow removal of detergent enables the sample to remain at an optimum LDR for longer periods of time than would be possible with faster detergent removal rates, i.e., fusion events are much more likely to take place and PVSG processes are favoured. Bio-Beads are compatible with a whole range of detergents (Rigaud et al., 1997) including Triton X-100 and DDM (dodecyl-beta-D-maltopyranoside). When using Bio-Beads for detergent removal, one should be aware that they also adsorb lipids. This lipid adsorption can, however, be overcome by pre-incubation with sonicated liposomes (Rigaud et al., 1997). This means in practice that flow-dialysis may be advantageous as it allows a precise adjustment of the detergent removal parameters. The speed of detergent removal depends, however, not only on the dialysis parameters but also on the detergent used, e.g., cholate removal proceeds faster than the removal of CHAPS (3-[(3-cholamidopropyl)-dimethylammonio]-1-propanesulfonate). Higher cmc detergents can be more rapidly (hours) removed than lower cmc detergents (days to weeks). Furthermore, the speed of PVSG processes is detergent-dependent, e.g., CHAPS causes a faster size growth than cholate. Fast PVSG processes have also been reported for OG (octyl-beta-D-glucopyranoside) which is also a detergent for which the micelle-to-vesicle transition occurs at a lower LDR than for cholates. The increase in hydrodynamic radius of aggregates formed by lipid-OG mixtures beyond the transition region (hydrodynamic radius of mixed micelles < 10 nm, hydrodynamic radius of vesicles > 10 nm) is orders of magnitude larger than with cholates. The detergents C8E5 (octyl-pentaoxyethylene) and 8-POE (octyl-(polydisperse)oligooxyethylene) fall into the same category as OG, and the resulting vesicles are much larger than with cholates. Assuming that above considerations for binary lipid-detergent systems are, within reason, transferable to ternary lipid-detergent-protein systems, the screening for an optimum detergent for reconstitution experiments should ideally include each of the above three detergents (C8E5, 8-POE, and OG). Having said this, it is often not possible to use either of these three detergents because they may not keep the protein in solution, inactivate the protein, are not compatible with the purification etc. Then, DDM or Triton X-100 may be the detergent of choice. As mentioned above, detergent mixtures (e.g., DDM/CHAPS) may also prove extremely useful in cases where one single detergent has a too low cmc (see also Zhuang et al., 1999).

During the course of dialysis, as the LDR increases, vesicles will form and the protein may or may not have been incorporated into the bilayer, depending on how the chosen parameters govern the interactions between protein, lipid and detergent. The exact sequence and nature of events are still not entirely understood. However, it has been proposed that vesicle formation may precede incorporation of protein into the lipid bilayer when detergent is removed slowly, and that proteins are incorporated during micelle growth when detergent is removed quickly. This phenomenon is modulated according to the nature of the detergent used. Where exactly the borderline is located between fast and slow is a contentious issue, but it is certainly true that the protein alters the micelle-to-vesicle phase transition point (Dolder et al., 1996). The same authors have called this transition the critical step in crystal assembly and it is therefore possible that protein incorporation commences close to this transition point or not at all. In the latter case, one is left with perhaps large but empty vesicles. Whatever the exact situation, if incorporation takes place, it seems to do so long before the cmc of the detergent (mixture) is reached and constitutes the prerequisite for the subsequent assembly of the protein molecules into ordered lattices. At a certain stage during detergent removal, the vesicles start to grow more rapidly. This rapid growth may be concomitant with lattice assembly which is mainly governed by protein-lipid and protein-protein interactions (Dolder et al., 1996). Later on, after the vesicle growth processes have ceased, a further removal of detergent can improve crystal order. While in the earlier stages during dialysis some detergent is needed for size growth processes, a complete removal of the detergent should be striven for towards the end of the reconstitution process in order to obtain the best crystalline order possible.

Monitoring: The most straightforward way to prepare specimens for the TEM is by negative staining (see above) which one could view as the method of choice for monitoring reconstitution experiments. It is easy to judge from negatively stained specimens whether, e.g., vesicles have formed or not, formed vesicles are devoid of protein molecules or not, or the formation of 2-D crystals has occurred (but be aware of lipid phases that appear like 2-D crystals, see above).

Trouble-shooting: If non-satisfactory results have been obtained using the protocol in Table 2.34, attempt to optimise the parameter configuration using the foregoing rationales, and bearing in mind that parameters are easily changed – but maintained with difficulty. Therefore, every time one thinks about changing a parameter, one should also consider whether it will be possible to keep the changed parameter at a constant value. Because of the complexity of reconstitution experiments, all parameters need to be controlled most thoroughly, and it is advisable to formulate a record sheet to keep all data in a standardised, easily reviewable format. Furthermore, always use the electron microscope to analyse what happened during the course of dialysis. Some of the principal indicators for a lack of reconstitution and/or formation of 2-D crystals are given below with possible explanations. It should also be mentioned that as image processing procedures aimed at retrieving high-resolution information from even small and mosaic crystalline specimens have been established (e.g., Stoylova et al., 1999), not all is lost in case only poorly ordered crystals can be obtained:

- No vesicle formation: detergent not dialysable, defective dialysis apparatus, too small an amount of lipid present, lipid does not support bilayer formation (e.g., PE).
- No protein inside the vesicles: wrong lipid environment, wrong buffer components, protein aggregated, dialysis membrane exclusion limit too large, too rapid a removal of detergent, unsuitable temperature, ternary micelles were not allowed sufficient time to equilibrate.

Tab. 2.34 Reconstitution of membrane proteins towards the formation of 2-D crystals

Lipids and detergents can generally be used without further purification.
Be aware of the possibility that the mixed micelle solutions prepared can have properties which are never again reproduced (!). For this reason, one should prepare a batch of sufficient size (e.g., 25 ml of a solution containing 2 mg lipid, e.g., soy bean PC/ ml) to last for a large number of experiments and freeze the whole batch in small aliquots (~ 250 μl). This way the starting conditions can always be standardised.
While the following protocol is taylored to a dialysis-type of reconstitution, it should be mentioned that removing the detergent by adsorption is a viable alternative. The use of Bio-Beads is very straight forward and described, together with some examples, by Rigaud et al. (1997).

Prepare the necessary equipment and reagents:
- flow-dialysis apparatus including programmable water bath, peristaltic pump or any other suitable dialysis device (this may obviate the preparation of dialysis membranes)
- dialysis membranes (boiled in 0.35 M sodium hydrogen carbonate and 1 mM EDTA prior to use)
- suitable lipid
- chloroform-methanol (2:1) or just chloroform
- suitable detergent dissolved in buffer (detergent-buffer; variation-1) or chloroform methanol-(lipid) (variation-2)
- detergent-solubilised protein solution

1.) Dissolve 50 mg of lipid in 1 ml chloroform-methanol (2:1) or chloroform alone in a small round bottomed flask, preferably under nitrogen or argon to avoid oxidation.
2.) Either evaporate the solvent in a rotary evaporator under a fine stream of nitrogen or argon to produce a thin film of lipid, and then add detergent-buffer to resuspend and solubilise the lipid. Adjust the detergent concentration as required (variation-1) or add the detergent to the resuspended lipid, or dissolve it in chloroform-methanol prior to its addition to the lipid. Then evaporate the solvent, vacuum-dry, and subsequently add buffer (variation-2).*) **)
3.) The next step will involve the addition of the detergent-solubilised protein solution to the mixed micelles. Adjust the temperature to the value at which the dialysis will be (initially) carried out prior to the addition of the protein. The detergent concentration in the detergent-solubilised protein solution should be about 1.5 x cmc or whatever is required to keep the protein in solution. Check the sample in the TEM for homogeneity, precipitates etc. before proceeding.
4.) Mix the detergent-solubilised protein with the detergent-lipid mixed micelles; the final detergent concentration should be about 3 x cmc with high cmc detergents or as low as possible for low cmc detergents.

N.B. Allow some time for the ternary micelles to equilibrate before you proceed. Depending on the exact properties of the protein, binary micelles or liposomes etc., this may take up to several hours. The equilibration process is generally faster when starting with liposomes + detergent-solubilised protein (~1 h) as opposed to detergent-lipid mixed micelles + detergent-solubilised protein (≤ 12 h).

5.) Load a flow-dialysis well – or other suitable dialysis devices, like microdialysers, dialysis cassettes, dialysis buttons, capillaries etc. (all available from Pierce Inc.) – making sure that the volume used entirely fills the well.
6.) Start the dialysis and note down all relevant parameters and changes of them (e.g., temperature, flow rate, composition of dialysis buffer used).
7.) Take samples after, e.g., 0.5, 2, 4, 6, 12, 24, and 48 h, and prepare negatively stained specimens. Document the results by taking or via an on-line digital image acquisition system.
8.) Assess the specimens for incorporation of protein into bilayers and the formation of ordered patches.
9.) Check the quality of any ordered arrays using a laser diffractometer or on-line fast Fourier transforms. It is imperative to note every detail down from the specimen preparation method and support films used to the data obtained by Fourier transformation (see chapter 2.4.9.2).

*) Note that in reconstitution experiments starting from mixed micelles prepared according to variation-2, a longer period of dialysis may be required for vesicles to form compared to experiments starting from variation-1 micelles (S. Renfrey, personal communication).

**) It is of course also possible to use liposomes (prepared by e.g., dissolving lipids in chloroform-methanol, drying them down, adding water or buffer and sonicating the suspension) and mix these with the detergent-solubilised protein solution (step 4).

- Protein inserted but no formation of 2-D crystals: check buffer, salt, pH, too high a LPR, unsuitable detergent/lipid, inappropriate flow rate, protein denatured, protein of low intrinsic symmetry (higher intrinsic symmetry may be obtainable by oligomerisation or by limited digestion), protein may require to be conformationally locked, protein concentration too low, insufficient detergent removal.
- Formation of poorly ordered arrays: incomplete removal of detergent, too high a LPR, suboptimal detergent/lipid, non-ideal temperature or temperature ramps chosen, inadequate phospholipase treatment procedures employed (where applicable).

2.4.8.10 Emerging techniques and concepts for protein crystallization

Annular and neutral lipids. Sabra et al. (1998) propose that in order to promote the formation of 2-D crystals, two types of lipids should be present in the membrane, one of which should interact more strongly with the protein than the other. While the relative amount of the two lipid species depends on protein concentration and size, strongly interacting endogenous lipids such as those – hopefully – still present after detergent solubilisation may act as one of the lipid species (annular lipid) while the lipid added for reconstitution purposes would then represent the other lipid type (neutral lipid). In their report, Sabra et al. conclude that in the absence of annular lipids but presence of neutral lipids no arrays would form. This is quite important when dealing with extensively solubilised membrane preparations, as the questions may arise as to whether or not there is indeed still some of the endogenous annular lipid present around the protein. In the presence of neutral lipids only, there may be then less of a chance for the protein to form arrays. Perhaps, and this is a speculation, by utilising lipid mixtures one can circumnavigate the latter problem, as the two lipids species could be recruited from the overall lipid pool. Sabra et al.s extended model also includes a ground (low-energy, low-entropy, ordered hydrophobic chains) and an excited state (high-energy, high-entropy, chains disordered) for the annular lipids. If the right fraction of annular lipids is in the excited state, the lipids belonging to this fraction would show less attractive interactions with protein and as such behave like neutral lipids; array formation could take place even if all lipids were annular lipids. The distribution of annular lipids between the ground and excited state depends on the temperature with higher temperatures decreasing the number of lipids present in the ground state. This concept explains why increasing the temperature may improve the quality of 2-D crystals.

Precipitant-induced surface growth of 2-D crystals: The following approach, originally based on a 3-D crystallisation protocol, has been successfully used to obtain 2-D crystals of the plasma membrane proton-ATPase from *Neurospora crassa* (Cyrklaff et al., 1995). With this method, the crystals grow on the surface of a droplet containing precipitating agents (PEG, ammonium sulphate) and only low protein concentrations (and amounts) are required. The drawback is that from each droplet (i.e., one specific crystallisation condition) crystals can only be harvested once. With the reconstitution method, a good batch of crystals can be stored and many specimens (for the TEM and cryo-TEM) prepared and analysed.

In order to attempt a precipitant-induced surface growth, the detergent-protein solution is mixed with appropriate stock solutions (prepared in standard buffer) resulting in a solution containing

- 60 – 300 µg protein/ml buffer (standard buffer as per purification),
- detergent as used for the final purification step (concentration could be decreased towards the cmc as long as the protein remains in solution),
- PEG 4000 or 6000 (8 – 12%, w/v),
- ammonium sulphate (50 – 120 mM),
- glycerol (if normally used, e.g., at 30%, v/v).

The optimum composition needs to be determined by trial and error within the given range of concentrations. The solution is then allowed to stand for several hours up to 2 days on ice and spun in a microfuge. Then, 5 – 7 μl droplets of the supernatant are placed on a siliconised glass coverslip and incubated for 10 min to several hours (even overnight) at 4 °C prior to harvest by touching the surface with a freshly glow-discharged carbon-coated grid. Alternatively, a baked carbon-grid (baked for 1 h at 160 °C for added stability) can be placed onto the droplet and allowed to incubate for the same period of time. After retracting the grid, it is negatively stained or plunge-frozen for observation in the TEM or cryo-TEM, respectively. For cryo-TEM, the specimens remain regularly unstained. Adrian et al. (1998) have shown that a specimen preparation procedure which constitutes a hybrid between negative staining and a cryo-preparation as the grid is floated for 60 s on a droplet of 16% ammonium molybdate (pH 7 to 8) prior to blotting and plunge-freezing. This procedure (**Cryo-negative staining**) may be a valuable alternative to the observation of unstained cryo-specimens with their extremely low contrast. Spectacular electron micrographs are presented in their paper, but at the moment it remains to be established what level of molecular details can be extracted from such specimens. It should be noted that the last crystallisation step can either be carried out in an open system, upon which the precipitant concentration increases as water evaporates, or in a closed system using a hanging or sitting drop set-up (like the one used for 3-D crystallisation trials), where the increase in precipitant concentration depends on the solution present in the reservoir (e.g., 1 ml of 0.5 – 1.5 M magnesium chloride). The closed system is prone to require longer incubation times (more than 4 h) than the open system (10 min up to a few hours).

Lipid-layer technique. This powerful technique, originally designed by Fromherz (1971) and, later on, Uzgiris & Kornberg (1983), is based on the binding of a protein to a specific lipid ligand incorporated into a lipid monolayer at an air-water (buffer) interface. The protein solution (with usually not more than 200 μg protein/ml and as little as 10 μg/ml) is placed into a cylindrical Teflon well (4 mm wide and 1 mm deep) and overlaid with a lipid solution (in chloroform or chloroform/hexane) consisting of a lipid ligand and, in cases where the lipid ligand does not form fluid monolayers, a dilution lipid (Vénien-Bryan et al., 1998). The solvent evaporates and the lipids form a monolayer at the interface. 2-D crystallisation is thought to proceed via the following stages (Brisson et al., 1994):

(i) The protein recognises and interacts with the ligand.
(ii) Through this interaction the protein molecules are drawn to the lipid layer and concentrated up to 10,000 times.
(iii) The protein-lipid complexes can diffuse in the plane of the lipid layer.
(iv) Provided there are favourable interactions between the protein-lipid complexes, they can undergo self-oganisation and assemble into 2-D cystals.

While this method has, until recently, only successfully been applied to soluble and membrane-associated proteins (Vénien-Bryan et al., 1998; Brisson et al., 1994), it now emerges that this general approach could prove very useful with transmembrane proteins as Lévy et al. (1999) describe an appropriate modification of the technique. With the advent Ni-chelating lipids (a commercially available Ni-chelating lipid can be purchased from Avanti Polar Lipids, Inc.), the way has now been paved for a general 2-D crystallisation strategy of histidine-tagged proteins (e.g., Vénien-Bryan et al., 1997; Bischler, 1998). As a detailed methodolgical account of the lipid-layer technique does not fall within the scope of this volume, the reader is referred to Schultz et al. (2001). Finally, it should be mentioned that other useful functionalised lipids like biotinylated lipids, Fab'-coupled lipids, hapten lipids etc. have been described (Schultz et al., 2001, and Dietrich et al., 1995).

2.4.9
Imaging and Image Evaluation under the Transmission Electron Microscope

Specimens may be visualized in different imaging modes (Table 2.34), most of which are available with the majority of transmission electron microscopes. With all transmission electron microscopes, several basic features have to be considered in order to obtain images of satisfactory quality. The full resolving power of the electron microscope (around 0.1–0.3 nm) is not achieved with most biological objects. The actual resolution depends on specimen thickness and staining conditions and is – for biological objects – around ten times lower than the resolving power. Image quality is influenced by following factors:

Choice of accelerating voltage. For electron microscopes in biomedical research, a range of accelerating voltages between 20 and 100 kV may be selected. High accelerating voltages produce highly resolved images with low contrast. If necessary, contrast may be enhanced by use of low accelerating voltages, but chromatic aberration becomes more apparent at low voltages and impairs image quality. Most objects are taken at accelerating voltages of 60–80 kV.

Filament saturation. Illumination of the specimen and resolution of the final image is inferior below the saturation point of the filament. Saturated filaments provide a homogeneous illumination spot.

Apertures. Condenser lens apertures mainly influence beam intensity, but not the image contrast. Resolution can be improved by selecting a small condenser aperture. The use of a small objective aperture improves contrast in the final image; at high magnification it is necessary to check carefully for astigmatism.

Focusing. The focus level of the objective lens influences the appearance of structures dramatically. In non-ordered objects, out-of-focus images usually show a characteristic granularity with no linear pattern.

Stability of the specimen. Movement of the specimen in one direction is common and may be traced back to specimen stage instability or drift of the support film under the electron beam. These movements result in unidirectional blurred images. Drift often occurs only for a short time, before the object becomes stable. Drift becomes less apparent in micrographs taken with short exposure times.

Specimen stability in the electron beam. Hydrocarbon vapors, present as contaminants in the high vacuum, may become deposited around specimen structures. Anticontamination devices act as cold traps to reduce contamination at the specimen, but may produce the opposite effect, causing visible material loss of the specimen. Perhaps the most severe problem that affects high resolution of biological objects is radiation damage from the electron beam. The energy transfer from beam electrons to orbital electrons by inelastic scattering on interaction with the electron beam causes fragmentation at the molecular level as well as vibrational and rotational excitations (i.e., heat transfer). This leads to rearrangement of molecular structures and mass loss. Though in principle unavoidable, radiation damage may be reduced by working at the lowest possible electron dose, possibly by minimal dose focusing (Table 2.34), and by stabilization of the specimen. For all studies in which high resolution is required, specimens should routinely be inspected at low radiation dosages with the aid of minimal dose focusing. Analysis of images (see below) usually includes critical evaluation of specimen damage. Obvious specimen damage at medium and low resolutions is a less severe problem. Since stains and embedding resins differ in their stabilities, appropriate adaptation of the specimen preparation reduces obvious radiation damage.

2.4.9.1 Stereo images

Production of stereo images requires a goniometer stage and is useful for gaining insight into the third dimensions of samples. With conventional TEMs the extension of the third dimension is restricted to the relatively low specimen thickness allowed, which should not exceed 150–200 nm if a small objective aperture is used. Transmission of thicker sections is achieved by selection of higher accelerating voltages (100–500 eV) or by filtering of inelastically scattered electrons in energy-filtering transmission electron microscopes (EFTEMs). The energy filtering permits examination of 1 μm thick specimens – dimensions previously restricted to high voltage electron microscopes (see, e.g., Martin, 1996) – without appreciable loss in image quality. Three-dimensional reconstruction by electron microscopic tomography is facilitated by EFTEMs, because of the increase in image contrast (Olins et al., 1993, 1994).

Stereo images for simple illustration of three-dimensional structures do not need extensive adjustment procedures (see Table 2.35). Medium and high magnifications usually require a tilt angle of 12°, the images of the stereo pair being taken at –6° and at +6° relative to the untilted position. This technique is very useful for evaluation of marker (ferritin or colloidal gold) distribution after pre-embedding and post-embed-

Tab. 2.35 Imaging modes in standard transmission electron microscopy

Imaging/ operation modes*	Description	Comments
Bright-field	Standard imaging mode: scattered and unscattered electrons participate in the formation of the image; the total electron dose specimens are subjected to is typically in the range of 1000 electrons per Å^2	Routinely used for stained specimens (ultra-thin sections, negative stained, metal-shadowed)
Low-dose bright-field	Total electron dose reduced to max. 10 electrons per Å^2.	Used to reduce beam-induced damage of the specimen (e.g., with negatively stained protein) in the electron beam
Performed in connection with minimal dose focusing	Focussing is carried out on an area different to that of interest (but in close proximity to it) by deflection of the electron beam	
Dark-field	Only scattered electrons participate in the formation of the image; electron doses required are as much as ~30 times higher (!) than in bright-field mode (cf. Misell, 1978)	Visualization of, for example, nucleic acids in unstained form
Diffraction mode	Causes a magnified (magnification determined by the camera length) electron diffraction pattern, which is present at the back focal plane of the objective lens, to be projected onto the screen	Used to record electron diffraction patterns of crystalline specimens; contains high-resolution amplitude information which replaces the usually lower resolution amplitude information present in the image and is combined with the image phase information (used in materials science and biological high-resolution TEM; cf. Misell and Brown, 1987)
Tilt series	The specimen is tilted relative to the incident electron beam by use of a eucentric goniometer	Tilted projections of single molecules (Osmani et al., 1984), crystals and (thick) sections provide three-dimensional structural information (cf. Amos et al., 1982, and Engelhardt, 1988)

*) A comprehensive description of electron microscope operation is given in Agar et al., 1974.

ding labeling of specimens. Localization of marker positions relative to cellular components in thick sections requires a three-dimensional image. Stereological measurements require more precise adjustments (see Helmcke, 1980).

Three-dimensional image reconstruction of complex tissue from serial sections by application of conventional TEM is time-consuming, since sections of 80 nm in thickness have to be used for high-resolution images. Thick sections allow fast reconstruc-

Tab. 2.36 Performance of stereo imaging

1.) Select the desired area.
2.) If a eucentric gonimeter stage is not used, rotate the grid until the desired area is positioned in the tilt axis.
3.) Tilt the specimen, reposition the desired area, and readjust focus position. If an exact correction of magnification is necessary, report change in objective lens current. For a pair of stereo images, use tilt angles of –6° and +6°.
If parallactic shift exceeds 0.5 cm in the final image, smaller tilting angles must be used.

tion of three-dimensional views with only minimal loss in resolution when ESI is applied (Heppelmann et al., 1989).

In principle, tilted images obtained from serial sections may be analyzed and recalculated to produce a three-dimensional model of subcellular structures when appropriate image processing is applied (Olins et al., 1994).

The actual magnification obtained depends on the objective lens current. If the area is positioned in the tilt axis, refocusing is not necessary or, at least, kept to a minimum. If refocusing is necessary, the difference in the objective lens currents has to be determined and used for recalculation of the actual magnification (the formula depends on the instrument type and will be included in the respective instruction manuals).

Correction of different magnifications in a stereo pair may also be achieved in preparation of photographic prints. Since parallactic shift parallel to the tilting axis is zero, the distance between two points on a line parallel to the tilting axis does not change. Therefore, magnifications of reproductions should be adjusted in such a way that these points are the same distance apart in a stereo pair. In order to achieve the stereo effect, parallactic shifts in the final image pair should not exceed 0.5 cm.

2.4.9.2 **Fourier-transformed images**

Qualitative and quantitative assessment of electron micrographs provides information not readily obtainable from a mere description of the object. Prior to this assessment, one should make sure that only optically sound electron micrographs, depicting specimens free of obvious artifacts, should be subjected to further analysis.

Processing of electron microscopic images, either as digitized signals or as electron micrographs, in order to enhance or quantify signals is becoming increasingly important for interpretation of results. Numerous software tools for particle counting, densitometry, measurement of surface dimensions, and stereology are available. These tools do not significantly differ from the corresponding software tools for light microscopy. One common tool in image analysis, Fourier analysis, is of outstanding importance for electron microscopy and is applied in several steps for averaging and three-dimensional reconstruction of projections of biological macromolecules. Since Fourier analysis itself is easily applicable to electron micrographs, key features of its theoretical background and applications are briefly discussed here, while a comprehensive treatise on Fourier analysis can be found in Misell (1978).

The real-space image is transformed into the reciprocal (Fourier) space, allowing several procedures for assessment and correction both of image quality and of averaging. Fourier transformations are standard procedures for analysis of spectra, and are applied to images by a similar principle. Every image may be viewed as a collection of sine waves (differing in frequency and amplitude) or the reciprocal spatial frequencies. These are, put simply, the number of changes in brightness value per unit distance in any part of the image. Tonally smooth, gradual changes represent low frequencies and tonally rough, abrupt changes (small dots, sharp edges, etc.) are high frequencies. Fourier transformation produces a complex number valued data set containing amplitude information (information on the spatial frequencies) as well as information on the position of the sine function (phase, information on the position of a structural detail). The Fourier-transformed image is usually displayed as two images, containing the frequency and phase information, respectively. In practice, the frequency information in Fourier-transformed image is displayed in optical diffractograms of the real image or by calculation of quasi-optical diffractograms (power spectra). The diffractogram contains easily extractable information on image quality (focus, astigmatism, noise, resolution of the image) and the spatial frequencies from the crystal; the latter as diffraction spots arranged on a lattice in reciprocal space (Fig. 2.46). The phase information, however, is lost, and so the diffractogram cannot be used to reconstruct the image correctly. Diffractograms and power spectra are translationally invariant: the diffractogram of an ideal two-dimensional crystal does not change if it is shifted.

A diffractogram also allows filtering for frequencies. A standard procedure is the filtering of very small and very large structures that do not add relevant information to the real image (band-pass filtering). In a similar way, all the non-periodic information outside the reciprocal lattice diffraction spots may be filtered out with a filter mask.

Calculation of quasi-optical diffractograms is achieved by numerous software tools, parts of commercially available general-purpose image analysis software packages. Several general-purpose software packages, as well as highly sophisticated software

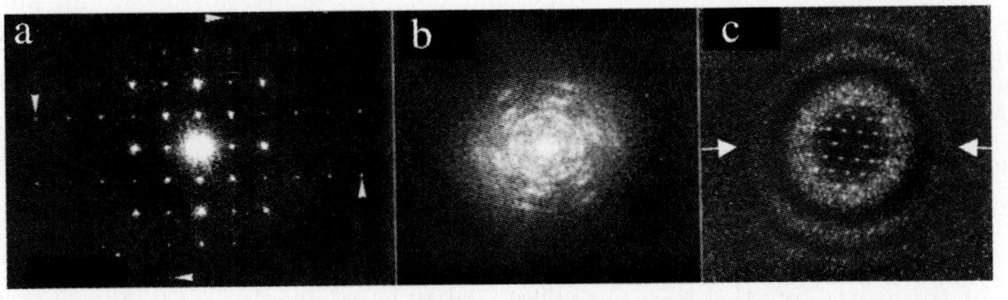

Fig. 2.46 Power spectra (diffractograms) of two-dimensional crystals.
(a) Apparently undisturbed crystal under appropriate focus conditions. Highest order diffraction spots are marked by arrowheads (see also Holzenburg and Mayer, 1989).
(b) Distorted crystal showing obvious disorder. The diffraction spots appear radially elongated (Engelhardt, 1988).
(c) Power spectrum indicating underfocus and specimen drift. The image shows concentric rings (Thon rings) indicating the inferior focus conditions; higher order diffraction spots – representing finer image details – are partially eliminated. The rings are not completely closed (arrows), indicating specimen drift (Engelhardt, 1988)

for two-dimensional and three-dimensional reconstruction of macromolecular structures, are also available free of charge.

2.4.9.3 **Electron diffraction**

Like photons, electrons also form diffraction patterns at the rear focal plane of the objective lens. Electron diffraction patterns are therefore monitored in all TEMs and may be used for creation of reciprocal lattices from two-dimensional crystalline specimens. The method is regularly applied for determination of inorganic crystal structures, as well as for analysis of crystalline biopolymers such as cellulose or chitosan. From the diffraction patterns, the sizes of, for instance, cellulose microfibrils, and the geometries of unit cells (if present), may be determined.

Large and highly ordered two-dimensional crystals of protein molecules have also been subjected to electron diffraction. The technique theoretically provides a resolution sufficient to determine atomic coordinates and might therefore be an alternative to X-ray crystallography. Preparation of the specimen by cryoelectron microscopy and low electron doses minimizes beam damage to the specimen, a prerequisite for near-atomic resolution.

2.4.9.4 **Determination of macromolecular structure by particle averaging**

Evaluation of the structures of large macromolecular assemblies has become a prominent field in electron microscopy. It is now possible to gain atomic resolution details of macromolecular assemblies by fitting the high-resolution X-ray and NMR structures of individual components with the subatomic resolved structure of the entire assembly obtained by electron microscopy. The tools of molecular modeling are therefore also made applicable to macromolecular structure. In order to decrease the positional uncertainty when modeling X-ray structures within them, structures recorded under the electron microscope need to be resolved at 0.7 nm (7 Å) or better. This is achievable by analysis of one-dimensional or, better, two-dimensional crystals (see chapter 2.4.8.9) and lies at the limit of the resolution attainable with single-particle analysis. Nevertheless, determination and optimization of crystallization conditions may be a very time-consuming task, so use of single particles for analysis may be an alternative.

Though single particles are easiest to obtain, by various methods, for spreading on a carbon film by negative staining or preparation of frozen-hydrated specimens, and electron microscopic images can be quickly recorded, evaluation of the data may be laborious. As a first step, determination of overall topology (shape, number of subunits, obvious symmetries) is often evident from the structure without further processing of the image. Antibodies directed against specific subunits reveal the positioning of the units in a protein complex and help evaluation of subunit stoichiometry. Together with biochemical data on subunit composition, quite satisfying models on subunit topology may often be constructed. Models of cellulosome structure and function have been made exclusively by this approach.

Rotational symmetries of particles may be confirmed by a rotational enforcement technique (Markham et al., 1963). The image of the particle is rotated around its center by n increments of a circle (by 120° and 240°, say) and the resulting images are superimposed. If n meets a periodicity of the structure (e.g., a threefold symmetry), the resulting image will reflect this periodicity by revealing enhanced contrast (by exhibiting three well defined units) arranged in rotational symmetry.

Superimposed structures, variability introduced by preparation, noise from any source, and low image contrast (especially of vitrified specimens), reduce the perceptibility of the structure of interest in individual molecules. Enhancement of the signal through averaging of many equivalent projections of the molecule – and three-dimensional reconstruction by comparing these projections – make better use of the theoretically, but still unattainable, near-atomic resolving power of the electron microscope. The outcome of such an approach strongly depends on the number of the particles used for averaging, the quality of the preparation and of the electron microscopic images, and already existing knowledge of the structure of the molecule. Complexes of high internal variability (such as cellulosomes) are, of course, unsuitable for averaging techniques. While analysis of single particles is software-assisted, an operator has to decide at crucial "branching points", which may still influence the final result. Several software tools for single-particle analysis and analysis of two-dimensional crystals are now available, including SPIDER (Frank et al., 1981; Frank et al., 1996), CRISP (Hovmöller, 1992), IMAGIC (van Heel and Keegstra, 1981; van Heel et al., 1996), SPARK (Lanzavecchia et al., 1993), and EMAN (Ludtke et al., 1999). All these tools are packages consisting of several individual programs that may be combined individually to resolve a specific structure.

The principal steps of single-particle analysis are briefly outlined here. As a first step in processing of digitized images, band-pass filters are applied to suppress the highest and lowest spatial frequencies. Very low spatial frequencies may be due to fluctuations of staining and very high frequencies include noise, but they may still also contain relevant information on object structure, so this information may be added to the image later during refinement processes.

The first alignment step is used to center single particles, creating a rotationally averaged total sum of the particles. The data set of single particles is compared to this average, which allows centering of the single particles. Alignment is achieved by cross-correlation, in which two images to be aligned are shifted relative to each other by defined vectors. Cross-correlation reaches a maximum where a motif present in both images overlaps. This step is iterated several times until good alignment is achieved. The aligned molecules now have to be analyzed, due to their individual variations, which is a prerequisite for classification into a limited number of characteristic projection forms. This is performed by correspondence analysis: a multivariate statistical analysis that allows the variations in a set of images to be quantified. An automated classification step is then used to generate a set of class averages. Selected class averages are used as references to search and align the entire data set. With this reference now available, each image is aligned to the class averages. After some iterative cycles of correspondence analysis and automatic classification, class averages with greatly improved signal-to-noise ratios can be achieved. The images are projections of the three-dimensional particles in a two-dimensional plane. Since they represent several views of the molecule, they contain all information necessary to obtain a three-dimensional image, but the spatial relationship between the projections has to be determined.

The first step to obtaining a three-dimensional model is to determine projection directions for the class averages. These orientations, expressed as Euler angles, are determined by use of the angular reconstitution method. This is based upon the common line projection theorem, which states that any two two-dimensional projections of a three-dimensional object share at least one line (one-dimensional) projection. All possible one-dimensional projections (sinograms) from the class averages are therefore determined. The sinograms of these projection averages are compared pairwise, which finally results in the definition of the common tilt axis between the class averages (see van Heel, 1987 for details). From the common tilt axis, it is possible to determine the Euler angles of the class averages. An initial three-dimensional reconstruction can be obtained from these data, but has to be subjected to several refinement procedures, essentially by reprojecting the three-dimensional reconstruction, from which the projection forms of the model are obtained. These may be used as references for the class averages. Class averages that do not fit to the model are discarded, and exceptional molecular views missed in the first round of alignment will become significant and be used to calculate a refined three-dimensional model. The alignment procedures, correspondence analysis and automatic classification, are the same as those used for the generation of class averages. Several iterative rounds are necessary, until the model shows no further improvement. The final reconstruction may be depicted in various ways, such as surface-rendered views. Figure 2.47 a shows a diagram illustrating the key steps of the three-dimensional reconstruction of a complex molecular assembly. The available software packages perform the iterative refinement process of the three-dimensional models in various ways, which also influence the processing time considerably.

When macromolecules are arranged in orderly fashion in a one-dimensional or two-dimensional crystal lattice, analysis and three-dimensional reconstruction are, in principle, simplified, because of the defined orientation of the individual molecule in the crystal lattice. Unlike in the case of preparation of single particles, optimization of crystallization parameters are therefore essential for facilitated analysis. Numerous filaments, such as polymerized actin, are naturally occurring one-dimensional crystals with helical symmetry. Some macromolecules, such as the gap junction protein and the bacterial S-layer, also occur in vivo as two-dimensional crystals. In most cases however, crystallization has to be optimized with purified protein in vitro. Two-dimensional crystals may be obtained with soluble and membrane proteins. It has been observed that their formation is easier to achieve with membrane proteins, which

Fig. 2.47 Three-dimensional reconstruction of macromolecules. ▶
(a) Key procedures for three-dimensional reconstruction of single particles depicting a selection of original views of a particle (calcium release channel of skeletal muscle) in the upper row, the class averages (left column, reprojected views of these averages (right column), and a stereoscopic view of the final three-dimensional model. See text for further explanations (Serysheva et al., 1995).
(b) Key procedures for three-dimensional reconstruction of two-dimensional crystals showing the original specimen (top; untilted view, several views at different tilt angels are necessary), a corresponding power spectrum (upper right) for determination of lattice parameters, a contour plot of the untilted view of a unit cell average (upper left), a scheme of lattice lines (lower left), and a final three-dimensional view (lower right). See text for further explanations (Engelhardt, 1988; Sosinsky and Perkins, 2000)

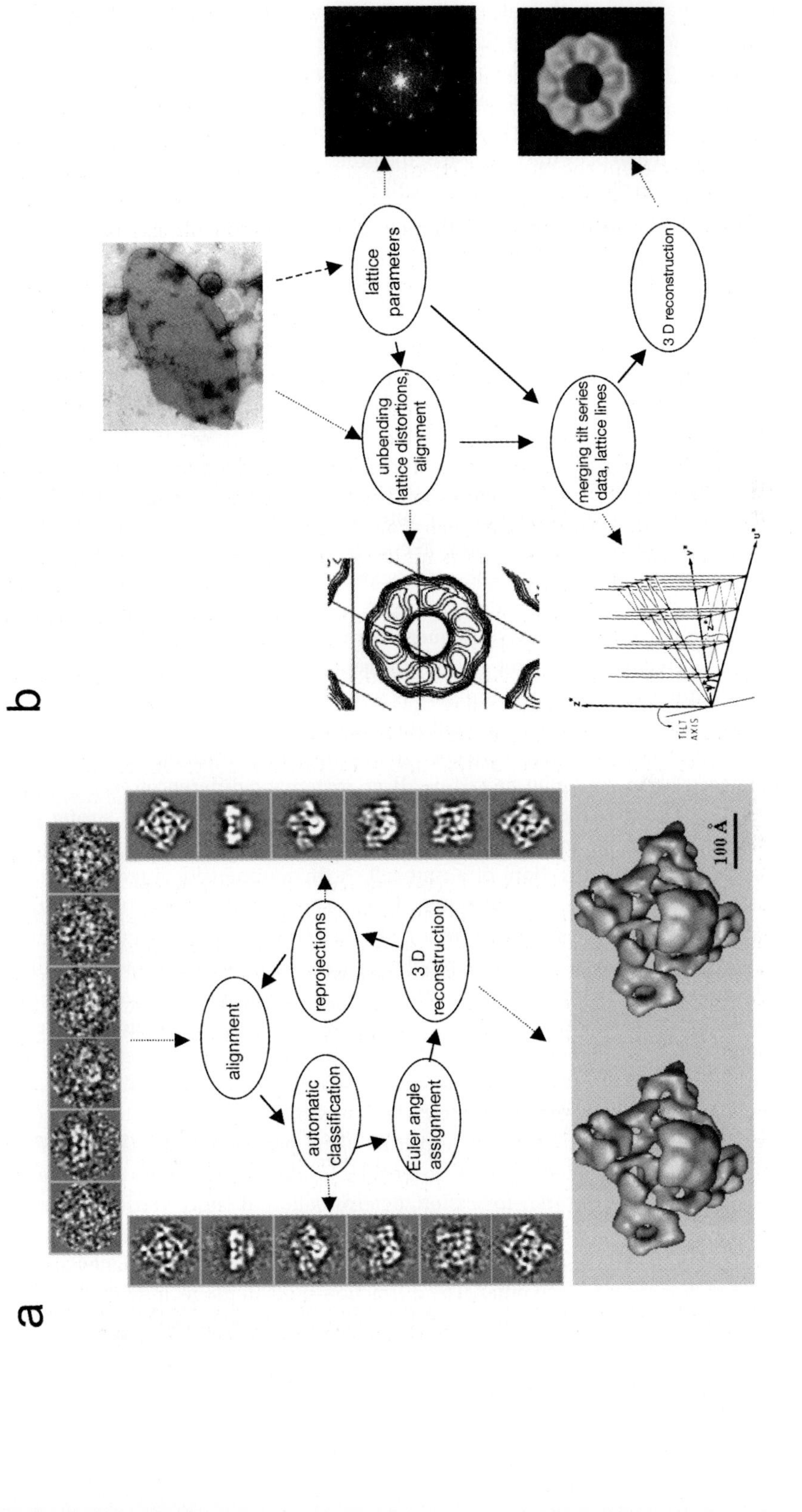
a
alignment
automatic classification
Euler angle assignment
3 D reconstruction
reprojections
100 Å
b
lattice parameters
unbending lattice distortions, alignment
merging tilt series data, lattice lines
3 D reconstruction
TILT AXIS

is reasonable, since they also occur naturally as two-dimensional sheets. Thus, a respective protocol has been included here (see chapter 2.4.8.9).

The individual molecules in two-dimensional crystals may be packed in various fashions. For biological molecules (which are all chiral), 17 plane groups are possible. In natural crystals such as membrane channel proteins, only those plane groups are realized that obey the function of the protein, which is regularly vectorial (transport or signal transduction). Molecules with inverted orientation with respect to the layer plane are therefore not encountered. In artificial two-dimensional crystals, any of the 17 plane groups may be observed.

Reconstruction of the molecular structure starts with a power spectrum of a (digitized) image of the crystal. Analysis of the regular arrangement of the diffraction pattern provides information on the crystal symmetry, the lattice constants, and the lattice base vectors, defining the shape and orientation of the lattice in the original image. This also takes account of (nearly unavoidable) distortions of the crystal. Together with information from the original image and the phase image, the plane group of the crystal may be identified. Peak filtering by masking out all information outside the reciprocal lattice of the diffractogram is applied to suppress noise in the retransformed real image (quasi-optically filtered image). A retransformed diffractogram or power spectrum is something like an image average over the entire crystal, due to the loss of phase information (see above). Because of the uncorrected lattice distortions, this average is not directly used for further reconstruction but may be an appropriate reference for subsequent correlation averaging. To correct the distortions, correlation averaging of digitally generated crystal patches is applied, using, for example, an area (minimum size is one unit cell) from the quasi-optically filtered image as reference. The alignment procedure is iteratively repeated. Similarly to single-particle analysis, correspondence analysis may be performed with the patches after the correlation averaging procedure. If no systematic differences exist, classification can be used to exclude from averaging those unit cells that exhibit strong variability with respect to the majority of the unit cells. The resulting average is, of course, the face-on view (x/y direction) of a unit cell. A three-dimensional reconstruction must also include data on the z-direction. For crystallized specimens, these data are obtained by tilting the specimen on a goniometer stage and recording the tilted view. Tilting poses several technical problems that do not allow recovery of all data in the z-direction. Firstly, strongly inclined specimens become very thick in the direction of the optical axis, which may have consequences for image resolution; recording at an angle of 90° is, of course, impossible. The fact that most goniometers only allow tilt angles of up to 60° (which leads to an information gap known as the "missing cone" problem) may be overcome by technical modification of the instrument (if this is technically possible), or, most simply, by using pre-bend grids. The tilt increments have to be determined in order to ensure to obtain maximum resolution (Engelhardt, 1998). Fourier transformation, determination of lattice constants, and correlation averaging is then performed essentially as described for the untilted image. For merging into a three-dimensional data set, the unit cells must be aligned and normalized with respect to each other. The diffraction spots in the reciprocal lattice are fitted to lattice lines. These data are then fitted and interpolated by least-squares techniques, finally to obtain a three-dimensional data set of layer slices. Both real space and Fourier space data are used in different software packages. The final reconstruction may be depicted in various ways, such as in surface-rendered view (Fig. 2.47 b).

2.4.10 Standard Specimen Preparation for Scanning Electron Microscopy

For visualization under a SEM, it is necessary that specimens emit a sufficient number of secondary electrons and develop as few surface charges as possible in the electron beam. Only a few biological samples – such as insect exoskeletons, bone, or diatom frustules – fulfil these prerequisites. In most cases, special preparation procedures are necessary. For further information on the preparation of bacteriological specimens, the reader is referred to Watson et al. (1980) and the fine selection of SEM-specific references quoted by Holt and Beveridge (1982). A low-temperature SEM setup is described by Read and Jeffree (1991).

Preparation of a specimen for SEM comprises:

1. Fixation

Fixation can be achieved either by conventional or by cryo methodology. Conventional (chemical) fixation of the sample is carried out in an isotonic medium (the media are described in more detail for fixation of samples to be embedded in resin; see above) containing 1–4% (v/v) glutardialdehyde. Since osmium tetroxide considerably increases the conductivity of the specimen (and thereby gives rise to stronger emission of electrons), a second fixation step is usually recommended: the sample is subjected to alternate incubation procedures in osmium tetroxide (1–2%, w/v) and tannic acid (1–7%, w/v) or thiocarbohydrazide (saturated solution). During this procedure a film of osmium is deposited on the surface of the specimen, so the objects often do not need further metal coating. A different strategy involving the fixation of single bacterial colonies in osmium tetroxide vapor has been devised by Springer and Roth (1972) and comprises the steps listed in Table 2.37.

The block is now ready to be processed further. In this particular procedure, the agar blocks are glued onto stubs for drying in air or in a desiccator. However, the gentlest and most efficient means of artifact-free fixation of samples is by plunging them into

Tab. 2.37 Osmium staining for SEM

1.)	Cut out a small block (~2 mm in height) of agar supporting a single colony.
2.)	Place the agar block in a Petri dish at the periphery and incline the dish so that the cube is located at the highest point.
3.)	Place a few drops of OsO_4 in the dish, cover with a lid, seal tightly, and incubate overnight.
4.)	With a pipette, remove the OsO_4 and carefully rinse the lower area with water.
5.)	Level the dish out on a bench and surround the block with water, always making sure that no water ever comes into contact with the upper surface of the block. Incubate for 15 min and repeat twice. This step is carried out to reduce the salt concentration in the agar block.
6.)	Excess liquid is removed by transferring the block onto filter paper.

liquid nitrogen (physical fixation; see above) after the cells have been adsorbed onto suitable specimen supports, such as carbon-coated Cu grids or carbon films of 0.1–0.2 mm thickness. The latter films are self-supporting and are purchased as larger sheets (e.g., from Goodfellow Cambridge Ltd., UK) from which smaller areas can be cut off to suit the sample. When adsorbing filamentous or network-forming microorganisms onto a support, one has to be aware that any shearing forces acting on the sample may introduce artifacts by disrupting the original organization of the colony. In cases in which this organization matters (such as cases in which organization into mats, fibers, etc. is macroscopically evident), one should therefore avoid centrifugation steps and transfer them directly from the medium onto the support.

2. Dehydration and drying
The chemically fixed specimens are then subjected to dehydration in a graded series of acetone or ethanol, as described for the dehydration of samples for resin-embedding and ultramicrotomy (see above). Physically fixed specimens are sensibly subjected to freeze-substitution (see above). Freeze-substitution with acetone can be carried out by using 2% osmium tetroxide as fixing additive for 3×8 h at –90, –60, and –30 °C, respectively (Rieger et al., 1997). After the water has been removed from the medium, the organic solvent also needs to be removed. In order to avoid drying artifacts, this is done by critical point drying or freeze-drying. In most cases, critical point drying is the preferred method since it is cheap and fast. In brief, the specimen is transferred from the dehydration fluid (ethanol) into the drying medium (gaseous CO_2) via a transitional fluid (liquid CO_2). Exchange of the transitional fluid for the drying medium is done under high pressure (critical pressure) and at the critical temperature of the transitional fluid (31 °C for CO_2, but 374.2 °C for water, which explains why it is not suitable). Above the critical point, the densities of the drying medium in its liquid and gas phase are identical. If the temperature of the system remains above the critical point, the gas phase can be vented off without subsequent recondensa-

Tab. 2.38 Principal steps involved in critical point drying

1.) Precool the apparatus to 5–10 °C below the ambient temperature by flushing with liquid CO_2 (modern pieces of apparatus are likely to be equipped with cooling coils, usually obviating the need for this flushing step).
2.) Load samples immersed in ethanol (or acetone) and seal the chamber.
3.) Open the CO_2 inlet valve and allow the chamber to fill with CO_2.
4.) Slightly open the exhaust valve to expel any ethanol.
5.) Close the exhaust valve and fill the chamber with CO_2 again.
6.) Close the inlet valve and allow the specimen to remain in liquid CO_2 for about 30 min, to allow time for the CO_2 to replace any dehydration fluid.
7.) Flush again with CO_2 until there is no ethanol left.
8.) Close both valves.
9.) Turn on the heater and allow the temperature to rise above 31 °C and the pressure to build up to 1100–1200 psi.
10.) Once the critical temperature and pressure are exceeded, slowly vent off the CO_2, while keeping the temperature elevated.
11.) The specimens are now ready to be mounted on stubs and sputter-coated.

tion. Because CO_2 is removed after its transition from the liquid to the gas phase at the critical point, the specimen is dried without structural damage. After removal of the gas and decompression of the chamber, the dried specimen is transferred to a desiccator (see Table 2.38 for details).

3. Sputtering

The conductivity of (biological) specimens needs to be increased prior to visualization under the SEM. The osmium fixation procedure may provide sufficient conductivity (see above), but normally sputtering (Au, W, or Pt, the latter two have a smaller particle size) or vacuum evaporation of Au or Au-Pd are necessary.

Sputtering of the metal is performed in a vacuum chamber equipped with a gold-plated cathode and a glow discharge unit. Glow-discharging results in the release of gold atoms, which subsequently cover the sample (anode).

In addition to this standard procedure, special preparation methods such as fracturing of specimens, elemental analysis, and immunocytochemical localisation can be used in conjunction with SEM (see Table 2.39).

Tab. 2.39 Special methodological approaches that utilize SEM

Approach	Description	Useful for:
Fracturing/cleaving of objects	Sectioned embedded material, freeze-fractured specimens, or mechanically disrupted material is prepared for visualization under the SEM	Visualization of internal structures of, for example, wood, leaves. or tissues (see, e.g., Hotta et al., 1990)
Elemental analysis	Measurement of cathodoluminescence or X-rays caused by interaction of the electron beam with the specimen	Recording of spectra or images depicting the elemental distribution in the sample
Immunocytochemical localization	Localization of antigens with the aid of colloidal gold, subsequent preparation for SEM	Localization of antigens exposed to the surface of the specimen (Hodges et al., 1987)

References

General Textbooks on Methods for Electron Microscopy

1 Aldrich, H.C., Todd, W.J. (**1986**). Ultrastructure techniques for Microorganisms. Plenum Press, New York.
2 Glauert, A.M. (**1973-1987**). Practical Methods in Electron Microscopy (vol. 1–12). Elsevier, Amsterdam.
3 Hayat, M.A. (**1978**). Introduction to Biological Scanning Electron Microscopy. University Park Press, Baltimore.
4 Kay, D. (**1965**). Techniques for Electron microscopy. Blackwell Scientific Publ., Oxford.
5 Maunsbach, A. B., Afzelius, B. A. (**1999**) Biomedical electron microscopy. Illustrated Methods and Interpretations. Academic Press, London
6 Mayer, F. (**1988**). Methods in Microbiology. Vol. 20: Electron Microscopy in Microbiology. Academic Press London.
7 Reimer, L. (**1997**). Transmission electron Microscopy: Physics of Image Formation and Microanalysis (Springer Series in Optical Sciences, Vol. 36). 4th ed. Springer, Heidelberg.
8 Robinson, D.G., Ehlers, U., Herken, R., Herrmann, B., Mayer, F., Schürmann, F.W. (**1987**). Methods of preparation for Electron microscopy. Springer-Verlag, Berlin.

References

1 Acker, G. (**1988**). Immunoelectron microscopy of surface antigens (polysaccharides) of Gram-negative bacteria using pre-and post-embedding techniques. In: *Methods in Microbiology*, Vol. 20, *Electron Microscopy in Microbiology* (ed. F. Mayer), pp. 147–174. Academic Press, London.
2 Adrian, M., Dubochet, J., Fuller, S. D., Harris, J. R. (**1998**). Cryo-negative staining. *Micron* 29, 145-160.
3 Adrian, M., Dubochet, J., Lepault, J., McDowall, A. W. (**1984**). Cryo-electron microscopy of viruses. *Nature* 308, 32–36.
4 Agar, A. W., Alderson, R. H., Chescoe, D. (**1987**). Principles and Practice of Electron Microscope Operation. In: *Practical Methods in Electron Microscopy*, Vol. 2, (ed. A. M. Glauert). Elsevier, Amsterdam.
5 Aldrich, H. C., McDowell, L., Barbosa, M. F., Yomano, L. P., Scopes, R. K., Ingram, L. O. (**1992**). Immunocytochemical localization of glycolytic and fermentative enzymes in *Zymomonas mobilis*. *J. Bacteriol.* 174, 4504–4508.
6 Allen, E. D., Wheatherbee, L. (**1979**). Ultrastructure of red cells frozen with hydroxyethyl starch. *J. Microsc.* 117, 381–394.
7 Almeida, J. D. (**1980**). Practical aspects of diagnostic electron microscopy. *Yale J. Biol. Med.* 53, 5–18.
8 Amos, L. A., Henderson, R., Unwin, P. N. T. (**1982**). 3-dimensional structure determination by electron microscopy of two-dimensional crystals. *Prog. Biophys. Molec. Biol.* 39, 183–231.
9 Bauer, R. (**1988**). Electron spectroscopic imaging: an advanced technique for imaging and analysis in transmission electron microscopy. In: *Methods in Microbiology*, Vol. 20, *Electron Microscopy in Microbiology* (ed. F. Mayer), pp. 113–146. Academic Press, London.
10 Beesly, J. E. (**1989**). Colloidal Gold: A new perspective for cytochemical marking (Microscopy Handbooks Vol. 17). Royal Microscopic Society and Oxford University Press, Oxford.
11 Bendayan, M. (**1985**). The enzyme-gold technique: a new cytochemical approach for the ultrastructural localization of macromolecules. In: *Techniques in Immunocytochemistry*, Vol. 3 (eds. G. R. Bullock, P. Petrusz), pp. 179–201. Academic Press, London.

12 Bendayan, M., Barth, R. F., Gingras, D., Londono, I., Robinson, P.T., Alam, F., Adams, D. M., Mattiazzi, L. (**1989**). Electron spectroscopic imaging for high-resolution immunocytochemistry: use of boronated Protein A. *J. Histochem. Cytochem.* 37, 573–580.

13 Bischler, N., Balavoine, F., Milkereit, P., Tschochner, H., Mioskowski, C., Schultz, P. (**1998**). Specific interaction and two-dimensional crystallization of histidine tagged yeast RNA polymerase I on nickel-chelating lipids. *Biophys. J.* 74, 1522-1532.

14 Brenner, S., Horne, R. W. (**1959**). A negative staining method for high resolution electron microscopy of viruses. *Biochim. Biophys. Acta* 34, 103–110.

15 Brisson, A., Olofsson, A., Ringler, P., Schmutz, M., Stoylova, S. (**1994**). Two-dimensional crystallization of proteins on planar lipid films and structure determination by electron crystallography. *Biol. Cell* 80, 221-228.

16 Carlemalm, E., Garavito, R. M., Villiger, W. (**1982**). Resin development for electron microscopy and an analysis of embedding at low temperature. *J. Microsc.* 126, 123–143.

17 Cevc, G., Marsh, D. (**1987**). *Phospolipid Bilayers.* Wiley-Interscience, New York.

18 Cyrklaff, M., Auer, M., Kühlbrandt, W., Scarborough, G. A. (**1995**). 2-D structure of the *Neurospora crassa* plasma membrane ATPase as determined by electron cryomicroscopy. *EMBO J.* 14, 1854-1857.

19 Däkena, P., Rohde, M., Dimroth, P., Mayer, F. (**1988**). Oxaloacetate decarboxylase from *Klebsiella pneumoniae*: Size and shape of the enzyme, and localization of its prosthetic group by electron microscopic affinity labeling. *FEMS Microbiol. Lett.* 55, 35–40.

20 Devereux, J., Haeberli, P., Smithies, O. (**1984**). A comprehensive set of sequence analysis programs for the VAX. *Nucl. Acids Res.* 12, 387–395.

21 Dietrich, C., Schmitt, L., Tampé, R. (**1995**). Molecular organization of histidine-tagged biomolecules at self-assembled lipid interfaces using a novel class of chelator lipids. *Proc. Natl. Acad. U.S.A.* 92, 9014–9018.

22 Dolder, M., Engel, A., Zulauf, M. (**1996**). The micelle to vesicle transition of lipids and detergents in the presence of a membrane protein: towards a rationale for 2D crystallization. *FEBS Lett.* 382, 203-208.

23 Dubochet, J., Adrian, M., Chang, J.-J., Homo, J.-C., Lepault, J., McDowall, A. W., Schultz, P. (**1988**). Cryo-electron microscopy of vitrified specimens. *Quart. Rev. Biophys.* 21, 129–228.

24 Dubochet, J., Ducommun, M., Zollinger, M., Kellenberger, E. (**1971**). A new preparation method for dark-field electron microscopy of biomacromolecules. *J. Ultrastruct. Res.* 35, 147–167.

25 Dubochet, J., Lepault, J., Freeman, R., Berriman, J. A., Homo, J.-C. (**1982**). Electron microscopy of frozen water and aqueous solutions. *J. Microsc.* 128, 219–237.

26 Egger, D., Troxler, M., Bienz, K. (**1994**). Light and electron microscopic in situ hybridization: non-radioactive labeling and detection, double hybridization, and combined hybridization-immunocytochemistry. *J. Histochem. Cytochem.* 42, 815–822.

27 Engelhardt, H. (**1988**). Correlation averaging and 3-D reconstruction of 2-D crystalline membranes and macromolecules. In: *Methods in Microbiology*, Vol. 20, *Electron microscopy in Microbiology* (ed. F. Mayer), pp. 357–413. Academic Press, London.

28 Estis, L. F., Haschemeyer, R. H., Wall, J. S. (**1981**). Uranyl sulphate: a negative stain for electron microscopy. *J. Microsc.*, 124, 313–318.

29 Faberge, A.C., Oliver, R.M. (**1974**). Methylamine tungstate, a new negative stain. *J. Microscopie (Paris)* 20, 241–246.

30 Frank, J. J., Radermacher, M., Penczek, P., Zhu, J., Li, Y., Ladjadj, M., Leith, A. (**1996**). SPIDER and WEB: Processing and visualization of images in 3D electron microscopy and related fields. *J. Struct. Biol.* 116, 190–199.

31 Frank, J., Shimkin, B., and Dowse, H. (**1981**). SPIDER – a modular software system for electron image processing. *Ultramicroscopy* 6, 343–358.

32 Frens, G. (**1973**). Preparation of gold dispersions of varying particle size: controlled nucleation for the regulation of the particle size in monodisperse gold suspensions. *Nature Phys. Sci.* 241, 20–22.

33 Freudenberg, W., Mayer, F., and Andreesen, J. R. (**1989**). Immunocytochemical localization of proteins P1, P2, P3 of glycine decarboxylase and of the selenoprotein P_A of glycine reductase, all involved in anaerobic

glycine metabolism of *Eubacterium acidaminophilum. Arch. Microbiol.* 152, 182–188.
34 Fromherz, P. (**1971**). Electron microscopic studies of lipid protein films. *Nature* 231, 267-268.
35 Frösch, D., Westphal, C. (**1989**). Melamine resins and their application in electron microscopy. *Electron Microsc. Rev. 2*, 231–255.
36 Geoghegan, W. D., Ackerman, G. A. (**1977**). Adsorption of horseradish peroxidase, ovomucoid and anti-immunoglobuline to colloidal gold for the indirect detection of concanavalin A, wheat germ agglutinin and goat antihuman immunoglobuline G on cell surfaces at the electron microscopic level: a new method, theory and application. *J. Histochem. Cytochem.* 25, 1187–1200.
37 Geymayer, W., Grasenick, F., Hodl, Y. (**1977**). Stabilising ultra-thin cryosections by freeze-drying. *J. Microsc.* 112, 39–46.
38 Glauert, A. M., Rogers, G. E., Glauert, R. H. (**1956**). A new embedding medium for electron microscopy. *Nature,* 178, 803.
39 Gregg, M., Reznik-Schüller, H. M. (**1984**). An improved method for rapid electron microscopic autoradiography. *J. Microsc.* 135, 115–118.
40 Griffiths, G., McDowall, A. F., Back, R., Dubochet, J. (**1984**). On the preparation of cryosections for immunochemistry. *J. Ultrastruct. Res.* 89, 65–78.
41 Griffiths, G., Simons, K., Warren, G., and Tokuyasu, K. T. (**1983**). Immunoelectron microscopy using thin, frozen sections: application to studies of the intracellular transport of semliki forest virus spike glycoproteins. *Meth. Enzymol.* 96, 435–450.
42 Hainfeld, J. F. (**1989**). Undecagold-antibody method. In: *Colloidal Gold: Principles, Methods and Applications.* Vol. 2 (ed. M. Hayat.), pp. 413–429. Academic Press, San Diego, CA.
43 Hainfeld, J. F. (**1996**). Gold liposomes. In Proc. 54th Ann. Mtg. Micros. Soc. Amer., G. W. Bailey, J. M. Corbett, R. V. W. Dimlich, J. R. Michael, N. J., Zaluzec (eds.). San Francisco Press, San Francisco, CA, pp. 898–899.
44 Hainfeld, J. F., Furuya, F. R. (**1995**). Silver enhancement of nanogold and undecagold. In: *Immunogold-Silver Staining: Principles, Methods and Applications.* (ed. M. A. Hayat), pp. 72–91. CRC Press, Boca Raton, FL.
45 Hainfeld, J. F., Liu, W., Halsey, C. M. R., Freimuth, P., Powell, R. D. (**1999**). Ni-NTA-gold clusters target his-tagged proteins. *J. Struct. Biol.* 127, 185–198.
46 Harris, J. R., Horne, R. W. (**1994**). Negative staining – a brief assessment of current technical benefits, limitations and future possibilities. *Micron* 25, 5–13.
47 Haschemeyer, R. H., Myers, R. J. (**1972**). Negative staining. In: *Principles and techniques of electron microscopy,* Vol. 2 (ed. M. Hayat), pp. 101–147. Van Nostrand Reinhold Company, New York.
48 Hayat, M. A. (**1973–1977**). Electron microscopy of enzymes: Principles and Methods. Vol. 1 (**1973**); Vols. 2 and 3 (**1974**); Vol. 4 (**1976**); Vol. 5 (**1977**). Van Nostrand Reinhold, New York.
49 Hawes, C., Martin, B. (**1995**). Freeze-fracture deep-etch methods. *Meth. Cell Biol.* 49, 33–43.
50 Hayat, M. A. (**1981**). Fixation for electron microscopy. Academic Press, London.
51 Helmcke, J. G. (**1980**). Die Erzeugung von Stereobildern und Stereogrammen. In: *Methodensammlung der Elektronenmikroskopie 4.4.2* (eds. G. Schimmel, W. Vogell) Wissenschaftliche Verlagsgesellschaft, Stuttgart.
52 Heppelmann, B., Messlinger, K., and Schmidt, R. F. (**1989**). Serial sectioning, electron microscopy, and three-dimensional reconstruction of fine nerve fibres and other extended objects. *J. Microsc.* 156, 163–172.
53 Hermann, R., Schwarz, H., Müller, M. (**1991**). High precision immunoscanning electron microscopy using Fab fragments coupled to ultra-small colloidal gold. *J. Struct. Biol.* 107, 38–47.
54 Hippe-Sanwald, S. (**1993**). Impact of freeze substitution on biological electron microscopy. *Microsc. Res. Tech.* 24, 400–422.
55 Hodges, G. M., Southgate, J., Toulson, E. C. (**1987**). Colloidal gold – a powerful tool in scanning electron microscope immunocytochemistry: an overview of bioapplications. *Scanning Microsc.* 1, 301–318.
56 Hohenberg, H., Mannweiler, K., Müller, M. (**1994**). High-pressure freezing of cell suspensions in cellulose capillary tubes. *J. Microsc.* 175, 34–43.
57 Holt, S. C., Beveridge, T. J. (**1982**). Electron microscopy: its development and application to microbiology. *Can. J. Microbiol.* 28, 1–53.
58 Holzenburg, A., Mayer, F. (**1989**). D-ribulose-1,5-bisphosphate carboxylase/oxyge-

nase: function-dependent structural changes. *Electron Microsc. Rev.* 2, 139–169.

59 Hoppert, M., Mayer, F. (**1995**). Electron microscopy technique for immunocytochemical localization of enzymes in methanogenic archaea. In: *Archaea – A Laboratory Manual* (eds. F. T. Robb, K. R. Sowers, S. DasSharma, A. R. Place, H. J. Schreier, E. M. Fleischmann), pp. 269–278. Cold Spring Harbor Press, Cold Spring Harbor, NY.

60 Horisberger, M. (**1985**). The gold method as applied to lectin cytochemistry in transmission and scanning electron microscopy. In: *Techniques in Immunocytochemistry*, Vol. 3 (eds. G. R. Bullock, P. Petrusz), pp. 155–178. Academic Press, London.

61 Hotta, Y., Kato, H., Watari, N. (**1990**). A simple and rapid maceration method for scanning electron microscopy using microwave. *J. Electron Microsc.* 39, 63–66.

62 Hovmöller, S. (**1992**). CRISP – crystallographic image processing on a personal computer. *Ultramicroscopy* 41, 121-135.

63 Kessels, M. M., Qualmann, B., Klobasa, F., Sierralta W. D. (**1996**). Immunocytochemistry by electron spectroscopic imaging using a homogeneously boronated peptide. *Cell Tissue Res.* 284, 239–245.

64 Kushida, H. (**1967**). A new embedding method employing DER 736 and Epon 812. *J. Electron Microsc.* 16, 278–283.

65 Lanzavecchia, S., Bellon, P. L., Scatturin, V. (**1993**). SPARK, a kernel of software programs for spatial reconstruction in electron microscopy. *J. Microsc.* 171, 255–266.

66 Lebermann, R. (**1965**). Use of uranyl formate as negative stain. *J. Mol. Biol.* 13, 606.

67 Leduc, E. H., Bernhard, W. (**1967**). Recent modifications of glycol methacrylate embedding procedure. *J. Ultrastruct. Res.* 19, 196–199.

68 Lévy, D., Mosser, G., Lambert, O., Mouk, S. G., Bald, D., Rigaud, J.-L. (**1999**). Two-dimensional crystallization on lipid layer: A successful approach for membrane proteins. *J. Struct. Biol.* 127, 44-52.

69 Li, M. J., Rogers, K., and Rust, C. A. (**1995**). Environmental scanning electron-microscopes. *Adv. Mater. Process.* 7, 24–25.

70 Ludtke, S. J., Baldwin, P. R., Chiu, W. (**1999**). EMAN: Semi-automated software for high resolution single particle reconstructions. *J. Struct. Biol.* 128, 82–97.

71 Luft, J. H. (**1956**). Permanganate – a new fixative for electron microscopy. *J. Biophys. Biochem. Cytol.* 2, 799–801.

72 Luft, J. H. (**1959**). The use of acrolein as a fixative for light and electron microscopy. *Anat. Rec.* 133, 305.

73 Mannella, C. A. (**1986**). Mitochondrial outer membrane channel (VDAC, porin) two-dimensional crystals from *Neurospora. Methods Enzymol.* 125, 595–610.

74 Markham, R., Frey, S., Hills, G. J. (**1963**). Methods for the enhancement of image detail and accentuation of structure in electron microscopy. *Virology* 20, 88–102.

75 Martin, R. (**1996**). The structure of the neurofilament cytoskeleton in the squid giant axon and synapse. *J. Neurocytol.* 25, 547–554.

76 Massover, W. H., Marsh, P. (**1997**). Unconventional negative stains: Heavy metals are not required for negative staining. *Ultramicroscopy* 69, 139–150.

77 Maunsbach, A. B., Afzelius, B. A. (**1999**). Biomedical electron microscopy. Illustrated methods and interpretations. Academic Press, London.

78 Mayer, G., Bendayan, M. (**1999**). Immunogold signal amplification: application of the CARD approach to electron microscopy. *J. Histochem. Cytochem.* 47, 421–429.

79 Mayer, F., Rohde, M. (**1988**). Analysis of dimensions and structural organization of proteoliposomes. In: *Methods in Microbiology* Vol. 20, *Electron Microscopy in Microbiology* (ed. F. Mayer), pp. 283–292. Academic Press, London.

80 Meek, G. A. (**1976**). Practical Electron Microscopy for Biologists (2nd ed.). John Wiley & Sons, London.

81 Mellema, J. E., van Bruggen, E. F. J., Gruber, M. (**1967**). Uranyl oxalate as negative stain for electron microscopy of proteins. *Biochim. Biophys. Acta* 140, 180–182.

82 Miller, S. E., Howell, D. N. (**1997**). Concerted use of immunologic and ultrastructural analyses in diagnostic medicine: immunoelectron microscopy and correlative microscopy. *Immunol. Invest.* 26, 29–38.

83 Meyer, H. W., Richter, W. (**2001**). Freeze-fracture studies on lipids and membranes. *Micron* 6, 615–644.

84 Millonig, G. (**1961**). Advantages of a phosphate buffer for OsO_4 solutions in fixation. *J. Appl. Phys.* 32, 1637.

85 Misell, D. L. (**1978**). Image Analysis, Enhancement and Interpretation. In: *Practical Methods in Electron Microscopy* Vol. 7, (ed. A. M. Glauert). Elsevier, Amsterdam.

86 Misell, D. L., Brown, E. B. (**1987**). Electron Diffraction: An Introduction for Biologists. In: *Practical Methods in Electron Microscopy* Vol. 12, (ed. A. M. Glauert). Elsevier, Amsterdam.

87 Monosov, E. Z., Wenzel, T. J., Luers, G. H., Heyman, J. A., Subramani, S. (**1996**). Labeling of peroxisomes with green fluorescent protein in living *P. pastoris* cells. *J. Histochem. Cytochem.* 44, 581–589.

88 Muscatello, U., Horne, R. W. (**1968**). Effect of the tonicity of some negative-staining solutions on elementary structure of membrane-bounded systems. *J. Ultrastruct. Res.* 25, 73–83.

89 Newman, G. R., Hobot, J. A. (**1987**). Modern acrylics for post-embedding immunostaining techniques. *J. Histochem. Cytochem.* 35, 971–981.

90 Olins, A. L., Olins, D. E., Levy, H. A., Shah, M. B., Bazett-Jones, D. P. (**1993**). Electron microscope tomography of Balbiani Ring hnRNP substructure. *Chromosoma* 102, 137–144.

91 Olins, A. L., Olins, D. E., Olman, V., Levy, H. A., Bazett-Jones, D. P. (**1994**). Modeling the 3-D RNA distribution in the Balbiani ring granule. *Chromosoma* 103, 302–310.

92 Osmani, S. A., Mayer, F., Marston, F. A. O., Selmes, I. P., Scrutton, M. C. (**1984**). Pyruvate carboxylase from *Aspergillus nidulans*: Effects of regulatory modifiers on the structure of the enzyme. *Eur. J. Biochem.* 139, 509–518.

93 Paul, T. R., Beveridge, T. J. (**1993**). Ultrastructure of mycobacterial surfaces by freeze-substitution. *Int. J. Med. Microbiol. Virol. Parasitol. Infect. Dis.* 279, 450–457.

94 Qualmann, B., Kessels, M. M., Klobasa, F., Jungblut, P. W., Sierralta, W. D. (**1996**). Electron spectroscopic imaging of antigens by reaction with boronated antibodies. *J. Microsc.* 183, 69–77.

95 Read, N. D., Jeffree, C. E. (**1991**). Low temperature scanning electron microscopy in biology. *J. Microsc.* 161, 59–72.

96 Reimer, L. (**1995**). Energy-filtering transmission electron microscopy (Springer Series in Optical Sciences, Vol. 71). Springer Publ., Heidelberg.

97 Reynolds, E. S. (**1963**). The use of lead citrate at high pH as an electron-opaque stain in electron microscopy. *J. Cell Biol.* 17, 208–213.

98 Rieger, G., Müller, K., Hermann, R., Stetter, K.-O., Rachel, R. (**1997**). Cultivation of hyperthermophilic archaea in capillary tubes resulting in improved preservation of fine structures. *Arch. Microbiol.* 168, 373–379.

99 Rigaud, J.-L., Mosser, G., Lacapere, J.-J., Olofsson, A., Levy, D., Ranck, J.-L. (**1997**). Bio-Beads: an efficient strategy for two-dimensional crystallization of membrane proteins *J. Struct. Biol.* 118, 226–235.

100 Robards, A. W., Sleytr, U. B. (**1985**). Low Temperature Methods in Biological Electron Microscopy. In: *Practical Methods in Electron Microscopy* Vol. 10 (ed. A. M. Glauert). Elsevier, Amsterdam.

101 Roberts, I. M. (**1975**). Tungsten coating – a method of improving glass microtome knives for cutting ultra-thin sections. *J. Microsc.* 103, 113–119.

102 Rohde, M., Gerberding, H., Mund, T., Kohring, G. W. (**1988**). Immunelectron microscopic localization of bacterial enzymes: pre- and postembedding labeling techniques on resin- embedded samples. In: *Methods in Microbiology* Vol. 20, *Electron Microscopy in Microbiology* (F. Mayer, ed.), pp. 175–210. Academic Press, London.

103 Roos, N., Morgan, A. J. (**1990**). Cryopreparation of thin biological specimens for electron microcopy: Methods and applications (Microscopy Handbooks Vol. 21). Royal Microscopic Society and Oxford University Press, Oxford.

104 Rotenberg, M., Lichtenberg, D. (**1991**). What determines the size of phospholipid-vesicles made by detergent-removal techniques? *J. Colloid Interface Sci.* 144, 591–594.

105 Roth, J., Bendayan, M., Orci, L. (**1978**). Ultrastructural localization of intracellular antigens by the use of Protein A-gold complex. *J. Histochem. Cytochem.* 26, 1074–1081.

106 Roth, J., Bendayan, M., Orci, L. (**1980**). FITC-Protein A-Gold complex for light and electron microscopic immunocytochemistry. *J. Histochem. Cytochem.* 28, 55–57.

107 Sabatini, D. D., Bensch, K., Barrnett, R. J. (**1963**). Cytochemistry and electron microscopy: The preservation of cellular structure and enzymatic activity by aldehyde fixation. *J. Cell Biol.* 17, 19–58.

108 Sabra, M. S., Uitdehaag, J. C. M., Watts, A. (**1998**). General model for lipid-mediated two-dimensional array formation of membrane proteins: application to bacteriorhodopsin. *Biophys. J.* 75, 1180–1188.

109 Schäfer, C., Weipoltshammer, K., Almeder, M., Wachtler, F. (**1997**). Signal amplification at the ultrastructural level using biotinylated tyramides and immunogold detection. *Histochem. Cell. Biol.* 108, 313–319.

110 Schultz, P., Bischler, N., Lebeau, L. (**2001**). Two-dimensional crystallization of soluble protein complexes. *Methods Mol. Biol.* 148, 557–568.

111 Serysheva, I. I., Orlova, E. V., Chiu, W., Sherman, M. B., Hamilton, S. L., van Heel, M. (**1995**). Electron cryomicroscopy and angular reconstitution used to visualize the skeletal muscle calcium release channel. *Nature Struct. Biol.* 2, 18–24.

112 Shires, M., Goode, N. P., Crellin, D. M., Davidson, A. M. (**1990**). Immunogold-silver staining of mesangial antigen in Lowicryl K4M- and LR gold-embedded renal tissue using epipolarization microscopy. *J. Histochem. Cytochem.* 38, 287–289.

113 Shotton, D. M., Severs, N. J. (**1995**). An introduction to freeze-fracture and deep etching. In: *Rapid Freezing, Freeze, Fracture and Deep Etching* (eds. N. J. Severs, D. M. Shotton), pp. 1–30. Wiley-Liss, New York.

114 Sitte, H. (**1984**). Equipment for cryofixation, cryoultramicrotomy and cryosubstitution in bio-medical TEM routines. *Zeiss Information (Magazine Electron Microscopy)* 3, 25–31.

115 Sjöstrand, F. S. (**1990**). Common sense in electron microscopy. *J. Struct. Biol.* 103, 135–139.

116 Skaer, H. (**1982**). Chemical cryoprotection for structural studies. *J. Microsc.* 125, 137–147.

117 Sleytr, U. B., Robards, A. W. (**1977**). Plastic deformation during freeze-cleaving: a review. *J. Microsc.* 110, 1–25.

118 Sleytr, U. B., Robards, A. W. (**1981**). Understanding the artefact problem in freeze-fracture replication: a review. *J. Microsc.* 101, 187–195.

119 Sosinsky, G. E., Perkins, G. A. (**2000**). Electron crystallographic methods for investigating gap junction structure. *Methods* 20, 140–155.

120 Springer, E. L. and Roth, I. L. (**1972**). Scanning electron microscopy of bacterial colonies. I. *Diplococcus pneumoniae* and *Streptococcus pyogenes. Can. J. Microbiol.* 18, 219–223.

121 Spurr, A. R. (1969). A low-viscosity epoxy resin embedding medium for electron microscopy. *J. Ultrastruct. Res.* 26, 31–43.

122 Stoylova, S. S., Ford, R. C., Holzenburg, A. (**1999**). Cryo-electron crystallography of small and mosaic 2-D crystals: An assessment of a procedure for high-resolution retrieval. *Ultramicroscopy* 77, 113–128.

123 Stuart, M. (**1991**). In: *Electron Microscopy in Biology – A practical approach* (ed. R. Harris), pp. 229–242. IRL Press, Oxford.

124 Studer, D., Michel, M., Müller, M. (**1989**). High pressure freezing comes of age. *Scanning Microsc. Suppl.* 3, 253–268.

125 Takizawa, T., Robinson, J. M. (**2000**). Freeze-fracture cytochemistry: a new fracture-labeling method for topological analysis of biomembrane molecules. *Histol. Histopathol.* 15, 515–522.

126 Tauschel, H.-D. (**1988**). Localization of bacterial enzymes by electron microscopic cytochemistry as demonstrated for the polar organelle. In: *Methods in Microbiology* Vol. 20, *Electron Microscopy in Microbiology* (ed. F. Mayer), pp. 237–259. Academic Press, London.

127 Timmins, P., Hauk, J., Wacker, T., Welte, W. (**1991**). The influence of heptane-1,2,3-triol on the size and shape of LDAO micelles. Implications for the crystallisation of membrane proteins. *FEBS Lett.* 280, 115-120.

128 Tokuyasu, K. T. (**1986**). A technique for ultramicrotomy of cell suspensions and tissues. *J. Cell Biol.* 57, 551–565.

129 Torrisi, M. R., Mancini, P. (**1996**). Freeze-fracture immunogold labeling. *Histochem. Cell Biol.* 106, 19–30.

130 Tvedt, K. E., Kopstadt, G., Haugen, O. A. (**1984**). A section press and low elemental support for enhanced preparation of freeze-dried cryosections. *J. Microsc.* 133, 285–290.

131 Umrath, W. (**1974**). Cooling bath for rapid freezing in electron microscopy. *J. Microsc.* 101, 103–105.

132 Uzgiris, E., Kornberg, R. (**1983**). Two-dimensional crystallization technique for imaging macromolecules, with application to antigen-antibody-complement complexes. *Nature* 301, 125–129.

133 Valentine, R. C., Shapiro, B. M., Stadtman, E. R. (**1968**). Regulation of glutamine synthetase. XII. Electron microscopy of the enzyme from *Escherichia coli. Biochemistry* 7, 2143–2152.

134 Van Bruggen, E. F. J., Wiebinga, E. H., Gruber, M. (**1960**). Negative-staining electron microscopy of proteins at pH values below their isoelectric points. Its application to hemocyanin. *Biochim. Biophys. Acta* 42, 171–172.

135 Van Bruggen, E. F. J., Wiebinga, E. H., Gruber, M. (**1962**). Structure and properties of hemocyanins. I. Electron micrographs of hemocyanin and apohemocyanin from *Helix pomatia* at different pH-values. *J. Mol. Biol.* 4, 1–7.

136 Van Harrefeld, A., Crowell, J. (**1964**). Electron microscopy after rapid freezing on a metal surface and substitution fixation. *Anat. Rec.* 149, 381–385.

137 Van Harrefeld, A., Trubatch J., Steiner J. (**1974**). Rapid freezing and electron microscopy for the arrest of physiological processes. *J. Microsc.* 100, 189–198.

138 van Heel, M., Harauz, G., Orlova, E. V., Schmidt, R., Schatz, M. (**1996**). A new generation of the IMAGIC image processing system. *J. Struct. Biol.* 116, 17–24.

139 van Heel, M, Keegstra, W. (**1981**). IMAGIC: A fast, flexible and friendly image analysis software system. *Ultramicroscopy* 7, 113–130.

140 van Heel, M. (**1987**). Angular reconstitution: a posteriori assignment of projection directions for 3D reconstitution. *Ultramicroscopy* 21, 111–124.

141 Varga, A. R., Staehelin L. A. (**1983**). Spatial differentiation in photosynthetic and non-photosynthetic membranes of *Rhodopseudomonas palustris. J. Bacteriol.* 154, 1414–1430.

142 Venable, J. H., Coggeshall, R. (**1965**). A simplified lead citrate stain for use in electron microscopy. *J. Cell. Biol.* 25, 407–408.

143 Vénien-Bryan, C., Balavoine, F., Toussaint, B., Mioskowski, C., Hewat, E. A., Helme, B., Vignais, P. M. (**1997**). Structural study of the response regulator HupR from *Rhodobacter capsulatus.* Electron microscopy of two-dimensional crystals on a nickel-chelating lipid. *J. Mol. Biol.* 274, 687–692.

144 Vénien-Bryan, C., Lenne, P.-F., Zakri, C., Renault, A., Brisson, A., Legrand, J.-F., Berge, B. (**1998**). Characterization of the growth of 2D protein crystals on a lipid monolayer by ellipsometry and rigidity measurements coupled to electron microscopy. *Biophys. J.* 74, 2649–2657.

145 Vogt, B., Berker, R., Mayer, F. (**1995**). Improved contrast by a simplified post-staining procedure for ultra-thin sections of resin-embedded bacterial cells: Application of Ruthenium Red. *J. Basic Microbiol.* 35, 349–355.

146 Watson, L. P., McKee, A. E., Merrell, B. R. (**1980**). Preparation of microbial specimens for electron microscopy. In: *Scanning Electron Microscopy* Vol. **1980**/II, (ed. E. D. Johari), pp. 45–56. Scanning Electron Microscopy Inc., Chicago.

147 Welte, W., Wacker, T. (**1991**). Protein-detergent micellar solutions for the crystallization of membrane proteins: Some general approaches and experiences with the crystallization of pigment-protein complexes from purple bacteria. In: *Crystallization of Membrane Proteins* (ed. H. Michel), pp. 107-123. CRC Press Inc., Boca Raton, FL.

148 Wohlrab, F., Gossrau, R. (**1992**). Katalytische Enzymhistochemie. G. Fischer Verlag, Jena.

149 Zhuang, J., Privé, G. G., Werner, G. E., Ringler, P., Kaback, H. R., Engel, A. (**1999**). Two-dimensional crystallization of *Escherichia coli* lactose permease. *J. Struct. Biol.* 125, 63–75.

3
Other Microscopical Methods and their Application Potential

This chapter presents the basic principles of microscopic instrumentation and techniques that are not routinely used for biotechnological applications, but which are, on the other hand, very versatile tools in more specialized fields. Ordinary laboratories are not usually equipped with these instruments for a multitude of reasons. Their operation and maintenance may be too difficult, their application range may be limited to a small number of object types, or the instrumentation may be too expensive. Some techniques are under development, or still await recognition as readily applicable alternatives to established techniques. With some examples, the application range is still limited by a lack of appropriate preparation techniques. The equipment – and the expertise – is readily available at larger research facilities and may be used in collaborations or offered as a service.

3.1
Total Internal Reflection Microscopy

Total internal reflection (fluorescence) microscopy (TIR[F]M) has been applied in biophysical research for years. It is a highly versatile alternative to laser scanning microscopy, when exclusively one very thin plane at an optical interface (particles adhering to a glass slide, for example) has to be excited in a fluorescence experiment. It is very useful for the study of phenomena occurring at interfaces, especially membranes, down to the level of single molecules. In the past, TIRFM was difficult to perform, due to the complexity of the microscope setup and problems in achieving acceptable image brightness. The technique profited from the development of laser light sources (for other microscopes), new fluorescent dyes, and more sensitive image recording techniques (Tokunaga and Yanagida, 1997; Axelrod, 2001)

Total internal reflection is an optical phenomenon that occurs at boundaries of media with different refractive indices (optical interfaces). When light strikes the interface between these media at an angle greater than the critical angle it is totally reflected. The electromagnetic field of the incoming/reflected light, however, builds up a surface wave field at the optical interface that still extends into the z direction (that is, into the medium of lower refractive index; Fig. 3.1). The penetration depth of the exponen-

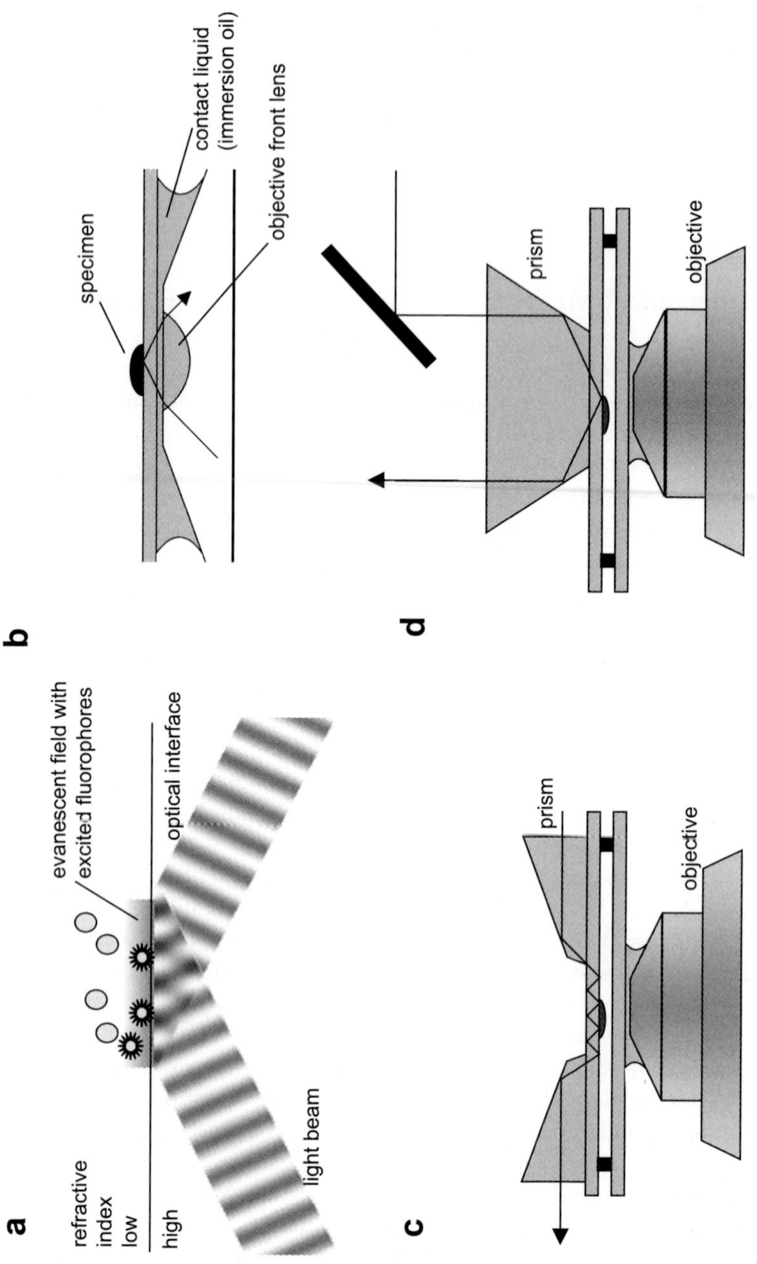

Fig. 3.1 Total internal reflection fluorescence microscopy.
(a) Schematic representation of evanescent wave generation upon reflection of a light wave at an optical interface. Excitation of fluorophores is restricted to a narrow plane near the surface.
(b–d) Principles of several TIRFM setups (only specimen stage and objective lens of inverted microscopes are depicted). (b) illumination through the objective lens (c, d). The beam from an external source is directed by prisms to the specimen

tially decaying field (also called the evanescent field) depends on the incident angle and on the wavelength of the light, as well as the refractive indices of the media. It usually ranges between 30 and 300 nm into the medium of lower refractive index. Within this region, the evanescent field excites fluorophores. Conversely, fluorophores in the bulk solution are not excited. This is the key advantage of TIRFM – only fluorophore molecules very close to the surface of a specimen are induced to emit, creating an extremely thin optical section. Outside the evanescent field, fluorescence is minimal. This makes the technique highly versatile for study of the binding, motion, aggregation, etc. of macromolecules at surfaces and cell membranes adhering to the interface between the two media.

In principle it is possible to use conventional illumination sources for TIRFM. Because of the low intensities of the evanescent field, though, laser light is widely preferred as a source of sufficient light intensities. High intensity laser sources are required for techniques as fluorescence recovery after photobleaching (FRAP; see below).

Several TIRFM setups have been developed; those commonly available are the types with illumination from below the specimen (the "condenser side") or through the objective lens. Two variations are shown in Fig. 3.1 b–d. A conventional microscope will often be upgradeable with a prism and a laser below the specimen plane (in upright microscopes). A fixed specimen stage is suitable to maintain the alignment of the laser beam. Maintaining a focused laser illumination source within a range of a few microns over long periods is difficult, however, so the microscope, laser source, and supporting optical components are secured in proper mounts and firmly fastened to the bench or microscope stage. The specimen chamber consists of a slide with the adhered specimen on the side facing the objective and the prism in optical contact with the reverse side. The optical contact is provided by a thin layer of immersion oil. The coverslip is usually supported by a Teflon ring, which provides a defined distance between coverslip and slide. This specimen chamber may also be constructed as a flow-through cell. When properly aligned, the collimated and focused laser enters the prism and is refracted in such a way that it strikes the interface between the slide and the specimen chamber at an incident angle exceeding the critical minimum. Several configurations with different prism setups are available.

For better manipulation (especially for applications in cell biology), illumination through a high-numerical aperture objective of an inverted microscope is more practical. The specimen must be positioned on the bottom surface of the cover glass or specimen chamber. The setup permits the use of other instrumentation aids such as micromanipulators.

TIRFM is mostly performed in conjunction with other imaging techniques. All instrumental setups also allow conventional bright-field illumination of fluorescence. Devices that oscillate rapidly between TIRFM and conventional fluorescence, resulting in practically simultaneous, but separate, recording of the two imaging modes, are also possible.

3.2
Multiphoton Excitation Microscopy

At high photon densities, a fluorophore will often be hit by two photons simultaneously, and these can combine their energies to excite the target molecule. For example, two photons of 640 nm wavelength (red light) can combine to excite an ultraviolet-absorbing fluorophore in the 320 nm (ultraviolet) region, which will result in fluorescence emission of longer (blue or green) wavelengths.

For technical application of this effect in microscopy, photon densities must be a million times higher than those used for excitation by a single photon. Excitation therefore only occurs in a very small volume of a tightly focused light spot (Fig. 3.2 a). The high photon densities are achieved with the aid of (expensive) high-power mode-locked pulsed lasers focused by the microscope objective lens. The lasers generate a significant amount of power during pulse peaks, but have an average power low enough not to damage the specimen. This minimization of the average excitation power level reduces the amount of single-photon absorption, which is important for reducing heating and some of the photodamage during fluorescence experiments. Pulsed laser beam intensity drops as the square of the distance from the focal

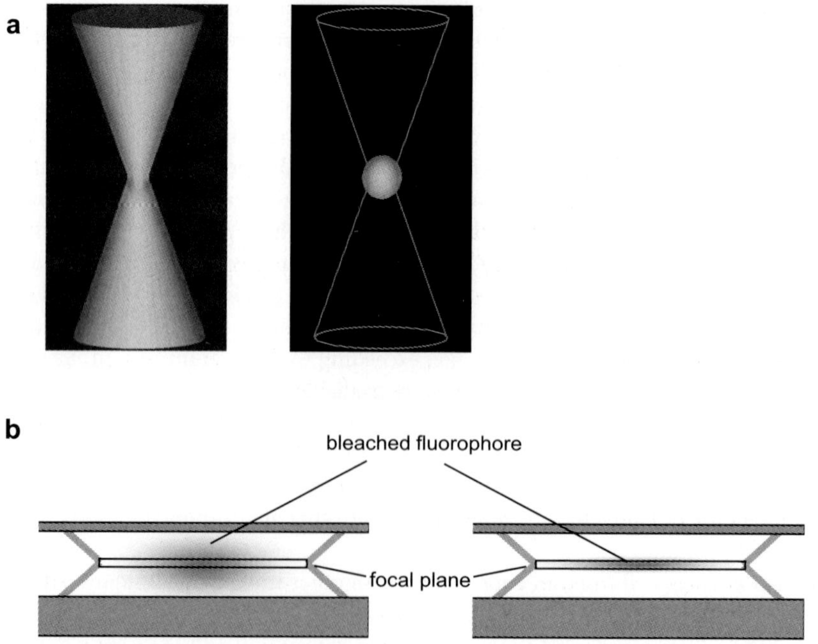

Fig. 3.2 Fluorescent light cones and bleaching patterns in two-photon microscopy.
(a) Schematic drawing of light cones in confocal and two-photon microscopy. In confocal microscopes, fluorophores are excited throughout the light cone as well as in the focal spot (left, light gray shade). In two-photon microscopes, only the focal spot is excited (light gray spot).
(b) Respective bleaching patterns in a schematic side view in confocal and two-photon microscopes. Darker gray represents stronger bleaching, which in two-photon microscopes is restricted to the focus

plane, so the excitation probability of a fluorophore decreases as the fourth power of its distance from the focal plane. Multiphoton microscopy thus generates fluorescence excitation (and subsequently, fluorescence emission) only at the focal spot. In confocal microscopes, fluorescence is excited in a comparably large volume centered at the focal spot in the specimen. Out-of-focus background fluorescence is filtered out by the confocal pinhole. In multiphoton microscopy the background signal is avoided altogether, also making the confocal pinhole unnecessary. As in confocal microscopy, raster scanning in two or three dimensions is applied to generate a two- or three-dimensional final image (So et al., 2000; Bryers, 2001; Gratton et al., 2001).

Two-photon microscopy does not improve optical resolution. In fact, resolution is about 30 % lower than in confocal microscopes. In numerous experiments, however, it is useful to avoid the excitation of fluorophores outside the focal spot.

For some experiments, excitation by three simultaneously absorbed photons is useful. For this type of excitation, the required photon concentration is only ten times greater than that needed for two-photon absorption. Compared with two-photon excitation, the higher proton densities are restricted to an even smaller volume in the center of the focal spot. As an example, a laser emitting infrared light at 1050 nm may excite a fluorophore that absorbs in the ultraviolet region (approximately 350 nm, one third of the excitation wavelength). The same laser can simultaneously excite another fluorophore at half the wavelength (525 nm), a useful combination in dual-labeled biological experiments. By utilizing shorter near-infrared wavelengths (down to 720 nm), three-photon fluorescence can extend the useful fluorescence imaging range into the deep ultraviolet.

Since photobleaching and photodamage are minimized in multiphoton excitation, and occur only in the immediate region surrounding the focal volume (see Fig. 3.2 b) multiphoton fluorescence microscopy is becoming a method of choice for dynamic imaging of living cells and tissues. The technique is also useful in specimens that require deeper penetration into the tissue. This is due to the use of longer excitation wavelengths and the fact that less light is absorbed above the focal plane. At present, due to the expense of appropriate lasers, confocal microscopes are still more attractive for most applications. The development of cheaper laser sources and a wider selection of appropriate fluorescence markers is to be expected in the next few years.

3.3 Analytical Features of Fluorescence Microscopes

In close conjunction with fluorescence microscopes or LSM, several analytical methods geared towards quantification of the fluorescence signal in time and space have been developed. Spectroscopy at (sub-) microscopic levels now allows determination of molecular interactions in vivo, in combination with a bright-field or fluorescent image (conventional or confocal) of the specimen. The techniques have already been briefly mentioned above (see Table 1.21; Hovius et al., 2000), and are described here in more detail. Specimen preparation is often quite similar to that used for standard fluorescence procedures. However, instrumentation and evaluation of raw data require at least instrumental upgrades on standard fluorescent microscopes as well as computer hard- and software for data evaluation. Applications of the related FRAP and FLIP are discussed in White and Stelzer, 1999, and Bastiaens and Pepperkok, 2000.

3.3.1 Fluorescence Recovery After Photobleaching (FRAP)

Fluorescence recovery after photobleaching is the recovery of fluorescence in a defined region of a specimen after a bleaching event. FRAP is the result of the movement of unbleached fluorophores from the surrounding zone into the bleached area. FRAP is used to measure the dynamics of two- or three-dimensional molecular mobility, such as diffusion, transport, or any other kind of movement of fluorescently labeled molecules in membranes or in living cells (Fig. 3.3 a). In a FRAP experiment, an area – such as a fluorescently labeled cell surface – is imaged with low laser intensity. An excitation light pulse of high intensity is then used to strongly bleach a defined region. Finally, the time course of recovery in the bleached region is monitored with a dim excitation laser beam As a result, FRAP indicates any kind of movement (passive, such as diffusion, or active, such as transport) of fluorescent molecules. The recovery time (half-recovery time) indicates the speed of this mobility (e.g., diffusion time).

3.3.2 Fluorescence Loss in Photobleaching (FLIP)

FLIP is the decrease/disappearance of fluorescence in a defined region adjacent to a repetitively bleached region. Like FRAP, FLIP is used to measure the dynamics of molecular mobility in membranes or in living cells.

3.3.3 Fluorescence Resonance Energy Transfer (FRET)

FRET (see, e.g., Clegg, 1995; Truong and Ikura, 2001) occurs when energy from an excited donor fluorophore is transferred over short distances to an acceptor without emission of a photon (see Fig. 3.4.). Selected pairs of fluorophores – GFP/rhodamine is one – are suitable for FRET experiments since, in addition to other prerequisites (such as dipole orientation, sufficient fluorescence lifetime), the donor emission spectrum has to overlap the excitation spectrum of the acceptor. In this way it is possible to measure changes in the relative distance between, or orientation of, two labeled proteins, or also two domains in a single polypeptide chain. FRET detection is also possible by examination of quenching of the donor emission by acceptor photobleaching.

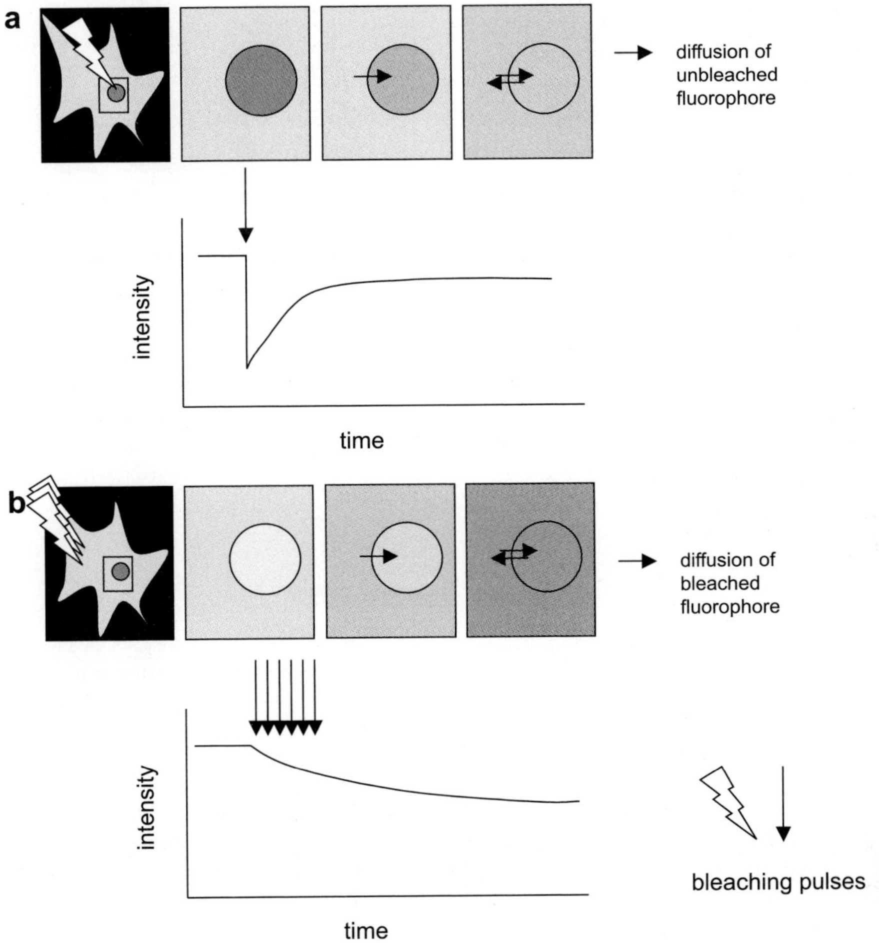

Fig. 3.3 FRAP and FLIP.
(a) Fluorescence recovery after photobleaching due to diffusion of fluorescent molecules into a bleached spot. In the schematic drawing of the object, the encircled region is bleached. The curve depicts the respective fluorescence intensity in the bleached spot.
(b) The related fluorescence loss in photobleaching due to diffusion of bleached molecules into a defined spot. Repetitive bleaching pulses are applied adjacent to the measured spot

3.4
Fluorescence Correlation Spectroscopy

The fluctuation of a fluorescence signal over time, as opposed to the averaged signal, carries important additional information relating to intermolecular interactions. The principle behind this method, fluorescence correlation spectroscopy, has been improved over the years and now allows measurement of molecular interactions in small (femtoliter) volumes. In a nanomolar solution, this small volume (one cubic micro-

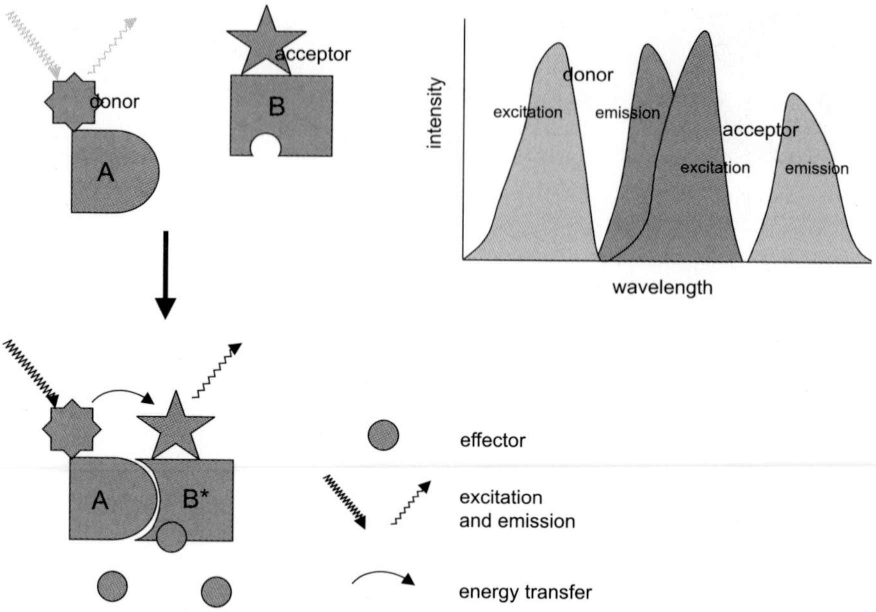

Fig. 3.4 Fluorescence resonance energy transfer between closely adjacent fluorophore pairs. In a typical experiment, binding between two molecules based upon effector-induced conformational change of one molecule (B → B*) is measured. The overlap of donor emission and acceptor excitation spectrum is a prerequisite for signal emission

meter) would contain, on average, 0.6 molecules. The actual number of molecules in the volume at any one time is determined by diffusion processes of the molecule. Fluorescence fluctuations in the small volume may thus be correlated with these diffusion properties.

By detection of fluorescence fluctuations, the interactions between two molecules – such as a fluorescence-tagged ligand and a receptor as target molecule – become examinable. Numerous fluorescent dyes are also suitable for FCS, including several Alexa fluor dyes, Cy3, and Texas Red. For labeling of molecules, the dyes may be covalently coupled through reactive groups such as isothiocyanates or carbodiimide.

The instrumentation is available in the form of the ConfoCor product series, manufactured by Zeiss. The essential parts of the optical unit comprise a laser light source, a dichroic mirror to split excitation and fluorescence light, and a photon detection unit. The laser light is focused by the objective into a cavity 3 μm in diameter. Fluorescent probes in the cavity are excited by the laser light and emit fluorescent light, which is captured by a microscope objective. Filters behind the objective separate excitation and fluorescence light. The light then passes through a variable pinhole, which blocks out photons from outside the illuminated spot and is directed to a single-photon-counting photodiode. Movement of fluorescent probes through the cavity causes fluctuations in fluorescence (see Fig. 3.5), from which the average residence time, average number of molecules, and the diffusion constant can be calculated after autocorrelation of the

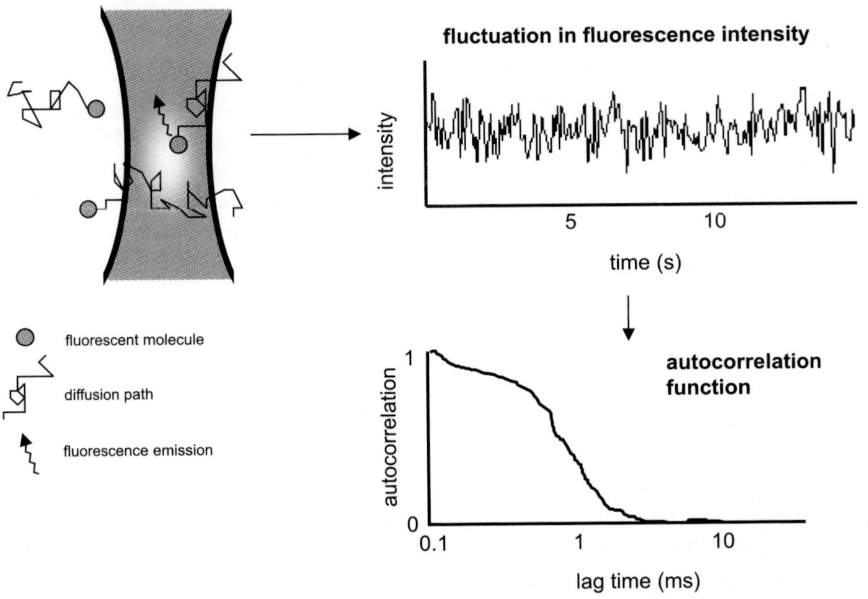

Fig. 3.5 Principle of fluorescence correlation spectroscopy (one-channel system) for measurement of particle diffusion parameters. Fluorescent molecules move in and out the detection volume. The autocorrelated function of the fluctuation signal decays from its initial value with a time dependence determined by molecular diffusion rates (faster moving molecules exhibit faster autocorrelation decay)

signal. Different diffusion characteristics, such as diffusion in a volume or membrane, may be distinguished. All changes in the diffusion coefficient (i.e., large changes in the hydrodynamic radius of a molecule, caused by conformational changes) are detectable. In this way it is also possible to determine the kinetics of reactions in which fluorescence characteristics change as a function of the reaction species. If a second pathway for emission light is available (two-channel system), two different fluorescent species may be recorded simultaneously, by cross-correlation of the fluctuating fluorescence signals from two channels originating from two target-bound fluorophores with different emission spectra. In this way, all reactions leading either to either separation or to association of the two fluorophores can be monitored. Measurement of kinetics with higher sensitivity than achievable by the autocorrelation approach is possible. Medina and Schwille (2002), and Hess et al. (2002) review recent applications.

3.5
Soft X-ray Microscopy

For a long time, the most logical approach to improvement of the resolution of microscopes appeared to be the development of microscopes operating with electromagnetic waves of shorter wavelength than visible light. However, the use of (electromagnetic) rays for imaging requires a way to refract or focus the ray with appropriate lenses. This is most easily achieved with glass lenses and visible light, but for other electromagnetic waves focusing may not be a trivial task.

Low-energy or "soft" X-rays have wavelengths of 1 – 10 nm, corresponding to photon energies between 100 and 1000 eV. Microscopy with soft X-rays provides resolutions between those of visible light and of electron microscopy with hydrated specimens several μm thick. The first big advantage over electron microscopy is that the specimen can be visualized under ambient conditions, so dehydration, embedding, drying, or freezing – unavoidable in electron microscopy – may not be necessary. However, only rigid specimens can withstand the photon exposure doses required to provide higher resolutions. Chemical fixation, cryotechniques, or "low-dose" exposure techniques therefore have to be applied to obtain images of specimens in a (near) living state. The microscopes operate in the "water window" spectral region between the K shell X-ray absorption edges of carbon (290 eV) and oxygen (540 eV), in which organic matter strongly absorbs X-rays, while water is relatively non-absorbing. Focusing of soft X-rays is impossible with conventional lenses. Concave lenses would be necessary (and are, in fact, now used for hard X-rays; see below) since the refractive index of X-rays in matter is less than 1. For soft X-rays, these lenses would have a very low numerical aperture, resulting in resolutions of only several μm. Higher aperture lenses would absorb the X-rays. Focusing of soft X-rays is therefore brought about by zone plates with fine diffraction gratings. The plates consist of series of concentric rings that become narrower at larger radii, the spatial resolution being determined by the width of the finest, outermost zones (100 nm down to 30 nm). A typical zone plate (60 μm in diameter) has a focal length of about 500 μm. The numerical apertures of zone plates optics are still low, but this also results in a large depth of field.

Two soft X-ray microscope combinations are commonly used (Fig. 3.6). Transmission X-ray microscopes have the same components as conventional optical and electron microscopes, with zone plates used as condensers and objectives. In scanning transmission X-ray microscopes, the zone plates are used as objectives to form focused spots. Scanning transmission X-ray microscopes in particular use electron synchrotrons as radiation sources, providing beams up to 10^{10} times brighter than those from conventional X-ray sources. Soft X-ray microscopy is therefore only possible in selected research facilities. The Göttingen X-ray microscope, for instance, is located at the BESSY storage ring in Berlin and operates with rays of 2.4 nm wavelength, with possible theoretical resolution down to 25 nm. In future, smaller instruments with alternative powerful X-ray sources may be made commercially available, but for the near future, X-ray microscopy remains a technique that requires very considerable instrumentation.

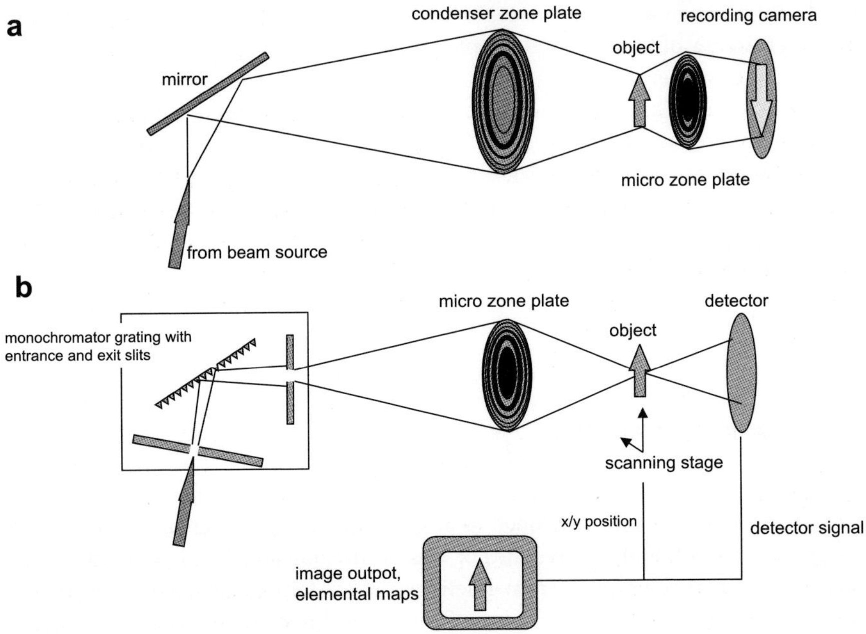

Fig. 3.6 Instrumental setups of X-ray microscopes
(a) Transmission instrument with two zone plates as condenser and objective (micro zone plate).
(b) Scanning instrument. For sensitive elemental mapping, a monochromator filters X-rays of defined wavelengths. A micro zone plate focuses the beam, which is scanned over the object. Only one detector for transmitted signals is depicted. Other detectors may record back-scattered electrons or luminescence

Soft X-ray microscopy is usually combined with spectromicroscopic applications and trace element mapping. When the energy of an X-ray is increased to or above the binding energy of an inner shell electron of a specimen atom, the electron can be completely removed from the atom and the absorption shows a step-like increase. Near this "edge" the electron can be promoted to weakly bound states that give rise to X-ray absorption near-edge structure or XANES resonances, which depend on the chemical binding state of the atom. The C, N, O, Ca, and K edges are particularly readily accessible to high-resolution soft X-ray microscopes. At the carbon edge, different amino acids have distinct spectroscopic fingerprints.

Specific labeling of macromolecular structures is also possible. This may be performed either with luminescent labels (e.g., phosphorus grains) that emit visible light under X-ray irradiation, or with antibody-conjugated gold labels. These are especially easy to visualize without the background of the specimen in the dark-field mode of the microscope and may be combined with a bright-field image of the specimen.

A different lens type for refraction of hard X-rays has recently been constructed. In order to obtain sufficient refraction, a refractive lens for hard X-rays, fabricated from Be or Al, consists of 20 to 300 single lenses arranged in a row along the optical axis. At present, with aluminum lenses, a resolution of 300 nm is achievable. The advantage of this type of X-ray microscope is the penetration of thick, strongly light-absorbing objects. In conjunction with spectroscopic methods, such as the measurement of fluorescence emission of elements, element-specific microtomograpy of relatively thick objects may be achieved.

X-ray microscopes are not yet available commercially, being at the stages of advanced (soft X-ray microscopes) or early (hard X-ray microscopes) development. The instruments are currently invaluable devices for high-resolution elemental mapping. In future, they should offer interesting application potential in the resolution range between light and electron microscopes, for real-time imaging of fully hydrated and living objects (see Jacobson, 1999; Osanna et al., 1996; Abraham-Peskir, 2000; Maser et al., 2000).

3.6 Microtomographic Techniques

Improvements in radiation sources and detectors have allowed the construction of instruments not originally developed as microscopes in the strict sense, but which now provide resolution down to scales of tens of micrometers. X-ray computer tomographs, and in particular (nuclear) magnetic resonance tomographs, have proven to be very versatile instruments for imaging of intact specimens down to micro-scales.

Thanks to the unique absorption characteristics of X-rays in soft tissue, the technique is most widely applied for visualization of features in the interiors of opaque solid objects. In comparison with microscopic specimens, the objects may be large – in the range of decimeters rather than microns. X-ray computer tomographs (CTs) are widely used in medical diagnostics. In clinical CT scanners, spatial resolutions of $1\,mm^3$ volume elements are common. The development of microcomputer tomographs with appropriate X-ray tubes, megapixel CCD detectors, and resolutions lower than $100\,\mu m^3$ has been achieved. For high-resolution tomography of extremely dense objects (such as geological specimens), an intense beam produced in a storage ring (as in soft X-ray microscopy) may be necessary. For most micro-CT applications with biological specimens, though, a microfocus X-ray tube is used. Here, focusing of the X-ray is performed with the aid of a non-refracting "lens" consisting of hollow capillaries (polycapillary X-ray optics) as waveguides (Fig. 3.7).

When an X-ray strikes the reflecting surface of a capillary at an angle smaller than the critical angle, the X-ray is externally reflected. A large array of these capillaries is used to collect X-rays over a large angle from a conventional X-ray source and focus them in a spot. The typical micro-CT scanner produces a focal spot of approximately $10\,\mu m$ in diameter. From there, the beam is reflected onto a (rotating) specimen stage. As a cone beam of X-rays is produced, magnification is achieved by moving the object towards the spot. After passing through the specimen the beams are directed through a fiber optic lens towards a CCD array. As in all tomographic techniques, X-ray projection views are acquired at multiple angular positions around the object. A computer is used to steer the stepping motors and to synchronize rotation, axial shift, and data recording. An especially laborious task is the three-dimensional image reconstruction from several hundreds of single images taken over an angle of $180\,^{\circ}$ and processed in a

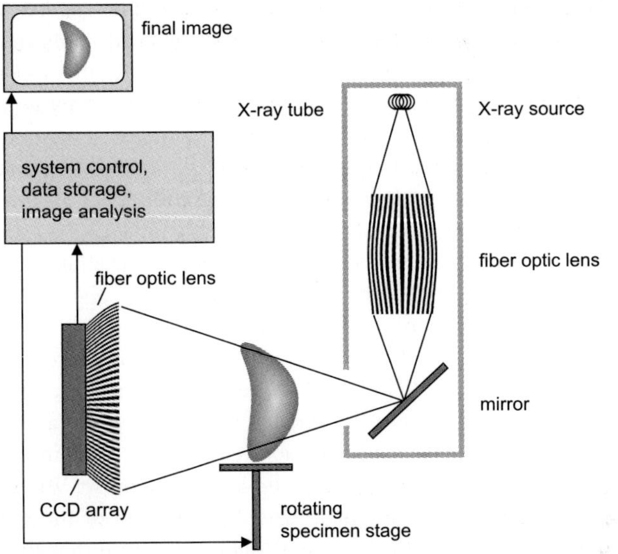

Fig. 3.7 Micro-CT scanner with non -refractive hollow fiber lenses as waveguides. See text for detailed description

back-projection algorithm. The number of volume elements (voxels) to be handled by the micro-CT computer typically corresponds to several gigabytes, requiring comparably large amounts of RAM and fast processors. For non-living specimens, a static microfocus X-ray tube and a rotating specimen stage are used. When optimized for maximum resolution, 15 – 50 μm resolution may be achieved, which is often sufficient for the determination of histological features. The technique is widely applicable for rigid, otherwise non-transparent materials such as bone, insect bodies, implants, other small technical devices, etc. Living animals can be inspected by micro-CT with rotating X-ray stages. Since only a reduced X-ray dose is tolerable for living organisms, the resolution limit is typically at 50 – 100 μm. See, for example, Hörnschemeier et al. (2002), in which an insect thorax is depicted with its internal histological features; see also Holdsworth and Thornton (2002), and Bentley et al. (2002) for other applications to small animals.

"Microscopic" resolution in the field of magnetic resonance imaging (MRI) has also been achieved (magnetic resonance microscopy; MRM). The magnetic resonance signal is mainly a product of spatial variations in water concentration in the specimen, due to the NMR signal of the hydrogen nuclei (protons). Other factors (other chemical components, temperature, magnetic susceptibility), though, also contribute to signal generation.

MRI is based upon nuclear magnetic resonance. As in conventional NMR, the specimens are placed in a magnetic field and then subjected to a radio frequency (RF) pulse. Since the positively charged nuclei spin on their axes, the (usually hydrogen) nuclei each have a miniature magnetic field associated with a magnetic moment. In an external magnetic field the otherwise randomly orientated nuclei align along the field lines (z-axis) in two possible orientations: nuclei at a lower energy level point in the "southward" direction of the field, those at the upper energy level in the opposite direction. At room temperature, the number of spins making up the lower energy level slightly outnumbers the number making up the upper level. As the temperature increases, this ratio approaches 1. The magnetic moments of the spinning protons precess around the direction of the external magnetic field. The frequency of precession (Larmor frequency) depends on a fixed term for the nucleus under investigation (the gyro-magnetic ratio) and the strength of the external magnetic field. When the sample is exposed to a radio frequency pulse at the same frequency as the precessing protons, "resonance" will occur, the protons taking up the energy and altering their orientation. The change in energy levels gives a net magnetization vector at 90° to the direction of the magnetic field. It takes some tens to some hundreds of milliseconds until the net magnetization vector returns to the state of being parallel with the magnetic field (spin-spin relaxation time, T2). The radio frequency pulse also increases the angle of precession of the nuclei, which affects the strength (but not the direction) of the net magnetization vector, since the spin axes of the nuclei now deviate more from the z-axis than they did before the pulse. When, over time, the precession angle returns to its original value, the strength of the net magnetization vector increases, returning to its normal value. The time that this takes is termed the spin lattice relaxation time (T1), and is usually from hundreds of milliseconds to seconds Since hydrogen atoms (the nuclei predominantly used for NMR and MRI) have different T1 and T2 relaxation properties in various tissues, it is possible to differentiate between tissue types in the body. Fat versus water content contributes most significantly to signal variation.

To obtain a tomographic image from a living animal or a specimen, the resonance signals are collected and processed in a different way than in CT. To "slice" the object, gradients of magnetic fields within the external magnetic field are created (see Fig 3.8). Upon pulsing with RF of a single frequency, only a defined "slice" (in the volume) of those nuclei, those situated in the appropriate field strength, will resonate. By variation of the slope of the magnetic field gradient, the thickness of the slice may also be varied. One slice now has to be resolved in two dimensions: the first dimension of the image is extracted with the aid of a "phase-encoding" gradient. Immediately after the RF pulse, all nuclei in the excited slice are "in phase" (i.e., their magnetic vectors all point in the same direction). When a phase-encoding gradient is applied at an angle of 90° to the frequency-encoding gradient and is then turned off, the frequency of precession remains constant, but the phases of the nuclei differ from each other. They are now "marked" in the first dimension. To resolve the second dimension, the phase-encoding gradient has to be applied in increments of increasing gradient strengths. The final image is recombined from this raw data by Fourier analysis.

Although "slicing" by electromagnetic gradients may be performed very precisely, voxels that are too small do not provide sufficient signal strength (are not resolvable). Resolution is further limited by background noise, magnetic susceptibility, motion of or inside the object, and diffusion. In the development process from the low-resolution (3 mm^3 voxel) instruments used for clinical applications to MRM, radio frequency coils with enhanced sensitivity, stronger magnetic fields, and improved systems for data handling and storage have been used. Unlike those instruments used for clinical applications, however, the high-resolution instruments are applied for small animals (insects, mice), plant seeds, etc.

It is a special feature of MRI and MRM that functional imaginary in living organisms and in dynamic systems (complex flow cells, etc.) is possible. Examples of applications include the imaging of lung function, fluid diffusion, and organ development. Paramagnetic compounds such as the rare earth element gadolinium may be used as additional stains. In principle, specific labelling techniques with gadolinium as a marker system are also possible. Though signal intensity is lower than with other techniques, gene expression studies may also be performed with MRM. The technique has great potential for application, especially in biomedical research, because it complements and often replaces (or will replace) conventional histological work. The low resolution relative to the light microscope notwithstanding, its biggest advantage is its ability to handle living systems, allowing processes to be observed in one and the same living organism without the need to sacrifice and prepare an experimental animal or to un-

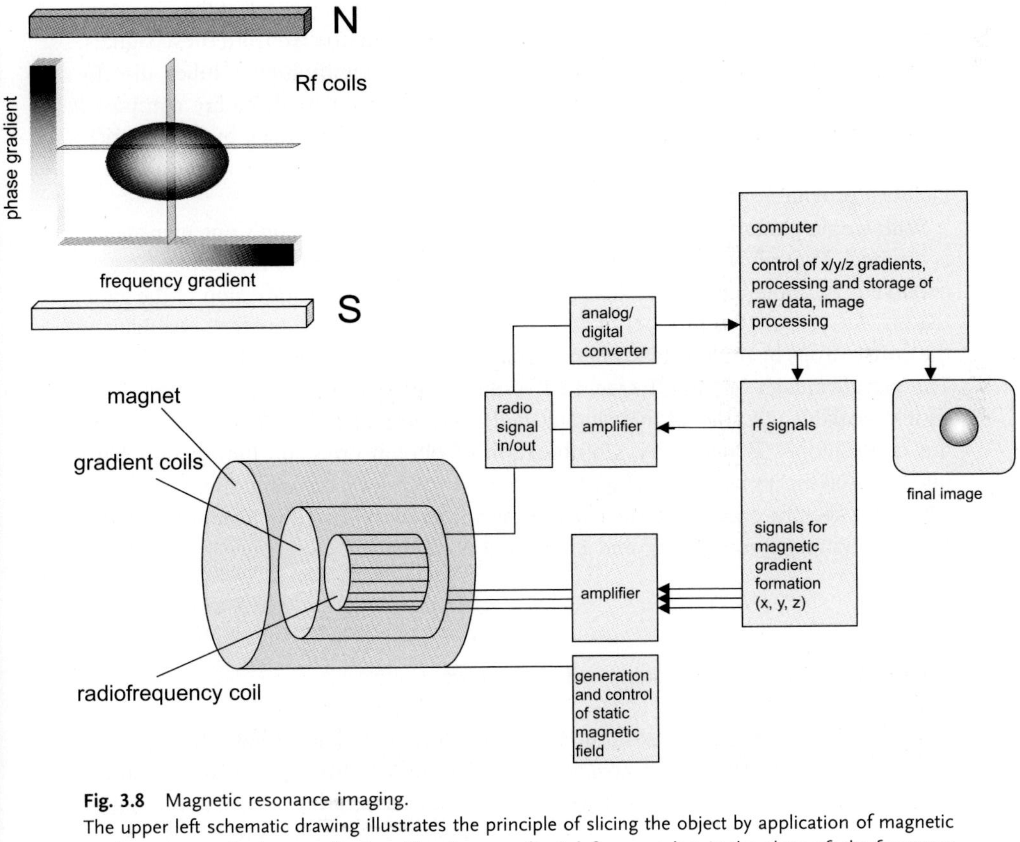

Fig. 3.8 Magnetic resonance imaging.
The upper left schematic drawing illustrates the principle of slicing the object by application of magnetic gradients perpendicular to each other. The phase gradient defines one line in the plane oft the frequency gradient slice. See text for detailed description.
Setup of the MRI or MRM scanner, representative of the clinical instrument as well as for the "desktop" instruments used for small objects

dertake a small biopsy. In particular, histological work in three dimensions – serial sectioning of a whole organ or an organism, for example – is a very laborious task by conventional methods, but may easily be performed with MRM. Potential applications are illustrated by Appel and Bauman (2002), Bulte et al. (2002), and Burg et al. (2002). Though instrumentation for low resolution (providing a magnetic field strength of around 1.5 Tesla) is readily available, high-resolution facilities (7 Tesla and higher) are often installed in larger research units.

Improved PET (positron emission tomography) imaging is also sometimes termed "microscopic", though it has yet to reach the resolutions of CT and MRM. The PET imaging process involves radiolabeling a molecule with a positron-emitting nuclide (^{18}F, ^{11}C) marker produced by a cyclotron. Generally, these isotopes are short-lived, which minimizes radiation exposure to the organism under investigation but also requires access to a cyclotron on site. When a positron meets an electron, the collision produces two gamma rays of equal energy, which are emitted in opposite directions. From the trajectories of those gamma rays, the line along which the positron was emitted can be defined. Detector modules surrounding the object detect these gamma rays, and the distribution of the radio tracer may be reconstructed from these signals. A conventional PET detector consists of a ring of photomultiplier tubes directly coupled to scintillating crystals. Higher-resolution detector modules are composed of luminous scintillation crystals coupled to photomultiplier tubes by fiber optics. This permits closer packing of the crystals and therefore higher resolution. These facilities provide linear resolution of 1–2 mm.

With single-photon emission computed tomography (SPECT), only one gamma ray is emitted. Though the device cannot indicate where the gamma ray comes from, a collimator (pinhole) in front of the detector is used to define the direction of the gamma ray. As a result, the number of detectable events is reduced, and the sensitivity of the device drops by roughly two orders of magnitude relative to micro-PETs sensitivity. The big advantage of SPECT over PET is that single gamma ray radioisotopes are widely available, because of their longer half lives, so no on-site cyclotron to provide the radioisotopes is necessary. Combination of photon emission tomographs with higher-resolving tomographs (e.g., CT) allows a more precise assignment of a PET signal to a specific organ or tissue type. Applications are reviewed in Reader and Zweit (2001), Chatziioannou (2002), and Sharma et al. (2002).

3.7
Microscopic Imaging with Protons and Heavy Ions

Accelerated ions are also a useful imaging medium, with some similarities to bombardment with electrons or X-rays, especially in conjunction with analysis of trace elements in specimens. The most widely used ions are protons, accelerated to energies of several MeV in linear accelerators or cyclotrons, which means that this technique is only possible in research centers providing these large facilities (see Roberts et al. (1999), and Hinderer et al. (1997) for descriptions of facilities at Lawrence Livermore National Laboratory and the accelerator laboratory at the Munich universities).

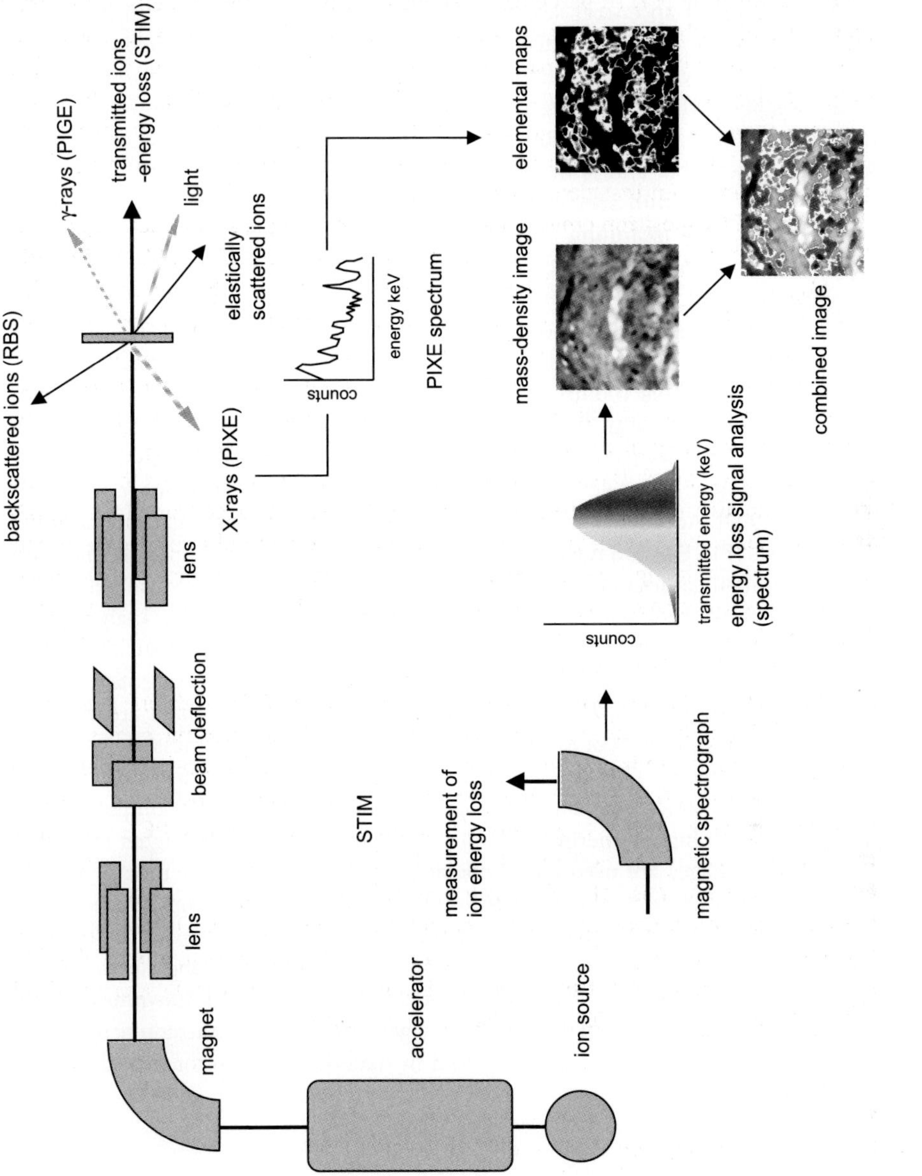

Fig. 3.9 Scanning transmission ion microscope.
After interaction with the specimen, several signals may be recorded for further analysis. For STIM, the transmitted ions are analyzed in a magnetic spectrograph, situated below the specimen plane. An energy loss image of the transmitted ion beam is combined with the elemental map generated from the X-ray signal

After passage of an energy-selective magnetic spectrograph (to obtain monoenergetic ions, which are all deflected in the same way in a magnetic field), the beam is focused by a system consisting of one or two quadrupole lenses and deflected by a "sweeper", which allows scanning – or "sweeping" – of a specimen in the x/y directions at high frequency. Resolution of this technique mainly depends on the size of the focal spot and is in the range of 1–2 μm down to 100 nm. The strong interaction of ions with specimen atoms leads to several readily detectable signals. In this way, STIM (scanning transmission ion microscopy) is applied analogously to scanning transmission electron microscopy (STEM), in conjunction with elemental mapping (Fig. 3.9).

The much higher expense necessary for accelerating ions instead of electrons is rewarded by better signal-to-noise ratios for element detection, which leads to detection limits of elements 100 times lower than achievable with electron-based spectroscopic techniques: here in the range of parts per million. The most widely applied analytical method, particle-induced X-ray emission analysis (PIXE analysis), is suitable for detection of heavy elements such as traces of metals in specimens (e.g., Schneider et al., 1999). Additional analytical features are particle-induced gamma ray emission analysis (PIGEA) and nuclear reaction analysis (NRA) – based upon the measurement of emitted gamma rays (useful for detection of light elements) – as well as analysis of elastically back-scattered particles (Rutherford back-scattering analysis, RBS) and X-ray fluorescence. As in electron microscopes, ion microscopes also allow tomographic analysis by scanning the beam over the specimen at many orientations (see, for example, Reinert et al., 2002). In addition, relatively thick specimens are transmissible (in the range of tens of μm for STIM). To obtain images of mass-density distribution (analogous to STEM or TEM images) the energy losses of transmitted ions are analyzed in a second magnetic spectrograph. The results may be composited into images in which higher energy losses of ions represent areas with higher mass densities. These images are used for combination with the scanned elemental maps.

Ion beam analysis should not be confused with field ion microscopy. Here, a sharp specimen tip is placed inside a chamber containing a small amount of hydrogen or helium gas, and pointed towards an imaging screen. When a strong electric field is applied to the specimen, gas atoms near the tips are ionized and drawn towards the screen by the electric field. In this way, an image of the tip surface at atomic resolution is formed. The method is exclusively used in materials science for investigation of metal surfaces.

3.8
Ultrasound Biomicroscopy and Scanning Acoustic Microscopy

Ultrasonography is one of the standard non-invasive tools for medical diagnosis. Diagnostic ultrasound scanners produce high-frequency sound waves, typically in the 2.5–10 MHz frequency range, to image the soft tissues of the body. The ultrasound wave is produced by a piezoelectric ceramic (transducer) that vibrates at a frequency defined by the frequency of an applied alternating electric current. When a pulse of sound passes through the body tissues, it produces a reflected pulse (an echo) whenever the main pulse passes from one tissue to another. The reflected pulses are received by the same transducer that produced the main pulse, The mechanical energy is converted back into an electric signal. Because the velocity of the ultrasound wave in soft tissue is (nearly) constant at 1540 m/s, the time taken for the wave to return to the probe can be used to determine the depth of the object causing the reflection. The amplitude of the detected echoes is shown by the brightness of areas, represented in shades of gray from black for no echo to white for a very strong echo. The transducer emits pulses in such a way that sound waves "illuminate" a section of tissue and display that section in real time, so that an image representing the structural changes across a plane of the body tissues is produced. Many ultrasound scanners incorporate two additional imaging techniques that allow rapid dynamic changes within the tissues to be investigated. If an ultrasound pulse is directed across a moving reflector (such as flowing blood), the frequency of the received ultrasound wave will be increased if the reflector is moving towards the transmitted probe and vice versa (Doppler effect). There is a constant relationship between this change and the velocity of the moving reflector, which can be used to calculate the velocity of the reflector. The resulting signal may be used to generate a wave form or a color-coded image superimposed with the black and white echo image. The color represents the frequency shift, which roughly translates to the velocity of the moving structure that caused the shift.

As a "microscopic" derivative of ultrasound imaging, ultrasound biomicroscopy achieves a maximum resolution of 50 μm. At the high ultrasound frequencies (40–50 MHz) necessary for this resolution, ultrasound penetration in soft tissues is limited to 5–10 mm, which may be sufficient for, say, study of a small animal. Unlike other techniques, especially MRI, the technique provides near real-time images (around 10 per second) of a specimen, so movements of a living animal or an organ are less disruptive and/or may be studied directly.

Acoustic microscopes also use ultrasound waves as their imaging medium, but with a different instrumental setup. As in other ultrasound techniques, contrast derives from inhomogeneities in the density or elasticity in the specimen. The most widespread application is in the form of the scanning acoustic microscope (SAM), equipped with acoustic lenses focusing ultrasound waves. Two acoustic microscope instrument types have been developed. With a through-transmission instrument, one transducer sends the signal, which traverses through the specimen, and another transducer receives it. Most instruments, however, uses the reflection technique, by which a single transducer both sends and receives the signal. This circumvents the problem of aligning two transducers.

Since sound waves behave analogously to light waves (with respect to reflection, refraction, and interference) when interacting with matter, they may be described by the same optical laws. However, the occurrence of longitudinal and transversal waves (penetrating the solid specimen or propagating at the surface) complicates the description of wave models and hence makes interpretation of an ultrasound image more difficult. Unlike light waves, acoustic waves do not exhibit considerable aberration, which simplifies the construction of lens systems. A single spherical acoustic lens surface focuses the acoustic wave sufficiently (Fig. 3.10). The acoustic lens is coupled to a transducer, which emits (or receives) ultrasound waves with frequencies around 100 MHz. The transducer is made up of a piezoelectric element, connected to a pulse generator and a receiver unit by a high frequency switch. The generator emits short pulses that excite the transducer. The reflected ultrasound pulses are registered by the transducer and directed to a receiver unit, the switch allowing distinction between transmitted and received signals. The lens is made of single-crystal sapphire or silicon, in both of which the velocity of sound is higher than in water, the coupling liquid. A surface image is produced when the acoustic beam is focused on the specimen surface. In the simplest case, reflected longitudinal waves are subjected to constructive or destructive interference. By moving the specimen towards the lens, the focal point is moved below the surface. Upon defocusing, surface waves and longitudinal waves interfere with each other either destructively or constructively, contributing to contrast generation in SAM. Such a "defocused" image contains information from both surface and subsurface.

The specimen is scanned by moving the lens and/or the specimen in the x/y-directions, ultimately allowing an image to be reconstructed. Quantification of rapid changes of mechanical parameters is also possible, by subtraction of consecutive images. Besides amplitude contrast (due to different reflective properties in a specimen), phase contrast, interference contrast, and other contrast phenomena also contribute to the final image. The interpretation of the final image is therefore a difficult task requiring digital analysis of the acoustic parameters – as well as considerable experience.

Though SAM is predominantly used in materials analysis, the technique is also used to investigate the mechanical properties of biological specimens (especially mineralized specimens such as bone), but also of soft tissue. An extended review of SAM applications in cell biology is presented in Bereiter-Hahn (1995).

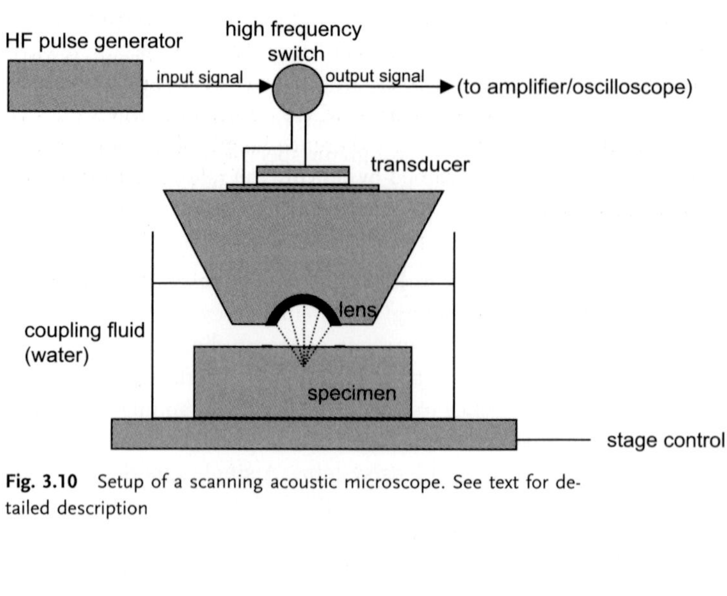

Fig. 3.10 Setup of a scanning acoustic microscope. See text for detailed description

3.9
Optical Coherence Microscopy/Tomography

Optical coherence microscopy was originally developed as a method to depict retinal tissue layers in ophthalmology. The method allows optical sectioning of a multitude of opaque objects, instruments with resolution down to the level of tissue layers now being readily availablc. Instruments with 1 μm resolution, enabling resolution down to the cellular level, have also been developed. Originally used for transparent tissue, imaging of non-transparent samples is now also possible. Since the technique can produce real-time images and has also been adapted as a micro-invasive endoscopic technique, it is also an interesting diagnostic tool in conjunction with endoscopy. Optical coherence microscopy or tomography (OCT) corresponds to ultrasonography with light instead of ultrasound as imaging medium, using photons back-scattered from structures inside a specimen for imaging. Although ultrasound scanning provides images from depths far beyond the limits of OCT, the resolution of OCT is superior and makes the technique an important non-invasive imaging technique of small specimens and a viable alternative to MRI or micro-CT. Photons (if not absorbed or transmitted) can be scattered in any direction, a single time or multiple times before leaving the specimen. OCT preferentially selects singly scattered photons. These are photons that have traveled a path length specified by a specimen structure (i.e., a discontinuity of the refractive index between two tissue layers). The optical coherence of the travelling light beams is analyzed by low-coherence interferometry in a Michelson interferometer (see Brezinski and Fujimoto, 1999, and Schmitt, 1999, for review; see Fig. 3.11), rather than by comparative measurements of signal traveling times (as in ultrasonograpy). The axial resolution of OCT is determined by the coherence length (inverse spectral width) of the light source. Superluminescent diodes are most widely used, thanks to their high radiation output and relatively low cost, but are insufficient for resolution of individual cells. Broad-spectrum laser light sources, which provide higher resolution, are not yet used in commercially available instruments, due to the higher complexity (and cost) of the system (see Kowalevicz et al., 2002). The light beam passing through the interferometer is split into a beam directed at the specimen and a reference beam. The reference beam is reflected by a movable reference mirror and brought to interference with the scattered beam – which occurs only when the two optical path lengths are matched to within the coherence length of the source. The depth-resolved image of the sample is obtained by moving the reference mirror. Optical sections (normally taken perpendicular to the specimen surface) or three-dimensional images are produced by scanning the beam across a small area of the specimen. Optical Doppler tomography, corresponding to the Doppler effect in ultrasound, allows measurement of flow velocities in small vessels.

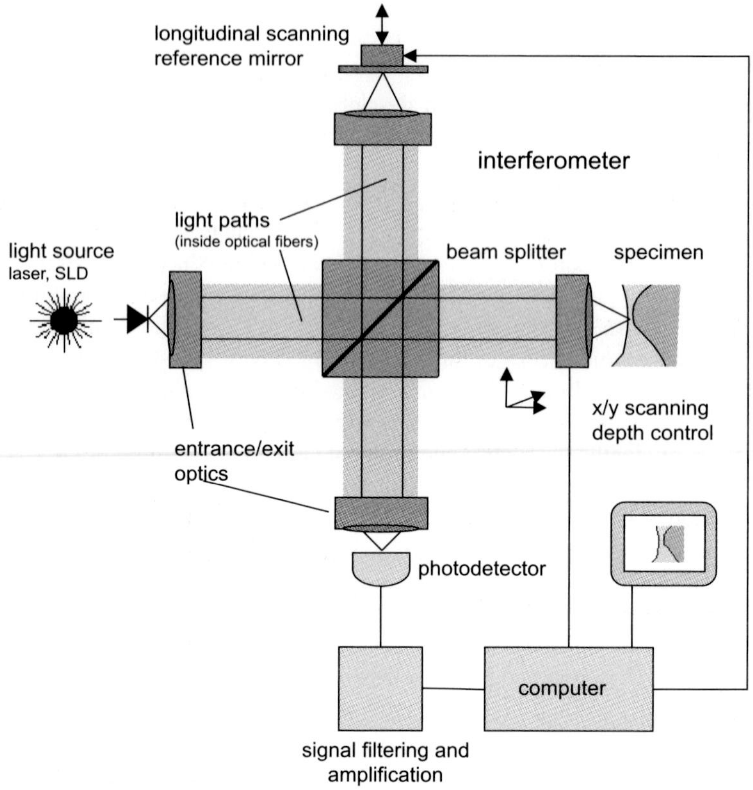

Fig. 3.11 Optical coherence microscope. A Michelson interferometer is used to detect the differences in the optical pathlength between the beams directed towards/reflected from the reference mirror and the specimen

3.10
Scanning Probe Microscopy

Since their development during the 1970s and early 1980s, scanning probe microscopes (SPMs) have developed into instruments suitable for analysis of surface topography (and other surface properties) down to near-atomic resolutions. All instruments belonging to this family contain essential components as depicted in Fig. 3.12. In the scanning process, surface material properties are mapped in all three directions, the essential parts of the system being the sharp scanning probe tip, which moves in the z direction, a piezoelectric scanner that moves the specimen in the x/y directions, and a sensor for tip positioning. Sensor and tip are connected in a feedback loop for correction of z positioning. The x/y/z positions are used to generate a topographic image of the specimen (TM microscopes, 2001).

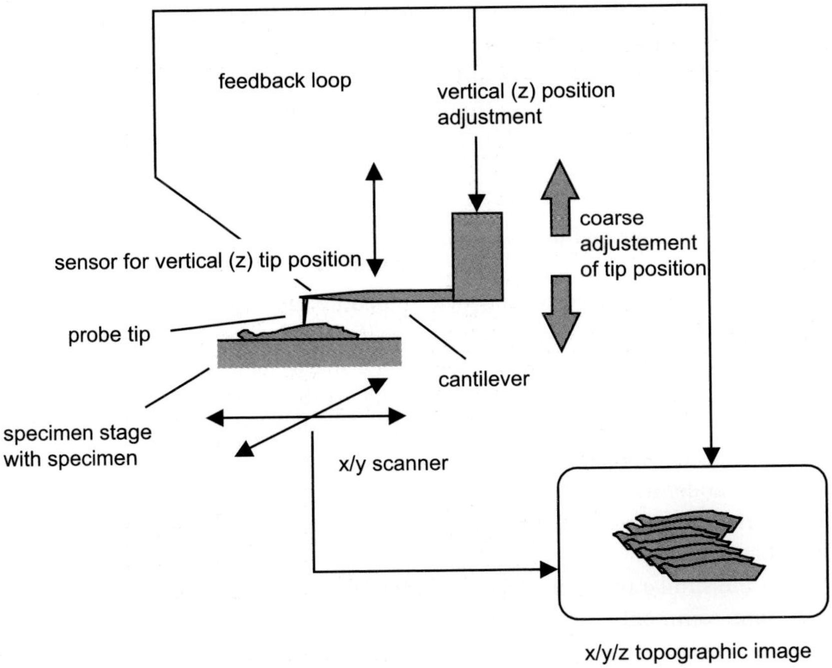

Fig. 3.12 Scanning force microscopy. The generation of the x/y/z topographic signal as indicated in the figure is achieved in several instrumental setups with different probe tips and imaging modes

3.10.1
The Scanning Tunneling Microscope (STM)

The scanning tunneling microscope is the ancestor of all scanning probe microscopes. The specimen needs to be conductive or semi-conductive and is predominantly used in materials sciences and chemistry. Though not ideally suited for biological specimens – the more recently developed SPM techniques are more applicable – STM has also been applied to biological specimens (see, e.g., Guckenberger et al., 1988). The instrument provides a surface topography with atomic resolution.

STMs use a sharpened conducting tip with a bias voltage (mV to 3 V) applied between the tip and the sample. If the distance between the tip and the sample is large, no current flows. When, however, the tip is brought very close (1 nm), but without physical contact, a small current (pico- to nano-Amperes) flows across the gap between the tip and the sample. This current is the result of the overlapping of wavefunctions between the tip atoms and the surface atoms. Electrons can "tunnel" across the vacuum barrier separating the tip and sample in the presence of a small bias voltage. The magnitude of the tunneling current increases exponentially when the tip approaches the specimen surface. This results in high system sensitivity: when the separation between the tip and the specimen changes by 10 %, the tunneling current changes by an order of magnitude. This ultimately provides surface images with lateral atomic resolution. To a first approximation, an image of the tunneling current depicts the topography of the sample. More accurately, the tunneling current corresponds to the electronic density of states at the surface. Rather than physical topography, it measures a surface of constant electron tunneling probability.

3.10.2
Atomic Force Microscopy and Related Techniques

Image formation in AFM is generally determined by sensing of interatomic forces, making the technique applicable to all solid specimens. The atomic force microscope (AFM) also probes the surface of a sample with a sharp tip, less than 10 nm in diameter. The tip is located at the free end of a cantilever that is 100 to 200 μm long. The cantilever is deflected by forces between the tip and the sample surface, a detector measuring the cantilever deflection as the tip is scanned over the sample (or the sample is scanned under the tip). The measured cantilever deflections are used to generate a map of surface topography. Conductors, semiconductors, and insulators may all be scanned (Fig. 3.12).

In AFMs and similar instrumental setups (see below) the minute movements of the cantilever are commonly detected by enhancement with laser-optical techniques. When a laser beam is projected onto the cantilever, the position of the reflected beam is shifted even upon a displacement of the cantilever as small as 1 nm, which results in detection of even sub-nm movement at the tip surface. Other methods are based upon optical interference (e.g., by a Michelson interferometer). A scanning tunneling microscope tip may also be used to read the cantilever deflection. If the cantilever is made from a piezoresistive material, its deflection can be detected electrically. When the cantilever deflection is detected, this can be used to generate the topographic data set directly. This mode (constant height) is often used for producing atomic-scale images of atomically flat surfaces, for which the cantilever deflections and variations in applied force are small. This mode is also essential for recording real-time images of changing surfaces, which requires high scan speeds. Alternatively, the deflection of the cantilever can be used in a feedback circuit that moves the scanner up and down in the z direction, responding to the topography by keeping the cantilever deflection constant. In this case, the image is generated from the scanners motion. With the cantilever deflection held constant, the total force applied to the sample is constant. This mode (constant force) is generally preferred for most applications.

Figure 3.13 depicts the typical force-distance curve for AFM, where two commonly used modes referring to the distance between tip and specimen distance are labeled. As the atoms of the tip and the specimen surface are gradually brought together, they first weakly attract each other by van der Waals forces. This attraction increases until the electron clouds of the atoms are so close together that their electrostatic repulsion progressively weakens the attractive force. The force becomes zero when the distance between the atoms reaches a couple of ngstroms, about the length of a chemical bond. In the **contact mode**, the cantilever is held less than a few Ångstroms from the sample surface, and the interatomic force between the cantilever and the sample is repulsive.

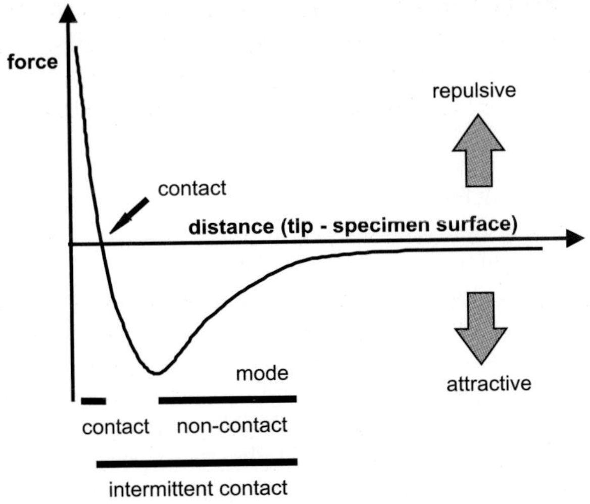

Fig. 3.13 Force-distance curve in scanning force microscopy. See text for detailed description

The repulsive force balances almost any force that attempts to push the atoms closer together: if the cantilever pushes the tip against the sample, the cantilever – with a low spring constant (lower than the effective spring constant holding the atoms of the sample together) – bends rather than forcing the tip atoms closer to the sample atoms. The total force applied to the tip is the sum of the capillary and the cantilever forces, and must be balanced by the repulsive van der Waals force. The magnitude of the total force exerted on the sample varies from 10^{-8} N to 10^{-7}–10^{-6} N. For biological (i.e., soft and wet) objects, the contact mode may induce several artifacts in the final image. This is avoided by application of modes in which less force is applied to the specimen. However, smaller forces also have to be detected, which in practice will often reduce the final resolution. In the **non-contact mode**, the cantilever is held of the order of tens to hundreds of Ångstroms from the sample surface, and the interatomic force between the cantilever and sample is attractive (largely as a result of long-range van der Waals interactions). In this mode, the stiff cantilever vibrates with a frequency of 100 to 400 kHz (the resonant frequency of the cantilever) with an amplitude of a few tens of nanometres. The resonant frequency of the cantilever varies with the force gradient acting on the cantilever, so changes in the resonant frequency of the cantilever can be used as a measure of changes in the force gradient, which in turn reflects changes in the tip-to-sample distance, or sample topography. The resonant frequency or vibrational amplitude of the cantilever is monitored and kept constant with the aid of a feedback loop that moves the scanner up and down in the z direction. By keeping the resonant frequency or amplitude constant, the system also keeps the average tip-to-sample distance constant. As in contact AFM (in constant-force mode), the motion of the scanner is used to generate the topographic data set. It thus detects changes in the resonant frequency or vibration amplitude as the tip comes near the sample surface.

Intermittent-contact AFM is similar to non-contact AFM, but the vibrating tip is brought closer to the sample in such a way that the sample is intermittently contacted or "tapped". As in non-contact AFM, the oscillation amplitude of the tip also changes in response to tip-to-sample spacing in intermittent-contact AFM. Intermittent-contact AFM is also less likely to damage the sample than contact AFM. Though it touches the specimen, lateral forces (friction or drag) between the tip and the sample are minimal.

Scanning force microscopy is a highly variable technique, because forces between a specimen and a AFM tip may be caused by a multitude of underlying factors. A modified scanning tip may therefore be used to measure a variety of parameters other than tunneling current or van der Waals surface forces. For biologists, the detection of chemical properties on biological surfaces is of outstanding interest. Some examples of measurement of material properties by AFM are listed below, while near-field scanning optical microscopy (SNOM, NSOM), is described in the next section.

Magnetic force microscopy (MFM) images the spatial variation of magnetic forces on a sample surface. For MFM, the tip is coated with a thin ferromagnetic film. The system operates in non-contact mode, detecting changes in the resonant frequency of the cantilever induced by the magnetic fields dependence on tip-to-sample separation.

Electrostatic force microscopy (EFM) measures the forces originating from an electrical field (E-field) between the AFM tip and the sample. When a voltage is applied between the tip and the sample (the tip is not in contact with the surface), the cantilever deflects when it scans over charged particles on the sample surface. In this way, EFM maps locally charged domains on the sample surface.

Lateral force microscopy (LFM) measures lateral deflections (twisting) of the cantilever caused by forces on the cantilever parallel to the plane of the sample surface. LFM studies are used to detect surface friction arising from inhomogeneity in surface material; an artificial enhancement of edges in the image may also be achieved.

In **force modulation microscopy (FMM)**, the elastic properties of a specimen are mapped. The cantilever tip is scanned in contact with the sample surface. As in constant-force contact AFM, the feedback loop uses the cantilever deflection signal to maintain a constant force between the tip and the sample and to generate a topography image. When a periodic oscillation is applied to the tip or the sample, the sample surface resists and the cantilever bends according to the elastic properties of the sample surface. The system generates an image, a map of sample stiffness, from changes in the amplitude of cantilever modulation.

Scanning thermal microscopy (SThM) measures the specimen temperature and thermal conductivity of the sample surface. The first SThMs were developed from a conventional STM with a modified tip for scanning non-conductive material. By using the temperature – as sensed by a modified tip with a thin film thermocouple – as a feedback signal to maintain a constant tip-substrate gap, this scanning thermal profiler could be used to image electrically insulating surfaces with a lateral resolution of 100 nm. Since the feedback signal was based on maintaining a constant probe temperature, it could not be used to obtain true thermal images of surfaces. A true SThM tip may be made from a thermocouple temperature sensor: a pair of dissimilar metals joined at the tip, the temperature difference between the tip and the open end generating a proportional net voltage. In this way, thermal data and surface topography (with the conventional AFM force feedback mechanism) can be measured at the same time. Probes made of Wollaston wires also operate as active heat sources. The probe resistance is proportional to the probe temperature. Changes in the current required to keep the probe at constant temperature produce a thermal conductivity map. When the tip is held at a constant current, the specimen temperature is mapped by changes in the probe resistance.

Phase detection microscopy (PDM) is another variation for mapping of surface properties such as elasticity, adhesion, and friction. In this way it resembles force modulation microscopy. Unlike FMM, the technique operates in the non-contact mode, which is more suitable for less rigid (biological) samples. Phase detection images can be produced while an instrument is operating in any oscillating cantilever mode, such as non-contact AFM, intermittent-contact AFM, or MFM mode. The technique monitors the phase lag between the signal causing the cantilever to oscillate and the cantilever oscillation output signal. Changes in the phase lag reflect changes in the mechanical properties of the sample surface. The systems feedback loop operates in the usual manner, using changes in the cantilevers deflection or vibration amplitude to measure sample topography. The phase lag is monitored while the topographic image is being taken, so that images of topography and material properties can be collected simultaneously.

The **pulsed force mode (PFM)** is used to monitor similar specimen properties. Besides sample topography, stiffness and adhesion may be measured simultaneously. In

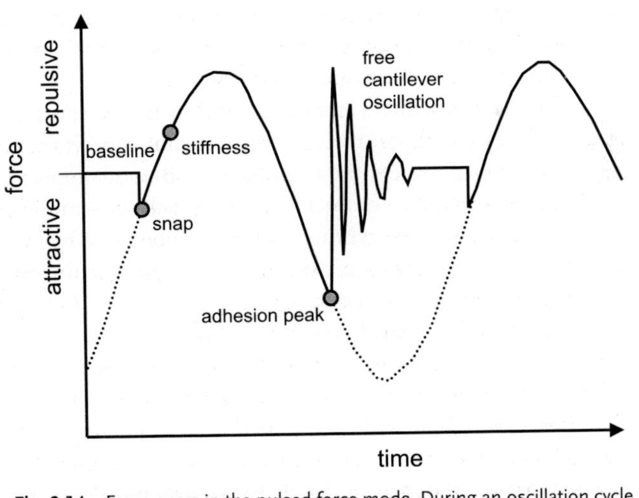

Fig. 3.14 Force curve in the pulsed force mode. During an oscillation cycle the probe tip is attracted by the surface and "snaps" into contact at a certain distance. The tip is then pushed closer to the sample, the force signal increases, the slope of the curve depending on the specimen stiffness. When the distance increases, the force signal decreases, and the tip loses contact (adhesion peak) and oscillates freely until the signal is damped to the baseline. Characteristic curve features are used to generate topographic, adhesion, and stiffness images

PFM, the probe scans the surface in contact mode. When a sinusoidal oscillation is applied to the cantilever, it is possible to analyze the force output as a function of the distance between the tip and the surface. The force output is depicted in Fig. 3.14, with the sections of the curve relevant to interpretation marked. Characteristic points of these curves are taken for processing to generate simultaneous topographic, stiffness (deformation of the specimen under an applied force), and adhesion maps.

Scanning near-field acoustic microscopy (SNAM) is the name given to a scanning force microscope operating in the non-contact mode, utilizing an ultrasonic quartz oscillator as resonator and at the same time as a sensor of the damping force exerted by the specimen. Damping increases when the tip approaches the surface. The instruments have been especially developed for scanning of larger areas, with a surface relief in the millimeter range. For low resolution, wide probing tips that are mainly damped by hydrodynamic friction forces are used. At resolutions of 100 nm, both hard and soft specimens are well suited for mapping. Higher resolution is obtained with sharp AFM tips (Murdfield et al., 1996).

With chemically modified tips it is possible to obtain modified force curves based on specific interactions between molecules on the specimen surface and a sensor molecule on the probe tip. This mode of imaging is sometimes referred to as **chemical force microscopy (CFM)**. The tip (normally made from silicon or silicon nitride) is modified by coating with a very thin metallic film (normally a 5 nm thick chromium layer followed by a 50 nm thick gold layer) onto the probe, followed by immersion of the probe in a solution of an organic thiol. One end of the thiol covalently bonds to the gold surface, forming a self-assembled monolayer, while the other end contains an appropriate functional group. Applications include mapping of the functional group microstructure in polymers and binding/recognition interactions (e.g., based upon the avidin/biotin system) in biological systems.

Lateral resolution in the atomic range, as achieved especially by contact-mode AFM, is not necessary for most biological specimens. Moreover, because of the large forces that act on the specimen, the technique is often not applicable. Thus, to provide only structural information, other techniques such as scanning electron microscopy or even light microscopic techniques are often more straightforward than AFM.

For biological specimens, AFM is especially useful for quantification of binding forces between receptors and ligands, proteins, and cells with the aid of chemically modified tips (Yip, 2001; Allison et al., 2002). Measurement of a distance-force curve between the RGD peptide and integrin has been performed in this way (Lehenkari and Horton, 1999). AFM is also used for study of the material properties of biological specimens (stiffness, plasticity, and viscosity; see, e.g., Rotsch et al., 1997). On the cellular level, the properties of the cytoskeleton and (induced) cytoskeletal changes become visible (see, e.g., Fung, 1993; Goldmann et al., 1998; and Wu et al., 1998, for examples of applications).

Although the probing tip has been described in this section as a surface sensor, it is also an interesting tool for manipulation of the specimen surface. Thanks to the flexibility of the tip in sensing surface properties, and in turn for use as a nanometric tool for manipulation of the surface, AFM is one of the most important tools for nanotechnology, also in conjunction with biological applications.

3.11 Near-Field Scanning Optical Microscopy

Fluorescence analysis can now provide insights into single-molecule events in vitro and in vivo. However, it is still not possible to localize these events accurately (i.e., below the resolution limit of the optical light microscope). For biologists, however, light is a nearly ideal imaging medium. Energy transfer to the specimen influences the molecular events under investigation only minimally, unlike with electromagnetic waves of shorter wavelengths. Light beams are easy to manipulate technically, and the microscope may be used under ambient conditions. Developments that use light as an imaging medium below its resolution limit in "non-optical" microscopes therefore have, a priori, great potential application for biologists (Gheber et al., 1998; de Lange et al., 2001; Edidin, 2001). Evanescent waves or near-field light as imaging media are described for TRIFM in a previous section (see Fig. 3.1). In addition, the ultra-fine AFM tip may be used for probing with near-field light at sub-wavelength resolution in near-field scanning optical microscopy (NSOM), so NSOM is based on the AFM scanning principle (Fig. 3.15). Though the instrument is, of course, not a true optical light microscope, the use of visible light in probing allows adaptation of fluorescence techniques for NSOM.

Near-field optical light microscopy works without the limits of numerical aperture. The light emitted by an object at its surface contains more information (or higher spatial frequencies) than is detectable by conventional lens optics operating at some distance (a large distance when compared with the wavelength of the imaging

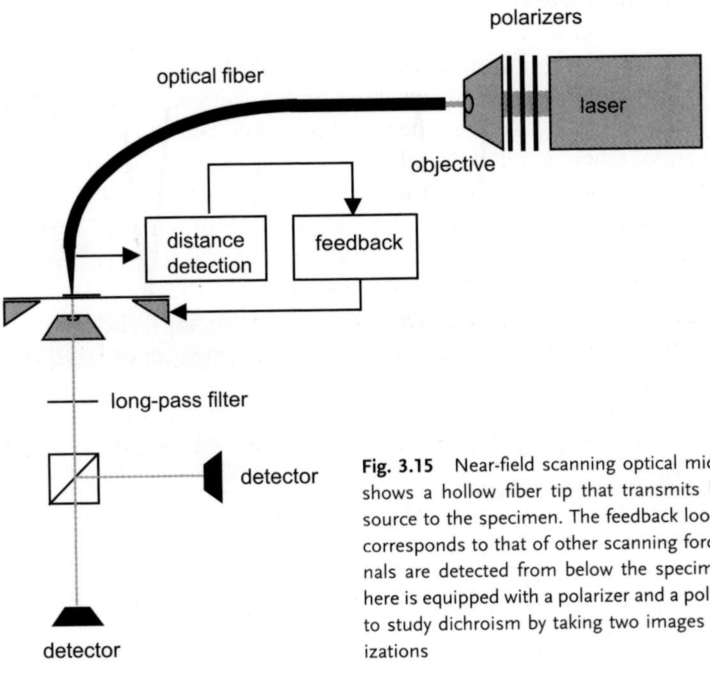

Fig. 3.15 Near-field scanning optical microscope. The setup shows a hollow fiber tip that transmits light from the laser source to the specimen. The feedback loop for tip positioning corresponds to that of other scanning force microscopes. Signals are detected from below the specimen. The instrument here is equipped with a polarizer and a polarizing beam splitter to study dichroism by taking two images at orthogonal polarizations

medium). A scanning probe similar to an AFM tip at a distance smaller than one wavelength (or in the "near field"), detects near-field (or evanescent) light waves that are not propagated, and so are lost for conventional microscopes. The specimen may be illuminated with conventional (or "far-field") light from below the specimen (in a total reflection mode) or from the top surface as depicted in Fig. 3.15. The near-field light on the top surface of the sample interacts with a sharp probe tip and scatters as far-field light. The scattered light is recordable with conventional light optical devices and originates from an area approximately equivalent to the size of the probe tip. If the tip size is smaller than the wavelength of light falling on the sample and is x/y-scanned over the sample at a correspondingly fine scale, an image with a resolution higher than the wavelength may be built up. In principle, resolution in NSOM is determined by the size of the probe tip. Theoretically, since atomically sharp tips may be generated, near-atomic resolution seems to be achievable. In practice, a probe tip of approximately 10

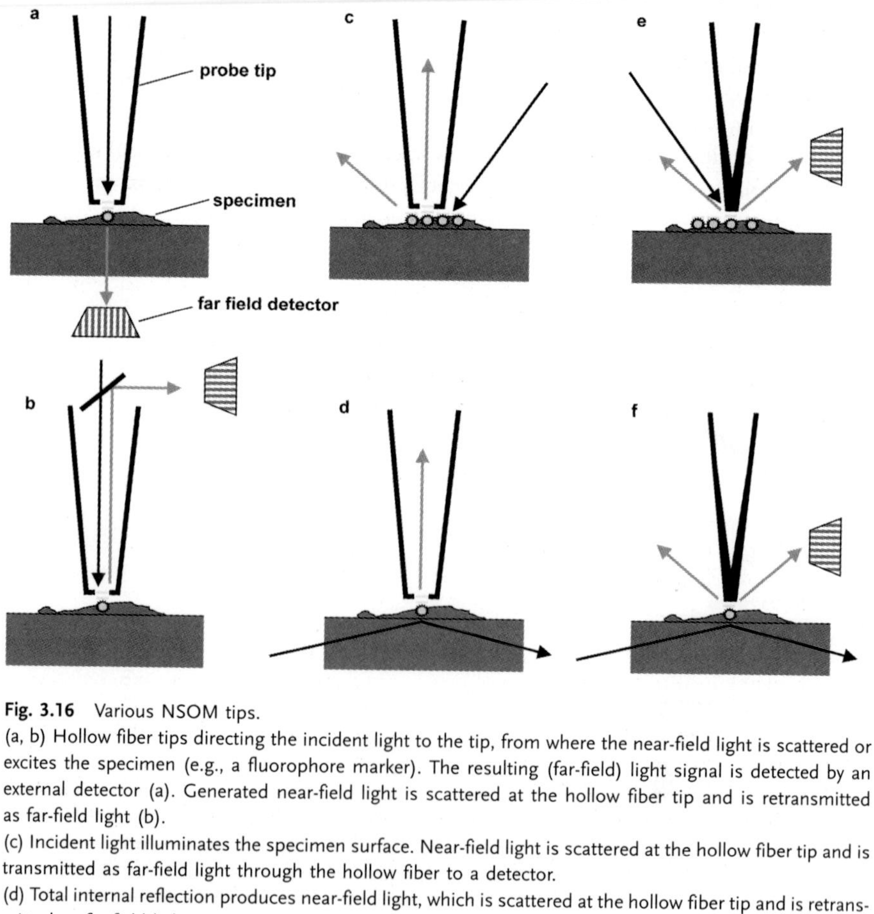

Fig. 3.16 Various NSOM tips.
(a, b) Hollow fiber tips directing the incident light to the tip, from where the near-field light is scattered or excites the specimen (e.g., a fluorophore marker). The resulting (far-field) light signal is detected by an external detector (a). Generated near-field light is scattered at the hollow fiber tip and is retransmitted as far-field light (b).
(c) Incident light illuminates the specimen surface. Near-field light is scattered at the hollow fiber tip and is transmitted as far-field light through the hollow fiber to a detector.
(d) Total internal reflection produces near-field light, which is scattered at the hollow fiber tip and is retransmitted as far-field light.
(e, f) Scattering of far-field light at apertureless tips generates near-field light that excites the specimen

nm would provide, in the end, a resolution of 100 nm. Rather than solid tips, most NSOMs currently work with hollow probe tips with very small apertures to illuminate the object in the near field as depicted in Figure 3.16 a–d. A probe consists of an optical fiber ending in a sub-wavelength sized aperture. This is placed close to the object surface (at a distance much less than one wavelength). When light is sent through the fiber, an optical near field is generated at the tip. The near field interacts with the sample surface. Far-field light emitted from the specimen will be scattered from the sample, but only from the near-field illuminated region, resulting in high resolvable detection. The same aperture as used to illuminate the sample can also collect the near-field light emitted from the specimen. This near-field light is scattered at the sub-wavelength aperture as far-field light and re-transmitted through the optical fiber. This technique is predominantly used in conjunction with spectroscopic techniques. The highest resolution may, in principle, be achieved with solid AFM tips (Fig. 3.16 e, f). Here, an incident light beam is scattered by the thin AFM tip, which generates the near field.

The NSOM tip operates like an AFM tip in the non-contact mode. In this way, the feedback signal itself is used to generate a surface topographic map of the sample. The typical setup for NSOM in cell biology is depicted in Fig. 3.15; the similarities to the setup of a conventional transmitted light fluorescence microscope are evident. The probe has a small aperture, generating the optical near field. Fluorescent (far-field) light is collected by a conventional optics below the specimen (i.e., an inverted microscope) and the topographical map is set in comparison with the fluorescence signal. Thanks to the extremely small excitation volume determined by the near-field light, together with the information on the coordinates of the probe, a high-resolution map of the fluorescent targets in the specimen is achieved.

With NSOM, the AFM topographic information is combined with optical properties of a specimen, such as refraction index, reflectivity, transparency, polarization Raman shift, and fluorescence.

References

1 Abraham-Peskir, J. V. (**2000**). X-ray microscopy with synchrotron radiation: applications to cellular biology. *Cell Mol. Biol.* 46, 1045–1052.

2 Allison, D. P., Hinterdorfer, P., Han, W. (**2002**). Biomolecular force measurements and the atomic force microscope. *Curr. Opin. Biotechnol.* 13, 47–51.

3 Appel, T. R., Baumann, M. A. (**2002**). Solid-state nuclear magnetic resonance microscopy demonstrating human dental anatomy. *Oral Surg. Oral Med. Oral Pathol. Oral Radiol. Endod.* 94, 256–261.

4 Axelrod, D. (**2001**). Total internal reflection fluorescence microscopy in cell biology. *Traffic* 2, 764–74.

5 Bastiaens, P. I. H., Pepperkok, R. (**2000**). Observing proteins in their natural habitat: the living cell. *Trends Biochem. Sci.* 12, 631–637.

6 Bentley, M. D, Ortiz, M. C., Ritman, E. L., Romero, J. C. (**2002**). The use of microcomputed tomography to study microvasculature in small rodents. *Am. J. Physiol. Regul. Integr. Comp. Physiol.* 282, R1267–1279.

7 Bereiter-Hahn, J. (**1995**). Probing biological cells and tissues with acoustic microscopy. In: *Advances in Acoustic Microscopy*, Vol. 1. Plenum Press, New York, pp. 79–115.

8 Brezinski, M. E., Fujimoto, J. G. (**1999**). Optical coherence tomography: High resolution imaging in nontransparent tissue.

IEEE J. Select. Topics Quant. Electron. 5, 1185–1192.

9 Bryers, J. D. (**2001**). Two-photon excitation microscopy for analyses of biofilm processes. *Meth. Enzymol.* 337, 259–269.

10 Bulte, J. W., van Zijl, P. C., Mori, S. (**2002**). Magnetic resonance microscopy and histology of the CNS. *Trends Biotechnol.* 20, S24–28.

11 Burg, K. J., Delnomdedieu, M., Beiler, R. J., Culberson, C. R., Greene, K. G., Halberstadt, C. R., Holder, W. D. Jr., Loebsack, A. B., Roland, W. D., Johnson, G. A. (**2002**). Application of magnetic resonance microscopy to tissue engineering: A polylactide model. *J. Biomed. Mater Res.* 61, 380–390.

12 Chatziioannou, A. F. (**2002**). Molecular imaging of small animals with dedicated PET tomographs. *Eur. J. Nucl. Med. Mol. Imaging* 29, 98–114.

13 Clegg, R. M. (**1995**). Fluorescence resonance energy transfer. *Curr. Opin. Biotechnol.* 6, 103–110.

14 de Lange, F., Cambi, A., Huijbens, R., de Bakker, B., Rensen, W., Garcia-Parajo, M., van Hulst, N., Figdor, C. G. (**2001**). Cell biology beyond the diffraction limit: Near-field scanning optical microscopy. *J. Cell Sci.* 114, 4153–4160.

15 Edidin, M. (**2001**). Near-field scanning optical microscopy, a siren call to biology. *Traffic* 2, 797–803.

16 Fung, Y. C. (**1993**). Biomechanics: mechanical properties of living tissues. Springer, New York.

17 Gheber, L., Hwang, J., Edidin, M. (**1998**). Design and optimization of a near-field scanning optical microscope for imaging biological samples in liquid. *Appl. Opt.* 7, 528.

18 Goldmann, W. H., Galneder, R., Ludwig, M., Xu, W., Adamson, E. D., Wang, N., Ezzell, R. M. (**1998**). Differences in elasticity of vinculin-deficient F9 cells measured by magnetometry and atomic force microscopy. *Exp. Cell Res.* 239, 235–242.

19 Gratton, E., Barry, N. P., Beretta, S., Celli, A. (**2001**). Multiphoton fluorescence microscopy. *Methods* 25, 103–110.

20 Guckenberger, R., Kösslinger, C., Gatz, R., Breu, H., Levai, W., Baumeister, W. (**1988**). A scanning tunneling microscope (STM) for biological applications. Design and performance. *Ultramicroscopy* 25, 795–802.

21 Hess, S. T., Huang, S., Heikal, A. A., Webb, W. W. (**2002**). Biological and chemical applications of fluorescence correlation spectroscopy: a review. *Biochemistry* 41, 697–705.

22 Hinderer, D., Dollinger, G., Datzmann, G., Körner, H. J. (**1997**). Design of the new superconducting microprobe system in Munich. *Nucl. Instr. Meth. Phys. Res. B* 130, 51–56.

23 Holdsworth, D. W., Thornton, M. M. (**2002**). Micro-CT in small animal and specimen imaging. *Trends Biotechnol.* 20, S34–39.

24 Hörnschemeyer, T., Beutel, R. G., Pasop, F. (**2002**). Head structures of *Priacma serrata* Leconte (Coeloptera, Archostemata) inferred form X-ray tomography. *J. Morphol.* 252, 298–314.

25 Hovius, R., Vallotton, P., Wohland, T., Vogel, H. (**2000**). Fluorescence techniques: shedding light on ligand–receptor interactions. *Trends Pharmacol. Sci.* 21, 266–273.

26 Jacobson, C. (**1999**). Soft X-ray microscopy. *Trends Cell Biol.* 9, 44–47.

27 Kowalevicz, A. M., Ko, T., Hartl., I., Fujimoto, J. G., Pollnau, M., Salathé, R. P. (**2002**). Ultrahigh resolution optical coherence tomography using a superluminescent light source. *Opt. Express* 10, 349–353.

28 Lehenkari, P. P., Horton, M. A. (**1999**). Single integrin molecule adhesion forces in intact cells measured by atomic force microscopy. *Biochem. Biophys. Res. Commun.* 259, 645–650.

29 Maser, J., Osanna, A., Wang, Y., Jacobsen, C., Kirz, J., Spector, S., Winn, B., Tennant, D. (**2000**). Soft X-ray microscopy with a cryo scanning transmission X-ray microscope: I. Instrumentation, imaging and spectroscopy. *J. Microsc.* 197 (Pt 1), 68–79.

30 Medina, M. A., Schwille, P. (**2002**). Fluorescence correlation spectroscopy for the detection and study of single molecules in biology. *Bioessays* 24, 758–64.

31 Murdfield, T., Fischer, U. C., Fuchs, H. (**1996**). Acoustic and dynamic force microscopy with ultrasonic probes. *J. Vacuum Sci. Technol.* B14, 877–881.

32 Osanna, A., Jacobsen, C., Kalinovsky, A., Kirz, J., Maser, J., Wang, S. (**1996**). X-ray microscopy: preparations for studies of frozen hydrated specimens. *Scanning Microsc. Suppl.* 10, 349–356.

33 Reader, A. J., Zweit, J. (**2001**). Developments in whole-body molecular imaging of live subjects. *Trends Pharmacol. Sci.* 22, 604–607.

34 Reinert, T., Sakellariou, A., Schwertner, M., Vogt, J., Butz, T. (**2002**). Scanning transmission ion microscopy tomography at the Leipzig nanoprobe LIPSION. *Nucl. Instr. Meth. Phys. Res. B* 190, 266–270.

35 Roberts, M. L., Grant, P. G., Bench, G. S., Brown, T. A., Frantz, B. R., Morse, D. H., Antolak, A. J. (**1999**). The stand-alone microprobe at Livermore. *Nucl. Instr. Meth. Phys. Res. B* 158, 24–30.

36 Rotsch, C., Braet, F., Wisse, E., Radmacher, M. (**1997**). AFM imaging and elasticity measurements on living rat liver macrophages. *Cell Biol. Int.* 21, 685–696.

37 Schmitt, J. M. (**1999**). Optical coherence tomography (OCT): A review. *IEEE J. Select. Topics Quant. Electron.* 5, 1205–1215.

38 Schneider, T., Haag-Kerwer, A., Maetz, M., Niecke, M., Povh, B., Rausch, T., Schüßler, A. (**1999**). Micro-Pixe studies of elemental distribution in Cd accumulating *Brassica juncea* L. *Nucl. Instr. Meth. Phys. Res. B* 158, 329–334.

39 Sharma, V., Luker, G. D., Piwnica-Worms, D. (**2002**). Molecular imaging of gene expression and protein function in vivo with PET and SPECT. *J. Magn. Reson. Imaging* 16, 336–351.

40 So, P. T., Dong, C. Y., Masters, B. R., Berland, K. M. (**2000**). Two-photon excitation fluorescence microscopy. *Annu. Rev. Biomed. Eng.* 2, 399–429.

41 TM microscopes (no authors listed) (**2001**). A Practical Guide to Scanning Probe Microscopy. TM microscopes Inc., Sunnyvale, CA.

42 Tokunaga, M., Yanagida, T. (**1997**). Single molecule imaging of fluorophores and enzymatic reactions achieved by objective-type total internal reflection fluorescence microscopy. *Biochem. Biophys. Res. Commun.* 235, 47–53.

43 Truong, K., Ikura, M. (**2001**). The use of FRET imaging microscopy to detect protein-protein interactions and protein conformational changes in vivo. *Curr. Opin. Struct. Biol.* 11, 573–578.

44 White, J., Stelzer, E. (**1999**). Photobleaching GFP reveals protein dynamics inside live cells. *Trends Cell Biol.* 9, 61–65.

45 Wu, H. W., Kuhn, T., Moy, V. T. (**1998**). Mechanical properties of L929 cells measured by atomic force microscopy: effects of anticytoskeletal drugs and membrane crosslinking. *Scanning* 20, 389–397.

46 Yip, C. M. (**2001**). Atomic force microscopy of macromolecular interactions. *Curr. Opin. Struct. Biol.* 11, 567–572.

Subject Index

d